T0329327

mmWave Massive MIMO

mmWave Massive MIMO
A Paradigm for 5G

Editors

Shahid Mumtaz
Instituto de Telecomunicações, Aveiro, Portugal

Jonathan Rodriguez
Instituto de Telecomunicações, Aveiro, Portugal

Linglong Dai
Tsinghua University, Beijing, China

AMSTERDAM • BOSTON • HEIDELBERG • LONDON
NEW YORK • OXFORD • PARIS • SAN DIEGO
SAN FRANCISCO • SINGAPORE • SYDNEY • TOKYO
Academic Press is an imprint of Elsevier

Academic Press is an imprint of Elsevier
125 London Wall, London EC2Y 5AS, United Kingdom
525 B Street, Suite 1800, San Diego, CA 92101-4495, United States
50 Hampshire Street, 5th Floor, Cambridge, MA 02139, United States
The Boulevard, Langford Lane, Kidlington, Oxford OX5 1GB, United Kingdom

Notices
Knowledge and best practice in this field are constantly changing. As new research and
experience broaden our understanding, changes in research methods, professional practices,
or medical treatment may become necessary.

Practitioners and researchers must always rely on their own experience and knowledge in
evaluating and using any information, methods, compounds, or experiments described herein.
In using such information or methods they should be mindful of their own safety and the safety
of others, including parties for whom they have a professional responsibility.

To the fullest extent of the law, neither the Publisher nor the authors, contributors, or editors,
assume any liability for any injury and/or damage to persons or property as a matter of products
liability, negligence or otherwise, or from any use or operation of any methods, products,
instructions, or ideas contained in the material herein.

Library of Congress Cataloging-in-Publication Data
A catalog record for this book is available from the Library of Congress

British Library Cataloguing-in-Publication Data
A catalogue record for this book is available from the British Library

ISBN 978-0-12-804418-6

For information on all Academic Press publications visit
our website at https://www.elsevier.com/

Working together
to grow libraries in
developing countries

www.elsevier.com • www.bookaid.org

Publisher: Joe Hayton
Acquisition Editor: Tim Pitts
Editorial Project Manager: Charlotte Kent
Production Project Manager: Julie-Ann Stansfield
Cover Designer: Greg Harris

Typeset by SPi Global, India

Contents

**CHAPTER 13 mmWave Cellular Networks: Stochastic Geometry
Modeling, Analysis, and Experimental Validation......313**

W. Lu and M. Di Renzo

Contributors

B. Ai
Beijing Jiaotong University, Beijing, China

T.E. Bogale
Western University, London, ON, Canada

L. Dai
Tsinghua University, Beijing, China

V. Dyadyuk
CSIRO, Sydney, NSW, Australia

Z. Gao
Beijing Institute of Technology, Beijing, China

X. Gao
Tsinghua University, Beijing, China

K. Guan
Beijing Jiaotong University, Beijing, China

S.A. Hassan
National University of Sciences & Technology (NUST), Islamabad, Pakistan

C. Hu
Tsinghua University, Beijing, China

X. Huang
University of Technology Sydney, Sydney, NSW, Australia

K.S. Huq
Instituto de Telecomunicações, Aveiro, Portugal

Y. Jay Guo
University of Technology Sydney, Sydney, NSW, Australia

M. Nasiri Khormuji
Huawei Technologies Sweden AB, Stockholm, Sweden

P.-H. Kuo
Industrial Technology Research Institute (ITRI), Hsinchu, Taiwan

L.B. Le
University of Quebec, Montreal, QC, Canada

G. Lee
Korea Advanced Institute of Science and Technology, Daejeon, South Korea

G. Li
Beijing Jiaotong University, Beijing, China

W. Lu
CNRS/Université Paris-Saclay, Gif-sur-Yvette, France

Z. Mulk
National University of Sciences & Technology (NUST), Islamabad, Pakistan

S. Mumtaz
Instituto de Telecomunicações, Aveiro, Portugal

S.A.R. Naqvi
National University of Sciences & Technology (NUST), Islamabad, Pakistan

M. Di Renzo
CNRS/Université Paris-Saclay, Gif-sur-Yvette, France

J. Rodriguez
Instituto de Telecomunicações, Aveiro, Portugal

M.Z. Shakir
University of the West of Scotland, Paisley, United Kingdom

B. Su
National Taiwan University, Taipei, Taiwan

Y. Sung
Korea Advanced Institute of Science and Technology, Daejeon, South Korea

X. Wang
Western University, London, ON, Canada

Z. Wang
Tsinghua University, Beijing, China

T. Xie
Tsinghua University, Beijing, China

C.-P. Yen
Industrial Technology Research Institute (ITRI), Hsinchu, Taiwan

D. Zhang
Waseda University, Shinjuku, Japan

J.A. Zhang
University of Technology Sydney, Sydney, NSW, Australia

Preface

Cellular communication systems historically have undergone a revolution about once every decade (e.g., an entirely new standard), driven by a combination of market demands and technology advances. We are now thinking 5G at the exploratory research phase, with industry consensus hinting toward commercialization around 2020 with widespread adoption by 2025. The market is demanding that 5G should support much higher system capacity ($100-1000\times$) than current 4G systems, which already are close to the Shannon limit in point-to-point communication systems.

To address the 5G design targets, the information theory suggests that there are predominantly three key approaches to achieve several orders of magnitude increase in system capacity: (i) *ultra-dense networks*: the network densification already has been adopted in existing 4G wireless cellular networks, which is essentially known as small cell technology, and a denser network can further boost the network capacity; (ii) *large quantities of new bandwidth*: Migrating toward higher frequencies will release a large amount of bandwidth available to support higher throughput transmission. In particular, the millimeter-wave ("mmWave," for carrier frequencies of 30–300 GHz) communications can be the promising candidate; and (iii) *high spectrum efficiency*: by using a large number of antennas (100 or more), massive MIMO can significantly improve the spectrum efficiency by extensively harnessing the available space resources.

Individually, each of these approaches is expected to offer an order of magnitude or more increase in wireless system throughput compared to current 4G systems. Fortunately, these three solutions share a symbiotic convergence in many respects: the very short wavelength of mmWave frequencies is attractive for massive MIMO since the physical size of the antenna array can be reduced significantly, smaller cell sizes are appealing for short-range mmWave communications, while the large antenna gains provided by massive MIMO is helpful to overcome the severe path loss of mmWave signals. Indeed, if there is way to harness each of these approaches, then one could expect to achieve the 1000-fold increase in capacity for 5G. Taking a step in this direction, we already have mmWave technology that takes the fundamental design blueprints of MIMO technology, and pushes up the operating frequency to the mmWave band. This not only takes a step toward significantly enhancing the MIMO gain of the system, but also is able to somewhat compensate the severe path loss of mmWave frequencies to allow realistic small cell sizes to exist within coverage areas of 200 m. Therefore, a natural step would be to combine mmWave communications and massive MIMO in synergy to harness the properties of wide area throughput coverage on demand and localized small cell hotspots through mmWave technology, leading to the notion of "mmWave massive MIMO," which is expected to provide a wireless networking platform constituting a wireless network of small cells, providing very high speed transmission data rate.

Many challenges spanning the breadth of communications theory and engineering, however, must be addressed before mmWave massive MIMO becomes a reality. This book aims to be the first one to systematically address the major research challenges of mmWave massive MIMO starting from antenna design, physical layer design, MAC layer design, network layer design, to experimental testing. It explains fundamental requirements for deploying mmWave massive MIMO from an architectural, technical, and testbed point of view. It also updates the research community on the mmWave massive MIMO roadmap as well as new features emerging for consideration in 3GPP/IEEE. Moreover, this book explores the issues surrounding the idea of mmWave massive MIMO, and offers some potential solutions and research directions in the near future.

In the compilation of this book, the editors have tried their best to make it at the forefront of mmWave massive MIMO research area and standardization. This book serves as the first one to comprehensively discuss the promising technology of mmWave massive MIMO for 5G, and hopefully it also serves as a useful reference not only for postgraduates students to learn more in this evolving field, but also to stimulate mobile communication researchers toward taking further innovative strides in this field and make their contributions in the 5G era.

Shahid Mumtaz
Jonathan Rodriguez
Linglong Dai

Acknowledgments

This book aims to be the first one to systematically address the major research challenges of mmWave massive MIMO for future 5G wireless communications, and the authors hope that it also will serve as a source of inspiration for researchers to drive new breakthroughs on this topic. The inspiration for this book stems from the editors' experiences at the forefront of European research on mmWave and massive MIMO for future wireless systems that includes the ECOOP project (UID/EEA/50008/2013), an interdisciplinary research initiative funded by the Instituto de Telecomunicações (Portugal), the European collaborative research project SPEED-5G with the grant agreement no. 671705, the National High Technology Research and Development Program of China (grant no. 2014AA01A704), the International Science & Technology Cooperation Program of China (grant no. 2015DFG12760), the National Natural Science Foundation of China (grant no. 61571270), and finally, the Fundação para a Ciência e a Tecnologia (FCT—Portugal) with the grant reference number: SFRH/BPD/100362/201, that supported this work.

This work would not be complete, however, if it were not for those who contributed along the way. The editors first would like to thank all the collaborators that have contributed with chapters toward the compilation of this book, providing complementary ideas toward building a complete vision of the mmWave massive MIMO. Moreover, a heartfelt acknowledgment is due to the members of the 4TELL Research Group at the Instituto de Telecomunicações and the members of the Broadband Communications & Signal Processing Laboratory of Tsinghua University who contributed with useful suggestions and revisions.

About the Editors

Shahid Mumtaz has more than 7 years of wireless industry experience and is currently working as a Senior Research Scientist and Technical Manager at Instituto de Telecomunicações Aveiro, Portugal under 4TELL group. Prior to his current position, he worked as a Research Intern at Ericsson and Huawei Research Labs in 2005 in Karlskrona, Sweden. He received his MSc and PhD degrees in Electrical & Electronic Engineering from Blekinge Institute of Technology (BTH) Karlskrona, Sweden and University of Aveiro, Portugal in 2006 and 2011, respectively. His MSc and PhD degrees were funded by the Swedish government and FCT Portugal. He has been involved in several EC R&D Projects (*5GPP-Speed-5G*, CoDIV, FUTON, C2POWER, GREENET, GREEN-T, ORCALE, ROMEO, FP6, and FP7) in the field of green communication and next generation wireless systems. In EC projects, he holds the position of technical manager, where he oversees the project from the scientific and technical side, managing all details of each work package, which gives the maximum impact to the project's results for further development of commercial solutions. He has been also involved in two Portuguese funded projects (SmartVision & Mobilia) in the area of networking coding and the development of a system level simulator for the 5G wireless system. He has several years of experience in 3GPP radio systems research with experience in HSPA/LTE/LTE-A and strong track-record in relevant technology fields, especially physical layer technologies, LTE cell planning and optimization, protocol stack, and system architecture.

Dr. Shahid research interests lie in the field of architectural enhancements to 3GPP networks (i.e., LTE-A user plan & control plan protocol stack, NAS, and EPC), 5G related technologies, green communications, cognitive radio, cooperative networking, radio resource management, cross-layer design, Backhaul/fronthaul, heterogeneous networks, M2M and D2D communication, and baseband digital signal processing. He has published more than 60 publications in international conferences, journal papers, and book chapters. He is serving as a *Vice-Chair of IEEE 5G Standardization*. In 2012, he was awarded an "Alain Bensoussan" fellowship by the European Research Consortium for Informatics and Mathematics (ERCIM) to pursue research in communication networks for 1 year at the VTT Technical Research Centre of Finland. He is also editor of three books and served as guest editor for special issue in IEEE Wireless Communications Magazine and IEEE Communication Magazine. Recently, he was appointed as permanent associate technical editor for IEEE Communication Magazine, IEEE Journal of IoT, and the Elsevier Journal of Digital Communication and Network. He has been on the technical program committee of different IEEE conferences, including Globecom, ICC, and VTC, and chaired some of their symposia. He was the workshop chair in many conferences and recipient of the 2006 IITA Scholarship, South Korea. He is a senior IEEE member.

Dr. Linglong Dai received a BS degree from Zhejiang University in 2003, an MS degree (with the highest honor) from the China Academy of Telecommunications Technology (CATT) in 2006, and a PhD degree (with the highest honor) from Tsinghua University, Beijing, China, in 2011. From 2011 to 2013, he was a Postdoctoral Research Fellow with the Department of Electronic Engineering, Tsinghua University, where he has been an Assistant Professor since Jul. 2013 and then an Associate Professor since Jun. 2016. His current research interests include massive MIMO, millimeter-wave communications, multiple access, and sparse signal processing. He has published over 50 IEEE journal papers and over 30 IEEE conference papers. He also holds 13 granted patents. He has received the Outstanding PhD Graduate of Tsinghua University Award in 2011, the Excellent Doctoral Dissertation of Beijing Award in 2012, the IEEE ICC Best Paper Award in 2013, the National Excellent Doctoral Dissertation Nomination Award in 2013, the IEEE ICC Best Paper Award in 2014, the URSI Young Scientist Award in 2014, the IEEE Transactions on Broadcasting Best Paper Award in 2015, the IEEE RADIO Young Scientist Award in 2015, the URSI AP-RASC Young Scientist Award in 2016. He currently serves as an editor of the IEEE Transactions on Communications, an editor of the IEEE Communications Letters, a guest editor of the IEEE Journal on Selected Areas in Communications (the Special Issue on Millimeter Wave Communications for Future Mobile Networks), and co-chair of the IEEE Special Interest Group (SIG) on Signal Processing Techniques in 5G Communication Systems. He is an IEEE senior member. He is particularly dedicated to reproducible research and has made a large amount of simulation code publicly available (http://oa.ee.tsinghua.edu.cn/dailinglong/).

Jonathan Rodriguez received his master's degree in Electronic and Electrical Engineering and PhD from the University of Surrey (UK), in 1998 and 2004, respectively. In 2005, he became a researcher at the Instituto de Telecomunicações (IT), Portugal, where he was a member of the Wireless Communications Scientific Area. In 2008, he became a senior researcher where he established the 4TELL Research Group (http://www.av.it.pt/4TELL/), which targeted next generation mobile networks, with key interests on green communications, radio resource management, security, and electronic circuit design. He has served as project coordinator for major international research projects that include Eureka LOOP and FP7 C2POWER, while serving as technical manager for FP7 COGEU and FP7 SALUS. In 2009, he became an invited assistant professor at the University of Aveiro (Portugal), and associate in 2015. He is author of more than 300 scientific works, which include 8 book editorials. His professional affiliations include: senior member of the IEEE and chartered engineer (CEng) since 2013, and fellow of the IET (2015).

Jonathan Rodriguez is affiliated to the Instituto de Telecomunicações, Aveiro, Portugal.

Introduction to mmWave massive MIMO

1

S. Mumtaz*, J. Rodriguez*, L. Dai[†]

Instituto de Telecomunicações, Aveiro, Portugal* Tsinghua University, Beijing, China[†]

CHAPTER OUTLINE

Wireless communication systems historically have undergone a revolution about once every decade (e.g., an entirely new standard), driven by a combination of market demands and technology advances. We are now thinking 5G at the exploratory research phase, with industry consensus hinting toward commercialization around 2020 with widespread adoption by 2025 [1]. The market is demanding that 5G should support much higher system capacity ($100-1000 \times$) than current 4G systems, which already are close to the Shannon limit in point-to-point communication systems [2].

To address the 5G design targets, the information theory suggests that there are predominantly three key approaches to achieve several orders of magnitude increase in system capacity [2,3]: (i) *ultra-dense networks (UDNs)*: the network densification already has been adopted in existing 4G wireless cellular networks, which is essentially known as small cell technology, and a denser network can further boost the network capacity [4–6]; (ii) *large quantities of new bandwidth*: migrating toward higher frequencies will release a large amount of bandwidth available to achieve higher capacity. In particular, the millimeter-wave ("mmWave," for carrier frequencies of 30–300 GHz) communications can be the promising candidate [7,8]; and (iii) *high spectrum efficiency*: by using a large number of antennas (100 or more), massive multiple-input multiple-output (MIMO) can significantly improve the spectrum efficiency by extensively harnessing the available space resources [9,10].

Individually, each of these approaches is expected to offer an order of magnitude or more increase in wireless system capacity compared to current 4G systems.

mmWave Massive MIMO. http://dx.doi.org/10.1016/B978-0-12-804418-6.00001-7

Fortunately, these three approaches share a symbiotic convergence in many respects [3,6]: the very short wavelength of mmWave frequencies is attractive for massive MIMO because the physical size of the antenna array can be reduced significantly, smaller cell sizes are appealing for short-range mmWave communications, while the large antenna gains provided by massive MIMO is helpful to overcome the severe path loss of mmWave signals. Indeed, if there is a judicious way to harness all of these three approaches, then one could expect to achieve the 1000-fold increase in capacity for 5G. Taking a step in this direction, we already have mmWave technology that takes the fundamental design blueprints of MIMO technology, and pushes up the operating frequency to the mmWave band. This not only takes a step toward significantly enhancing the MIMO gain of the system, but also is able to somewhat compensate for the severe path loss of mmWave frequencies to allow realistic small cell sizes to exist within coverage areas of 200 m [7,8]. Therefore, a natural step would be to combine mmWave communications and massive MIMO in synergy to harness the properties of wide area coverage on demand and localized small cell hotspots through mmWave technology, leading to the notion of "mmWave massive MIMO" [3], which is expected to provide a wireless networking platform constituting a wireless network of small cells, providing very high-speed data rate.

Although the potential of mmWave massive MIMO is exciting, many challenges spanning the breadth of communications theory and engineering must be addressed before mmWave massive MIMO becomes a reality, e.g., the large number of antennas inevitably introduce very high or even unaffordable hardware complexity and power consumption [11], the estimation and feedback of the large dimension channel involve high overhead that significantly reduces the expected gain in spectrum efficiency [12]. This book aims to systematically address the major challenges of mmWave massive MIMO starting from antenna design, physical layer design, medium access control (MAC) layer design, network layer design, to experimental testing.

As the introduction of this book, this chapter is organized as follows. In Section 1.1, we briefly introduce the requirements of key capabilities for future 5G recently defined by the International Telecommunication Union (ITU) [1]. Then, in Section 1.2 we describe the potential 5G network architecture based on mmWave massive MIMO to meet the 5G harsh requirements, and the corresponding challenges for realizing mmWave massive MIMO are discussed in Section 1.3. Finally, we summarize the structure and key contributions of this book in Section 1.4.

1.1 REQUIREMENTS OF KEY CAPABILITIES FOR 5G

The 5G wireless network has not yet been standardized. In Sep. 2015, however, ITU defined the requirements of key capabilities for 5G by eight key performance indicators as shown in Fig. 1.1, where the baselines of current 4G are also compared [1].

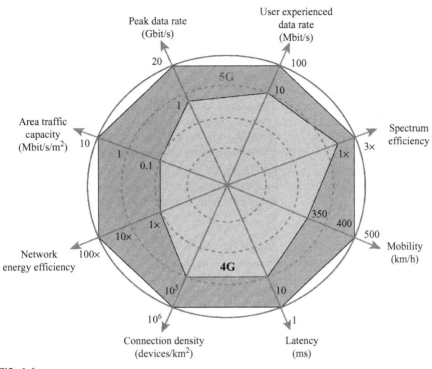

FIG. 1.1

Requirements of key capabilities for 5G [1].

The requirements of eight key capabilities for 5G are described below [1].

- *Peak data rate*: The peak data rate of 5G is expected to reach 10 Gbit/s, compared with 1 Gbit/s for current 4G. Under certain conditions and scenarios, 5G would support up to 20 Gbit/s peak data rate.
- *User experienced data rate*: 5G would support different user experienced data rates covering a variety of environments. For wide area coverage cases, e.g., in urban and suburban areas, a user experienced data rate of 100 Mbit/s is expected to be enabled, compared with 10 Mbit/s in current 4G systems. In hotspot cases, the user experienced data rate is expected to reach higher values (e.g., 1 Gbit/s in indoor scenarios).
- *Spectrum efficiency*: The spectrum efficiency of 5G is expected to be three times higher than that of 4G. The achievable increase in efficiency compared with 4G will vary between scenarios and could be higher in some scenarios (e.g., five times or higher in hot spots).
- *Mobility*: 5G is expected to enable high mobility up to 500 km/h with acceptable quality of service (QoS), while current 4G is mainly designed to support the mobility up to 350 km/h. This is envisioned in particular for high-speed trains.

- *Latency*: 5G would be able to shorten the over-the-air latency from 10 ms in current 4G systems to 1 ms, so 5G should be capable of supporting services with very low latency requirements.
- *Connection density*: 5G is expected to support a connection density about 10 times higher than that of 4G—up to $10^6/km^2$, for example, in massive machine-type communication scenarios.
- *Network energy efficiency*: The energy consumption for the radio access network of 5G should not be greater than 4G networks deployed today, while delivering the enhanced capabilities. Therefore, the network energy efficiency should be improved by a factor at least as great as the envisaged traffic capacity increase of 5G relative to 4G, e.g., about 100 times higher network energy efficiency.
- *Area traffic capacity*: 5G is expected to support 10 Mbit/s/m^2 area traffic capacity, for example, in hot spots, which is about 100 times higher than 0.1 Mbit/s/m^2 for 4G.

It should be pointed out that, while all key capabilities may to some extent be important for many use cases, the relevance of certain key capabilities might be significantly different, depending on the use cases/scenarios. For example, in the enhanced mobile broadband scenario, user experienced data rate, area traffic capacity, peak data rate, mobility, energy efficiency, and spectrum efficiency all have high importance, but mobility and the user experienced data rate would not have equal importance simultaneously in all use cases, e.g., a higher user experienced data rate in hotspots, but a lower mobility would be required than that in the wide area coverage case [1].

1.2 5G NETWORK ARCHITECTURE BASED ON mmWave MASSIVE MIMO

The information theory suggests that there are predominantly three key approaches [2,3] to address the 5G design targets presented above: UDNs, large quantities of new bandwidth (e.g., the mmWave band with large bandwidth), and high spectrum efficiency mainly contributed by massive MIMO.

Network densification via massive deployment of different types of cells such as macrocells, microcells, picocells, and femtocells is a key technique to enhance the network capacity, coverage performance, and energy efficiency. This cell densification approach already has been adopted in existing wireless cellular networks in particular LTE-Advanced (4G) systems, which essentially results in the multitier cellular heterogeneous networks (HetNets) [5]. Wireless HetNets also might comprise remote radio heads and wireless relays, which can further boost the network performance. It is anticipated that relaying and multihop communication will be among essential elements of the 5G wireless architecture, in contrast to the existing LTE-Advanced systems in which multihop communication is considered as an additional feature [4].

In general, radio resource management for HetNets plays a crucial role in achieving the benefits of this advanced network architecture [5]. Specifically, development of a resource allocation algorithm that efficiently uses radio resources, including bandwidth, power and antenna, while mitigating intercell and interuser interferences and ensuring acceptable QoS for active users, is one of the most critical issues. In addition, design and deployment of reliable backhaul networks that enable efficient resource management and coordination also are very important. It is believed that massive MIMO and mmWave technologies provide vital means to resolve many technical challenges of the future 5G HetNets [6], and they can be integrated seamlessly with the current networks and access technologies.

The deployment of the massive number of antennas at the transmitter and/or receiver can significantly enhance the spectrum and energy efficiency of the wireless network [9]. In a rich scattering environment, these performance gains can be achieved with simple beamforming strategies such as maximum ratio transmission or zero forcing [10]. Moreover, most of today's wireless systems operate at microwave frequencies below 6 GHz. The sheer capacity requirement of the next-generation wireless network inevitably would demand us to exploit the frequency bands above 6 GHz where the mmWave frequency ranging 30–300 GHz can offer huge spectrum, which is still underutilized [12]. Recent measurements at 28 and 38 GHz unveiled the potential for cellular communication at the lower end of the mmWave frequency: non-line-of-sight (LOS) communication is possible through reflections but with different path loss characteristics; rain losses are small for distances under 1 km [13]. Most importantly, as mmWave frequencies have extremely short wavelengths, it becomes possible to pack a large number of antenna elements in a small area, which consequently helps realize massive MIMO at both the base stations (BSs) and users.

In particular, mmWave frequencies can be used for outdoor point-to-point backhaul links or for supporting indoor high-speed wireless applications (e.g., high-resolution multimedia streaming). In fact, mmWave technologies already have been standardized for short-range services in IEEE 802.11ad. However, these frequencies have not been well-explored for cellular applications. Some potential reasons are the high propagation loss, penetration loss, rain fading, and the fact that these frequencies are absorbed easily or scattered by gases [13]. The massive deployment of small cells, such as pico and femto, in the future 5G HetNets renders the short-range mmWave technologies very useful. Therefore, the mmWave frequencies can be considered one of the potential technologies to meet the requirements of the 5G network. There are many possibilities to enable the 5G wireless HetNets based on mmWave massive MIMO. One such 5G network architecture is shown in Fig. 1.2, where we demonstrate how mmWave massive MIMO can be used in different parts of the future 5G wireless networks.

The architecture of Fig. 1.2 employs both mmWave and microwave frequencies. To determine the operating frequency bands of different communications in Fig. 1.2, several factors might need to be considered, such as the regulatory issues, application, channel, and path loss characteristics of various frequency bands. In general,

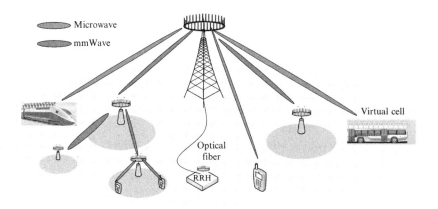

FIG. 1.2

Potential 5G network architecture based on mmWave massive MIMO.

path loss increases as the carrier frequency increases. This observation leads to the utilization of microwave frequencies for long-range outdoor communications. In mmWave frequency bands, different frequencies have distinct behaviors. For example, naturally occurring oxygen (O_2) absorbs electromagnetic energy to a much higher degree at 60 GHz than at 30–60 GHz. This absorption weakens (attenuates) 60 GHz signals over distance significantly; thus, the signals cannot reach faraway users. This makes the 60 GHz suitable for high data rate and secure indoor communications. Hence, selection of operating frequency depends on several factors such as application, different absorptions and blockages. Given these factors, however, there is a general consensus that mmWave frequency bands (30–300 GHz) can be useful for backhaul links, indoor, short range, and LOS communications.

Generally, the deployment of multiple antennas at the transmitter and/or receiver improves the overall performance of a wireless system. This performance improvement is achieved when the channel coefficients corresponding to different transmit-receive antennas experience independent fading. For a given carrier frequency, such independent fading channel is exhibited when the distance between two antennas is at least 0.5λ, where λ is the wavelength. Thus, for fixed spatial dimension, the number of deployed antennas increases as the carrier frequency increases, which consequently allows large number of antennas to be packed at mmWave frequencies. Moreover, the deployment of massive MIMO can be realized for different transportation systems such as trains and buses, even at microwave frequency bands because sufficient space is available to do so (i.e., as in Fig. 1.2). In recent years, three-dimensional (3D) and full-dimensional (FD) MIMO techniques have been promoted to increase overall network efficiency as they allow cellular systems to support a large number of users using multiuser MIMO techniques. Thus, massive MIMO and mmWave systems of the considered architecture also can be designed to be either 3D or FD. On the other hand, this architecture can support coordinated multipoint transmission where BSs are coordinated using either fiber or wireless backhauls. In addition, Fig. 1.2 incorporates a cell virtualization concept where the virtual cell can be defined either by the network (network centric) or the users (user centric), and also can be integrated as part of the cloud radio access network [14].

Moreover, massive MIMO is more likely to be standardized by 3GPP. To realize the advantages of using more antennas, the 3GPP RAN working groups recently have completed the standardization tasks of a new feature called FD-MIMO for LTE Release 13. In contrast to massive MIMO, mmWave is standardized by multiple international organizations, including ECMA, IEEE 802.15.3 Task Group 3c (TG3c), IEEE 802.11ad standardization task group, the Wireless HD consortium, and the Wireless Gigabit Alliance (WiGig) [15].

1.3 CHALLENGES FOR mmWave MASSIVE MIMO

The large available bandwidth and high spectrum efficiency certainly makes mmWave massive MIMO a promising choice to significantly improve overall system throughput for future 5G cellular networks. Such technology also would have advantages in terms of compact dimensions, energy efficiency, flexibility, and adaptivity, which make it ideal for a variety of picocell and femtocell applications. The following hurdles, however, must be considered carefully for the transition from microwave to mmWave frequencies.

- *Received power*: Let x be the transmitter and y be the receiver. Then, by neglecting small-scale variation, the received power $P_{xy}(d)$ in a microwave system is given by $P_{xy}(d) \approx P_x G_{xy} \chi \left(\frac{\lambda}{4\pi} \right)^2 d^{-\alpha}$, where $d = \|y - x\|$ is the distance between the transmitter and the receiver, λ denotes the wavelength, G_{xy} is the antenna gain and can be written as $G_x G_y$ for single input single output (SISO) case or point-to-point MIMO case, and shadowing is well modeled by an independent random variable (RV) χ. However, in a mmWave system, we have
  $P_{xy}(d) \approx P_x G_{xy} \chi(x, y, B) \left(\frac{\lambda}{4\pi} \right)^2 d^{-\alpha(x, y, B)}$, where G_{xy} can be at least two orders of magnitude smaller than in a microwave system, and the shadowing RV $\chi(x, y, B)$ is mainly caused by the blockage B, which is often very large and not independent of other RVs [13].
- *Total interference power*: Let I be the total interference power at the typical point (the origin o). For microwave, we have $I = \sum_x P_{ox}$ with each interferer at a distance $d = \|x\|$. Interference is dominated by a few nearby ones, and a large number of interferers also might cause a background interference floor. However, for mmWave, most interferers, including nearby ones, will be attenuated strongly by randomly aligned antenna gain patterns, or experienced

blockage. Thus, the interference distribution is not very distance dependent; instead it assumes an ON/OFF type of behavior.

- *Signal-to-interference-plus noise ratio (SINR)*: For microwave, the SINR changes slowly from cell center to cell edge. Thus, a user changes between "center" and "edge" status slowly. For mmWave, however, the SINR undergoes extremely rapid fluctuations, taking on an "On/Off" behavior mostly depending on the efficiency of the beamsteering, the presence/absence of blockage, and the random alignment of interfering beams [8].
- *Signal processing complexity*: Microwave usually has 1–8 antennas in a linear array, so channel state information (CSI) can be obtained easily at the receiver, which makes spatial multiplexing possible. However, mmWave usually adopts 8–256 antennas or even more in a 3D array, so CSI is very hard to get even just at the receiver [3,12]. For the same reasons, digital beamforming and spatial multiplexing require considerably higher hardware cost and power consumption [11].
- *Handover between BSs*: For microwave, handover between BSs usually is done at cell boundaries based on signal strength, and it might include load balancing considerations. For mmWave, however, cell "boundaries" are nearly irrelevant. Handovers will occur much more frequently because of blockage, beam alignments, and the high network density.

Based on these hurdles, great efforts must be endeavored to address the following challenges spanning the breadth of communications theory and engineering for mmWave massive MIMO.

- *Information theoretic issues of mmWave massive MIMO*

  The information theoretic analysis of mmWave massive MIMO systems can be facilitated significantly through random matrix theory (RMT), because the channel characteristics tend to become deterministic when the number of BS antennas goes to infinity, and the channels associated with different users tend to become orthogonal (which is termed as "asymptotic channel orthogonality") [9]. In addition, the impact of the system parameters on the system performance can be derived more easily on RMT. On the other hand, further developments in information theory for mmWave massive MIMO are needed in the regime of a large but finite number of BS antennas, and the trade-offs between spectral efficiency (bit/s/Hz) and energy efficiency (bit/Joule) should be quantified. Key differences from the previous analysis at cellular spectrum lie in the fact that fewer terminals might be supported per cell (because of the limited coverage area), and channel coherence time is smaller (because of more severe Doppler spread at higher frequencies). The effect of noisy CSI also has to be accounted for in various capacity bounds [16].
- *3D channel models for mmWave massive MIMO*

  Urban networks are definitely not flat, while the conventional approach from stochastic geometry treats all transmitters and receivers as living on a 2D plane [17]. Meanwhile, urban areas are projected to grow rapidly in population and density by 2050 according to a recent United Nations urbanization study [17], with about two-thirds of the world's population living in urban areas. At the same time, the number of wireless devices connected via the cellular network also is increasing rapidly and expected to accelerate, popularly referred to as the "internet of things" or machine-to-machine communications [18]. The implications on the communication environment of these two trends will be profound: more and more devices will be used in complicated urban environments. Although many principles of stochastic geometry can be extended to 3D urban areas, such extension is still challenging because of the nonhomogeneous distribution of users and infrastructure. The situation becomes more challenging at mmWave frequencies because of the sensitivity to blockages and the use of highly directional 3D beam patterns. Little is known currently, however, about the coverage and rates achieved in dense urban networks with planar deployments of infrastructure. New mathematical tools and models are required to analyze urban geometries and to realize the potential benefits of 3D beamforming in such environments.

- *Antenna and radio frequency transceiver architecture for mmWave massive MIMO*

    The assumptions underlying the theory of mmWave massive MIMO communications will drive many aspects of the antennas and radio frequency (RF) transceiver design, such as the need for high-efficiency antennas, low mutual coupling and RF channel crosstalk, stable and coherent local oscillator (LO) distribution, sharing of transceiver resources, modular and easily scalable architectures, tight RF and antenna integration, as well as the choice of carrier frequency, signal bandwidth, antenna directivity, antenna geometry, array size, and so on. Some initial work has investigated the impact of phase noise, mutual coupling, and unstructured statistical hardware errors, but such studies have been limited to models rather than actual transceiver implementations. No prior work has yet been reported about the design and fabrication of a complete mmWave massive MIMO transceiver, covering aspects such as antenna layout, array geometries, RF front-end architectures, local oscillator distribution, optimization of power dissipation, demodulation, baseband processing, sampling, and multichannel data aggregation [3].

- *Novel waveform and multiple access for mmWave massive MIMO*

    Radio access technologies for cellular mobile communications typically are characterized by multiple access schemes, e.g., frequency division multiple access (FDMA), time division multiple access (TDMA), code division multiple access (CDMA), and orthogonal frequency multiple access (OFDMA) for 1G, 2G, 3G, and 4G, respectively. OFDMA was a reasonable choice for achieving good system-level throughput performance for packet-domain services, but more advanced waveforms like filter-band multicarrier and multiple access schemes like nonorthogonal multiple access [19] could be more attractive for mmWave massive MIMO in future 5G systems.

- *Channel estimation techniques for mmWave massive MIMO*

    For very large number of BS antennas, channel estimation errors because of uncorrelated noise and interference are less problematic because the impact of such errors should vanish as the number of BS antennas goes to infinity [9,10]. The primary source of CSI errors is the limited channel coherence time, which limits the number of orthogonal training sequences that can be used and can lead to severe pilot contamination if the system is highly loaded with terminals. An interesting aspect of mmWave frequencies is the degree to which high path loss and near-LOS propagation would mitigate the pilot contamination effect [10,20]. Furthermore, for mmWave massive MIMO working in the frequency-division duplexing (FDD) mode, the downlink channel estimation is extremely challenging because of the prohibitively high pilot overhead, which is proportional to the number of BS antennas [21]. One might exploit the sparse nature or low-rank property of mmWave massive MIMO channels to design the compressive sensing (CS)-based channel estimation to significantly reduce the pilot overhead for channel estimation and feedback [21,22].

- *Modulation and energy efficiency for mmWave massive MIMO*

  Energy efficiency is one of the key advantages driving much of the interest in mmWave massive MIMO. The high peak-to-average power ratio (PAPR) in orthogonal frequency-division multiplexing, however, works against this advantage, and can impede good downlink performance. A recent study indicates that single-carrier modulation (SCM) with an equalization-free receiver [23] theoretically can achieve the near-optimal sum-rate performance in massive MIMO systems operating at low transmit-power-to-receiver-noise-power ratios, independent of the channel power delay profile. This is interesting for energy efficiency because SCM can be designed to have much better PAPR performance or even constant envelope waveforms. However, the results of Ref. [23] are based on the assumption of independent Rayleigh fading channels, which will not hold in the mmWave regime and could jeopardize the "equalization free" result. Furthermore, implementing SCM at mmWave frequencies implies very tight timing constraints on the order of a few nanoseconds or even less, which is nontrivial. Thus, the trade-offs involved with using SCM for mmWave massive MIMO need further study.

- *MAC layer design for mmWave massive MIMO*

  The ambition of mmWave massive MIMO entails many challenges on MAC because of the large number of antennas, special propagation features and hardware requirements. Therefore, there is need to design the proper MAC layer for mmWave frequencies that might differ from microwave networks in three main aspects: (1) control channel architecture; (2) initial access, mobility management and handover; and (3) resource allocation and interference management. Because the channel coherence time reduces with the carrier frequency, MAC layer decisions need to be made more frequently.

- *Interference management for mmWave massive MIMO*

  Several factors might mitigate interference management in a mmWave massive MIMO implementation [7]: (1) increased path loss at mmWave frequencies means limited coverage range and thus allows for higher frequency reuse (or even a unity frequency reuse factor); (2) shadowing effects because of

LOS or near-LOS propagation will reduce power leakage into adjacent cells; (3) sheer volume of spectrum available at mmWave frequencies leads to relaxation in frequency reuse constraints; and (4) beamforming with a massive MIMO array leads to narrow beam widths and high spatial selectivity, which limits the exposure of the signals to unintended receivers. Nonetheless, it is not difficult to envision scenarios in which small adjacent cells using mmWave massive MIMO have significant LOS overlap, and a unity frequency reuse factor is employed to maximize capacity. As such, there will be a need for interference mitigation in these networks. Potential approaches could exploit the large number of degrees of freedom available in a massive MIMO array to use subspace-based interference alignment methods [3].

- *New correlation sources for mmWave massive MIMO networks*

   In microwave networks, the spatial and temporal correlations of the interferences are introduced mostly by common locations of the transmitters and receivers [24]. In mmWave massive MIMO systems, physical blockage and high gain beamsteering introduces new sources of correlations, which might degrade the system performance and thus must be solved in an efficient way.

- *Mobility management in mmWave massive MIMO networks*

   Mobility is perhaps one of the most daunting problems for mmWave massive MIMO-based cellular networks, and the one most often raised by prominent skeptics, because the channel coherence time is much smaller at higher frequencies. As a user equipment (UE) moves around, the channel must be estimated frequently to adapt the narrow beamforming to its location. The feedback and latency requirements for this seem very difficult in FDD mode [8]; time-division duplexing (TDD) schemes also are limited by the channel coherence time. In the context of thrust, mobility can be viewed as the cause of decorrelation. Therefore, novel mobility models are needed for mmWave massive MIMO networks.

- *Backhaul transmissions for mmWave massive MIMO*

   MmWave massive MIMO can be beneficial in providing very high throughput backhaul in areas where it is too costly to install wire or fiber connections [3,8]. Massive antenna arrays could be arranged to relay information back and forth between cells or to nearby network hubs. Such an approach would have considerable advantages over microwave backhaul links that employ dish antennas and physical antenna alignment [25]. Cooperating massive antenna arrays could modify the beams adaptively to account for changes in the environment without a physical readjustment of the antenna arrays, and they could simultaneously communicate with multiple backhaul stations because their beams are electronically steerable.

- *System-level modeling in mmWave massive MIMO cellular networks*

   Overall, large available bandwidth and high spectrum efficiency certainly makes mmWave massive MIMO a tantalizing choice for future 5G cellular networks. Most of the existing works, however, focus on point-to-point channel models for mmWave. Therefore, there is need to consider system-level modeling in mmWave massive MIMO cellular networks.

- *Experimental demonstrations, tests and performance characterization for mmWave massive MIMO*

  The potential of mmWave massive MIMO wireless systems will be supported by real-time measurement campaigns. Ultimately, new channel models can be derived from measurement data. The impact of practical impairments (such as timing offset, frequency offset, and phase noise) on the overall system performance also has to be considered. Analytical techniques for the most important performance merits of mmWave massive MIMO systems (e.g., bit error rate (BER), outage probability, average rates, etc.) have to be derived.

- *Health and safety issues in mmWave communication*

  mmWave band is part of the RF spectrum, composed of frequencies between 30 and 300 GHz, corresponding to a wavelength from 1 cm to 1 mm. The photon energy of mmWave ranges from 0.1 to 1.2 millielectron volts (meV). Unlike ultraviolet, X-ray, and gamma radiation, mmWave radiation is nonionizing, and the main safety concern is heating of the eyes and skin caused by the absorption of mmWave energy in the human body [26]. Therefore, it is important to understanding how the propagation of mmWave affects the human body, as well as the inquiry of potential health effects related to mmWave exposures. Additionally, the current safety rules regarding RF exposure do not specify limits above 100 GHz [26], while the spectrum use of mmWave massive MIMO probably will move to these bands above 100 GHz, hence further investigations need to codify safety metrics at these mmWave frequencies.

- *Standardization and business model for mmWave massive MIMO*

  Despite the 3D channel modeling group in 3GPP, a separate group should be added to look at the inclusive aspects of mmWave massive MIMO and propose novel business models for operators to deploy this technology in their networks.

1.4 STRUCTURE AND CONTRIBUTIONS OF THIS BOOK

This book aims to address the major key research challenges of mmWave massive MIMO through 13 well-structured chapters starting from antenna design, physical layer design, MAC layer design, network layer design, to experimental testing. The structure of this book is shown in Fig. 1.3, and the contributions of this book can be summarized as follows.

Chapter 2 (SISO to mmWave massive MIMO): This chapter reviews the prior studies on SISO and MIMO and summarizes the emerging technology of mmWave massive MIMO. This chapter also describes the main aspects of channel modeling from microwave SISO to mmWave massive MIMO.

Chapter 3 (Hybrid antenna array for mmWave massive MIMO): This chapter introduces the massive hybrid array architecture, where antenna elements are grouped into multiple analog subarrays, and a single digital signal is received

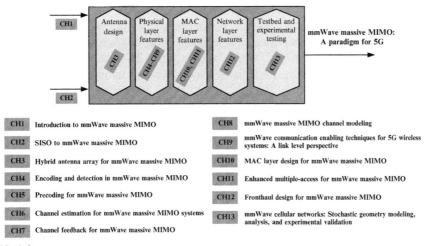

CH1   Introduction to mmWave massive MIMO

CH2   SISO to mmWave massive MIMO

CH3   Hybrid antenna array for mmWave massive MIMO

CH4   Encoding and detection in mmWave massive MIMO

CH5   Precoding for mmWave massive MIMO

CH6   Channel estimation for mmWave massive MIMO systems

CH7   Channel feedback for mmWave massive MIMO

CH8   mmWave massive MIMO channel modeling

CH9   mmWave communication enabling techniques for 5G wireless systems: A link level perspective

CH10 MAC layer design for mmWave massive MIMO

CH11 Enhanced multiple-access for mmWave massive MIMO

CH12 Fronthaul design for mmWave massive MIMO

CH13 mmWave cellular networks: Stochastic geometry modeling, analysis, and experimental validation

FIG. 1.3

Structure and contributions of this book.

from or sent to each subarray. It provides an affordable and spatially feasible solution for an mmWave massive array, and can achieve comparable performance with a fully digital array thanks to the temporal and spatial sparsity of mmWave propagation channels. This chapter starts by presenting the hybrid array architecture, highlighting two typical configurations of interleaved and localized arrays. It then presents four optional hardware implementations of this architecture. Array construction is discussed through quasi-Yagi antenna arrays and stacked patch antenna with perpendicular feed substrate. Two prototypes, developed by CSIRO and Samsung, also are introduced. After reviewing the hardware development, this chapter discusses signal processing techniques for hybrid arrays, with focus on angle-of-arrival estimation, single user LOS MIMO and its capacity, and spatial division multiple access techniques. Overall, this chapter shows that a massive hybrid array is a very promising technique for mmWave massive MIMO.

Chapter 4 (Encoding and detection for mmWave massive MIMO): This chapter considers a pilot reuse scheme for the uplink of a massive MIMO system. Subsequently, a lower bound on the throughput of the system has been derived, and is applicable for any number of antennas. In the supposed scenario, each user first transmits a training or pilot sequence to the BS, where a maximum ratio combining receiver decodes this message to determine the channel response, which then can be used to improve the average throughput. A hexagonal geometry of the system is considered wherein each cell contains uniformly distributed users and a fixed pilot sequence reuse pattern. The derived lower bound is shown to be limited by three types of interferences: intercell interference, intracell interference, and pilot contamination. These metrics have been used further to differentiate between the performance of ultra-high frequency and mmWave networks. Furthermore, a set of conditions, including the number of users, antennas, pilot reuse factor, and coherence period, is analyzed to achieve the lower bound of throughput. The results indicate that the reuse factor plays a critical role in the least achievable throughput. A minimum reuse factor is quantified for a given user density and coherence period.

Chapter 5 (Precoding for mmWave massive MIMO): This chapter briefly reviews the traditional digital precoding for MIMO systems and analog beamforming for mmWave communications. It then shows that they cannot be directly extended to mmWave massive MIMO systems. Thus, a novel precoding scheme called hybrid analog and digital precoding was investigated. The key idea of hybrid precoding is to divide the conventional digital precoder into a large-size analog precoder (realized by a large number of analog phase shifters) to increase the antenna array gain and a small digital precoder (realized by a small number of RF chains) to cancel interference. Thanks to the low-rank characteristics of mmWave channels in the spatial domain, a small digital precoder is sufficient to achieve the spatial multiplexing gain, which makes hybrid precoding enjoy the satisfying sum-rate performance with only a small number of RF chains. Finally, this chapter compares hybrid precoding with traditional digital precoding and analog beamforming, and highlights some other potential precoding schemes for mmWave massive MIMO.

Chapter 6 (Channel estimation for mmWave massive MIMO): This chapter reviews state-of-the-art channel estimation schemes for mmWave massive MIMO. At first, it introduces three key components in mmWave massive MIMO systems, including the sparse mmWave massive MIMO channels, hybrid MIMO transceiver structure with analog phase-shifter-network, and the receiver with one-bit analog-digital-converter. Then it discusses four kinds of channel estimation schemes for mmWave massive MIMO systems in detail, including the CS-based channel estimation, channel estimation with one-bit receiver, parametric channel estimation, and subspace estimation and decomposition-based channel estimation, and discusses their pros and cons. Finally, it briefly discusses how the existing channel estimation schemes initially proposed for conventional microwave massive MIMO can be tailored to mmWave massive MIMO.

Chapter 7 (Channel feedback for mmWave massive MIMO): This chapter discusses two different approaches of channel feedback for FDD mmWave massive MIMO, as well as a downlink precoding method that does not require CSI feedback. Firstly, by noting that a spatially correlated MIMO channel can have a sparse representation via certain linear transformations, a feedback load reduction method based on CS theory is introduced. Secondly, it describes a practical and efficient channel feedback mechanism utilizing multistage beamforming, where the pilot symbols are transmitted on angular-domain beams. Lastly, a downlink precoding technique based on angles-of-arrival knowledge of propagation paths instead of the full CSI is investigated in detail.

Chapter 8 (Channel models for mmWave massive MIMO): This chapter first surveys and highlights the main distinguished features of mmWave massive MIMO channels in terms of three layers: propagation mechanisms, static channel model, and dynamic channel model. This also chapter introduces and discusses the state-of-the-art of mmWave massive MIMO channel modeling and sounding, respectively. Even though the current available measurements or models for mmWave massive MIMO channels are still very limited, many researchers have been making efforts toward a general and standard channel model that can effectively guide system design for 5G and beyond.

Chapter 9 (MmWave communication enabling techniques for 5G wireless systems: A link level perspective): This chapter examines the performance of mmWave systems, which can be studied either at the network level or link level, with emphasis on link level performance. Link level performance of mmWave wireless system depends on a number of factors, including the transmission scheme (i.e., whether we employ beamforming, multiplexing or both), channel characteristics, and transmitted signal waveform structure.

Chapter 10 (MAC layer design for mmWave massive MIMO): This chapter provides an overview of user scheduling from SISO to massive MIMO based on the Rayleigh fading channel model suitable for rich scattering environments in the lower cellular band. Then, user scheduling for mmWave massive MIMO is introduced based on a channel model suitable for sparse mmWave propagation channels. The different properties between them are provided with basic theoretical results.

Chapter 11 (Enhanced multiple access for mmWave massive MIMO): This chapter presents a novel multiple access scheme for mmWave massive MIMO, hinging on semiorthogonal channel training and data transmission, such that in a given time-slot, some of the coordinated users appear orthogonal while the remaining users transmit nonorthogonally. The scheme is coined as semiorthogonal multiple-access (SOMA). SOMA allows transmitting additional data as well as scheduling more users for uplink transmission in contrast to the conventional TDD protocol, which results in a notable increase in the spectral efficiency of massive MIMO. The solution then is extended to generalized SOMA (GSOMA) by grouping the users, wherein the SOMA principle is applied per group and the pilot sequences within each group are mapped to the same time-frequency resources using code-division multiplexing. The proposed SOMA and GSOMA schemes are analyzed using information theoretic bounds to obtain the corresponding achievable aggregate throughput.

Chapter 12 (Fronthaul design for mmWave massive MIMO): This chapter discusses the fronthaul design for mmWave massive MIMO-based HetNets. The dense small-cell deployment requires the cost-effective fronthaul with high capacity to accommodate the 1000-fold incensement of cellular capacity for 5G. Against this background, the mmWave massive MIMO-based mesh fronthaul has attracted increasing interest in recent years. Compared with existing fronthaul solutions working at low frequency bands, the mmWave fronthaul is compatible with the ultradense deployment of small cells, because the fronthaul link can be short (typically 50–200 m) to mitigate the high path loss of mmWave signals and guarantee the LOS link. Moreover, by leveraging on the emerging mmWave massive MIMO technique, the mesh fronthaul topology can be facilitated easily to ease the installation and reduce the deployment cost. Additionally, the beamforming techniques of mmWave massive MIMO can make the mesh fronthaul more flexible and intelligent. This chapter presents a survey of existing fronthaul solutions, followed by the market requirement of the fronthaul network for future 5G HetNets. More importantly, it presents the concept of mmWave massive MIMO-based mesh network for fronthaul, where some issues, including antenna techniques, beamforming design, duplexing protocol, and in-band fronthaul, are discussed in detail.

Chapter 13 (MmWave cellular networks: Stochastic geometry modeling, analysis, and experimental validation): This chapter introduces a new mathematical framework for analysis of mmWave cellular networks. Its peculiarity lies in considering realistic path loss and blockage models, which are derived from the experimental data. The path loss model accounts for different distributions of LOS and non-LOS propagation conditions. The blockage model also includes an outage state that provides a better representation of the outage possibilities of mmWave transmission. By modeling the locations of the BSs as points of a Poisson point process, simple and exact integrals as well as approximated and closed-form formulas for computing the coverage probability and the average rate are obtained. With the aid of Monte Carlo simulations and using experimental data, the noise-limited approximation is shown to be sufficiently accurate for typical network densities. The noise-limited approximation, however, might not be sufficiently accurate for UDN deployments and for subgigahertz transmission bandwidths. For these such case, the analytical approach is generalized to take into account other-cell interference at the cost of increasing its computational complexity. The accuracy of the stochastic geometry modeling approach for mmWave cellular networks is investigated by explicitly taking into account realistic BS locations, building footprints, spatial blockages and channel propagations. We highlight that sufficiently dense mmWave cellular networks are capable of outperforming microwave cellular networks, in terms of coverage and rate.

REFERENCES

[1] ITU-R M.2083-0, IMT vision-framework and overall objectives of the future development of IMT for 2020 and beyond, 2015.
[2] R.W. Health, A. Lozano, T.L. Marzetta, P. Popovski, Five disruptive technology directions for 5G, IEEE Commun. Mag. 52 (2) (2014) 74–80.
[3] A.L. Swindlehurst, E. Ayanoglu, P. Heydasri, F. Capolino, Millimeter-wave massive MIMO: the next wireless revolution? IEEE Commun. Mag. 52 (9) (2014) 56–62.

[4] D. Feng, C. Jiang, G. Lim, L.J. Cimini, G. Feng, G.Y. Li, A survey of energy-efficient wireless communications, IEEE Commun. Surv. Tut. 15 (1) (2013) 167–178.

[5] E. Hossain, M. Rasti, H. Tabassum, A. Abdelnasser, Evolution towards 5G multi-tier cellular wireless networks: an interference management perspective, IEEE Wirel. Commun. Mag. 21 (3) (2014) 118–127.

[6] T.E. Bogale, L.B. Le, Massive MIMO and mmWave for 5G wireless HetNet: potential benefits and challenges, IEEE Veh. Technol. Mag. 11 (1) (2016) 64–75.

[7] L. Wei, R.Q. Hu, Y. Qian, G. Wu, Key elements to enable millimetre wave communications for 5G wireless systems, IEEE Wirel. Commun. 21 (6) (2014) 136–143.

[8] S. Hur, T. Kim, D. Love, J. Krogmeier, T. Thomas, A. Ghosh, Millimeter wave beamforming for wireless backhaul and access in small cell networks, IEEE Trans. Commun. 61 (10) (2013) 4391–4403.

[9] T.L. Marzetta, Noncooperative cellular wireless with unlimited numbers of base station antennas, IEEE Trans. Wirel. Commun. 9 (11) (2010) 3590–3600.

[10] F. Rusek, D. Persson, B.K. Lau, E.G. Larsson, T.L. Marzetta, O. Edfors, F. Tufvesson, Scaling up MIMO: opportunities and challenges with very large arrays, IEEE Signal Process. Mag. 30 (1) (2013) 40–60.

[11] X. Gao, L. Dai, S. Han, C.-L. I, R.W. Heath Jr., Energy-efficient hybrid analog and digital precoding for mmWave MIMO systems with large antenna arrays, IEEE J. Sel. Areas Commun. 34 (4) (2016) 998–1009.

[12] Z. Gao, L. Dai, Z. Wang, S. Chen, Spatially common sparsity based adaptive channel estimation and feedback for FDD massive MIMO, IEEE Trans. Signal Process. 63 (23) (2015) 6169–6183.

[13] T.S. Rappaport, S. Sun, R. Mayzus, H. Zhao, Y. Azar, et al., Millimeter wave mobile communications for 5G cellular: it will work! IEEE Access 1 (1) (2013) 335–349.

[14] C-RAN: The road towards green ran (white paper). China Mobile, December 2013, http://labs.chinamobile.com/cran/.

[15] E.G. Larsson, O. Edfors, F. Tufvesson, T.L. Marzetta, Massive MIMO for next generation wireless systems, IEEE Commun. Mag. 52 (2) (2014) 186–195.

[16] S. Furrer, D. Dahlhaus, Multiple-antenna signaling over fading channels with estimated channel state information, IEEE Trans. Inf. Theory 53 (6) (2007) 2028–2043.

[17] Q. Ye, B. Rong, Y. Chen, M. Al-Shalash, C. Caramanis, J.G. Andrews, User association for load balancing in heterogeneous cellular networks, IEEE Trans. Wirel. Commun. 12 (6) (2013) 2706–2716.

[18] S.-Y. Lien, K.-C. Chen, Y. Lin, Toward ubiquitous massive accesses in 3GPP machine-to-machine communications, IEEE Commun. Mag. 49 (4) (2011) 66–74.

[19] L. Dai, B. Wang, Y. Yuan, S. Han, C.-L. I, Z. Wang, Non-orthogonal multiple access for 5G: solutions, challenges, opportunities, and future research trends, IEEE Commun. Mag. 53 (9) (2015) 74–81.

[20] X. Zhu, L. Dai, Z. Wang, Graph coloring based pilot allocation to mitigate pilot contamination for multi-cell massive MIMO systems, IEEE Commun. Lett. 19 (10) (2015) 1842–1845.

[21] L. Dai, Z. Wang, Z. Yang, Spectrally efficient time-frequency training OFDM for mobile large-scale MIMO systems, IEEE J. Sel. Areas Commun. 31 (2) (2013) 251–263.

[22] W. Shen, L. Dai, B. Shim, S. Mumtaz, Z. Wang, Joint CSIT acquisition based on low-rank matrix completion for FDD massive MIMO systems, IEEE Commun. Lett. 19 (12) (2015) 2178–2181.

[23] A. Pitarokoilis, S. Mohammed, E. Larsson, On the optimality of single-carrier transmission in large-scale antenna systems, IEEE Wireless Commun. Lett. 1 (4) (2012) 276–279.

[24] R.K. Ganti, M. Haenggi, Spatial and temporal correlation of the interference in ALOHA ad hoc networks, IEEE Commun. Lett. 13 (9) (2009) 631–633.

[25] Z. Gao, L. Dai, D. Mi, Z. Wang, M.A. Imran, M.Z. Shakir, MmWave massive-MIMO-based wireless backhaul for 5G ultra-dense network, IEEE Wirel. Commun. 22 (5) (2015) 13–21.

[26] T. Wu, T.S. Rappaport, C.M. Collins, Safe for generations to come: considerations of safety for millimeter waves in wireless communications, IEEE Microw. Mag. 16 (2) (2015) 65–84.

SISO to mmWave massive MIMO

D. Zhang*, S. Mumtaz[†], K.S. Huq[†]

Waseda University, Shinjuku, Japan* Instituto de Telecomunicações, Aveiro, Portugal[†]

CHAPTER OUTLINE

2.1 OVERVIEW OF WIRELESS COMMUNICATION EVOLUTION

Wireless communications have experienced an evolution from analog communications systems (which is also called as 1G) to Global System for Mobile Communications (GSM, digital communications, also called 2G, where the Internet service is added in at the same time), third generation (3G, digital, supported data, packet switched, etc.), fourth generation (4G, wireless broadband, long-term evolution (LTE), and LTE-advanced (LTE-A)), and finally the fifth generation (5G), and so forth.

In wireless communications, the evolution of the transceiver has progressed from prior single-input signal-output (SISO) to multi-input multi-output (MIMO) and massive MIMO (mMIMO), and apart from the prior analog to digital communication

mmWave Massive MIMO. http://dx.doi.org/10.1016/B978-0-12-804418-6.00002-9

technologies, the channel modeling, channel coding, and also channel estimation technologies are proposed with the asymptotic sum rate or ergodic capacity analysis. As mmWave MIMO presents new challenges to the existing technologies, some new channel models are needed to cater to this potential technology in the near future.

In this chapter, we review the prior studies on SISO and MIMO and, finally, summarize the emerging technologies in mmWave mMIMO. As the main technology trends behind the evolution comprise the methods to boost the transmission rate and capacity using various versatile methods, in this chapter the main plotline is the channel modeling from SISO MIMO to mmWave mMIMO. As the knowledge of the authors and chapter structure are limited we maybe unable to convey the whole content. Therefore, if there is anything within this chapter that you want to discuss, further, please reach me at: di_zhang@fuji.waseda.jp.

2.2 THE CHANNEL MODELS BEHIND SISO, MIMO

In general, wireless channels can be summarized into the *large-scale propagation effects* [1] (*large-scale fading* [2]) and *small-scale propagation effects* [1] (*small-scale fading* [2]). Where the large-scale fading mainly comes from the path loss and shadowing effects related to large distances. In contrast, the small-scale fading originating from the variations over short distance, such as the constructive and destructive interferences from different paths.

2.2.1 WIRELESS PROPAGATION LOSS

Wireless propagation inherently consists of three different types, the ground wave propagation model, the sky wave propagation model, and the light of sight (LoS) propagation model [1, 3], which are shown in Fig. 2.1.

Mostly in cellular communications, we take the simple model of the LoS while doing the analysis [1, 3], although the reflection [1, 4], refraction [1, 4], and scattering [1, 4] effects (consist the shadowing) are also taken into consideration in different channel models. Whereas the ground wave propagation model is typically used for low and medium frequency radio communications [5], and the sky wave propagation model is mostly used for satellite and similar communications [4] that require transmission over thousands of kilometers.

The initial radio wave propagation research dates back to the early stages of work done by James Clerk Maxwell in 1864 and 1865, when, with his work titled "On Physical Lines of Force," "Electromagnetic Theory of Light," and "A Dynamical Theory of the Electromagnetic Field," he hypothesized on the existence of radio waves and that their transmission speed equaled the speed of light. Afterwards, the existence of radio waves was demonstrated by Heinrich Hertz in 1887, and in the following work of Oliver Lodge and Guglielmo Marconi who set the initial wireless communication systems. The first voice and music transmission was done by Feginald Fessenden in 1906 [1].

Although the exact electromagnetic wave propagation can be obtained by Maxwell's law due to different physical obstructions with reflections, reflections,

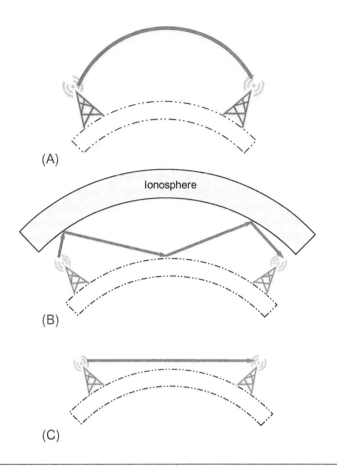

FIG. 2.1

The general wireless propagation models. (A) The ground wave model, (B) the sky wave model, and (C) the LoS model.

scattering, etc., this will present a greater challenge for computation complexity. Hence some approximative models are needed while studying the wave propagation. Generally, in wireless communications, particularly the cellular communications, the wave bands are restricted to 0.3–30 GHz (although 60 GHz is on the way with mmWave [6], yet it will result in different approximative models and this will be discussed in Section 2.4).

2.2.2 FREE SPACE PROPAGATION MODEL

The free space propagation model is based on the otherwise empty environment, which means that all of the other impacts, for example, reflection, reflection, etc., are empty, as shown in Fig. 2.2. While communication happens between the

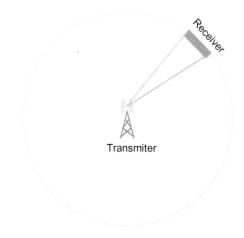

FIG. 2.2

The free space propagation model.

transmitter and receiver, the omnidirectional antenna wave transmission model can be adopted for the power attenuation, where the expression can be given as [1],

$$\frac{P_r}{P_t} = \left[\frac{\sqrt{G_l}\lambda}{4\pi d}\right]^2, \tag{2.1}$$

where $P_r, P_t, \sqrt{G_l}, \lambda, d$ represent the received power, transmitted power, product of the transmitter and receiver antenna field radiation patterns in free space transmission, wave length, and the distance between transmitter and receiver, respectively. As shown, the received power has a negative correlation with transmit distance square value d^2 and a positive correlation with wavelength λ. The received power can be further expressed in dBm as [1] while adopting the definition of dBm by,

$$x_{dBm} = 10\lg\left(\frac{x}{0.001}\right). \tag{2.2}$$

One can obtain the dBm expression as,

$$P_r = P_t + 10\lg\left(\left[\frac{\sqrt{G_l}\lambda}{4\pi d}\right]^2\right) = P_t + 10\lg(G_l) + 20\lg(\lambda) - 20\lg(4\pi) - 20\lg(d). \tag{2.3}$$

Additionally, the $20\log d$ pass loss are also used a lot for free space propagation loss in engineering, which is defined as,

$$P_r = P_t - 20\log\left(\frac{d}{d_\alpha}\right). \tag{2.4}$$

The standard experience expression of this is,

$$P_{loss} = 20\lg(f) + 20\lg(d) + 32.4, \tag{2.5}$$

where f yields the wave frequency. Take the assumption that the transmitted signal is a sinusoid wave with $\cos(2\pi ft)$, then the received signal at the receiver will be [2],

$$E(f,t,(r,\theta,\psi)) = \frac{\alpha_s(\theta,\psi,f)\cos\left(2\pi f\left(t-\frac{r}{c}\right)\right)}{r}, \tag{2.6}$$

where (θ, ψ) is the vertical and horizontal angles from the transmitter to the receiver and $\alpha_s(\theta, \psi, f)$ is the radiation pattern of the transmitting antenna at frequency f. Note that here both the transmitter and receiver antennas are assumed to be points whose size can be ignored. Further definition is given as,

$$H(f) = \frac{\alpha(\theta,\psi,f)e^{j2\pi f\frac{r}{c}}}{r}, \tag{2.7}$$

where $\alpha_s(\theta, \psi, f)$ is the radiation pattern of the transmitting antenna at frequency f as aforementioned. Then the received signal can be expressed as [2],

$$E(f,t,(r,\theta,\psi)) = \mathfrak{RE}[H(f)e^{j2\pi ft}]. \tag{2.8}$$

Generally, the free space propagation model is an ideal assumption that takes no shadowing effects. Mostly, it is used for calculation in a wide space transmission period with no obstructions. One can roughly estimate the received power at the receiver side or transmited power at the transmitter side while adopting the free space propagation model. But in the real world, especially the urban area with lots of building and tree obstructions, the moving conditions and other acting factors should be included, such as the shadowing, Doppler.

2.2.3 RAY TRACING

To step forward, the ray tracing model is used to approximate the urban or indoor environment that a wave transmitted in a space encounters multiple obstructions, which results in the reflection, diffraction, or scattering of the transmitted signal. Although the detail of the transmission period can be reflected by Maxwell's law, but due to its computational complexity, some approximation approach methods are provided here in the form of the two ray and general ray tracing models. Note that for the sake of compactness, we take the hypothesis that both the transmitter and receiver antennas are immobile.

Two ray tracing model

The two ray model is outlined in Fig. 2.3. As shown here, from the transmitter to receiver, there are two kinds of lights, the LoS and the reflected light, which is the transmitted wave reflected from the ground plane or other obstructions. The received signal from LoS can be given as aforementioned in Eq. (2.8), and the received signal from the reflection wave can be divided into two segments, x_1, x_2 here. While denoting $u(t)$ as the transmit signal at the transmitter, the received signal at the receiver side will be [1],

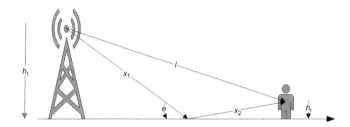

FIG. 2.3

The two ray model.

$$E_{2\text{-ray}} = \mathfrak{Re}\left\{\frac{\lambda}{4\pi}\left[\frac{\sqrt{G_1}u(t)e^{-j2\pi\frac{l}{\lambda}}}{l} + \frac{R\sqrt{G_r}u(t-\tau)e^{-j2\pi\frac{x_1+x_2}{\lambda}}}{x_1+x_2}\right]e^{j2\pi ft}\right\},\qquad(2.9)$$

where $\tau = \dfrac{x_1 + x_2 - l}{c}$ represents the time delay of the reflection, $\sqrt{G_1} = \sqrt{G_a G_b}$, $\sqrt{G_r} = \sqrt{G_c G_d}$ are the product of the transmit and receiver antenna field radiation patterns in the LoS section and reflection section relative to the x_1, x_2, respectively. In addition, R is the ground reflection coefficient. Whereas, as discussed in Tse and Viswanath [2], in the two ray model, the received signal is attenuated as r^2 with the received power attenuated as r^4. While denoting r as the distance from transmitter to the receiver. In particular, the difference between LoS and reflection paths goes to zero with r^{-1} as r increasing, and small relative goes to c/f with r approaching infinite. As demonstrated in the study by Goldsmith [1], the received power at the receiver side, while assuming a narrowband transmission, can be written as,

$$P_r = P_t\left[\frac{\lambda}{4\pi}\right]\left|\frac{\sqrt{G_1}}{l} + \frac{R\sqrt{G_r}e^{-j\Delta\phi}}{x_1+x_2}\right|^2,\qquad(2.10)$$

where $\Delta\phi = 2\pi(x_1 + x_2 - l)/\lambda$ yields the phase difference between LoS and reflection signals.

General ray tracing model
The general ray tracing model attempts to model the real transmission period in more detail. Therefore, the specific location properties (height, volume, etc.) of the transmitter and receiver, and also the obstructions should be included in detail. Specific expressions will vary in different environments of the general ray tracing model. Yet the LoS and reflection waves are still the dominant factors in the general ray tracing model overall, which is because the diffraction and scattering impacts are minor. While adopting the *Fresnel knife-edge diffraction model* [7] for diffraction wave lights and *Bistatic radar equation* for scattering lights in the general ray tracing model, adding the aforementioned LoS and reflection wave lights, the received signal at the receiver side reads [1],

$$E_{\text{genray}} = \mathfrak{Re}\left\{\left[\frac{\lambda}{4\pi}\right]\left[\frac{\sqrt{G_1}u(t)e^{j2\pi l/\lambda}}{l} + \sum_{i=1}^{N_r}\frac{R_{x_i}\sqrt{G_{x_i}}u(t-\tau_i)e^{-j2\pi x_i/\lambda}}{x_i}\right.\right.$$

$$+\sum_{j=1}^{N_d}\frac{4\pi}{\lambda}L_j(v)\sqrt{G_{d_j}}u(t-\tau_j)e^{-j2\pi(d_{j1}+d_{j2})/\lambda} \tag{2.11}$$

$$\left.\left.+\sum_{k=1}^{N_s}\frac{\sqrt{G_{sk}}\sigma_k u(t-\tau_k)e^{-j2\pi(s_{k1}+S_{k2})/\lambda}}{\sqrt{4\pi}s_{k1}S_{k2}}\right]e^{j2\pi ft}\right\}$$

where τ_i, τ_j, and τ_k represent the time delays of the given reflection, diffraction, and scattering lights with respect to the normalized LoS light delay. As previous studies show, $\sqrt{G_1}$, $\sqrt{G_{x_i}}$, $\sqrt{G_{d_j}}$, and $\sqrt{G_{sk}}$ yield the products of the transmit and receiver antenna field radiation patterns in the LoS, reflection, diffraction, and scattering sections, respectively.

2.2.4 EMPIRICAL MODELS

Okumura model

Some engineers have given various radio propagation models within different environments. The most common and a model that is still adopted a lot is the Okumura model [8], which suits the urban macrocells' radio propagation within 1–100 km, at frequency ranges from 150 to 1500 MHz, and for a transmit base station (BS) height 30–100 m. With these conditions in hand, its path loss after transmitting a distance d will be, by the Okumura model [1, 8, 9],

$$P_1(\text{okumura})\text{dB} = L(f,d) + A_\mu(f,d) - G(h_t) - G(h_r) - G_{\text{AREA}}, \tag{2.12}$$

where $L(f,d)$, $A_\mu(f,d)$, $G(h_t)$, $G(h_r)$, and G_{AREA} represent the free space path loss, median attenuation at distance d and frequency f, BS antenna high gain and receiver antenna height gain factors, and gain of the transmission environment, respectively.

Hata model

Similarly, Hata model [10, 11] is another widely used empirical propagation model in wireless communications that is also valid for the frequency range from 150 to 1500 MHz. Yet it provides a simpler closed-form expression of the urban environment. The expression of path loss with Hata model reads [1, 10, 11],

$$P_L(\text{Hata})_{\text{urban}}\text{dB} = 69.55 + 26.16\lg(f) - 13.82\lg(h_t) - \alpha(h_r) + (44.9 - 6.55\lg(h_t))\lg(d). \tag{2.13}$$

Note that here the parameters are the same as in the Okumura propagation model, but the $\alpha(h_r)$ takes different expressions in different environments. For small to medium sized cities, $\alpha(h_r)$ is given by,

$$\alpha(h_r)\text{dB} = 1.1\lg(f) - 0.7)h_r - (1.56\lg(f) - 0.8. \tag{2.14}$$

In contrast, for large cities with frequency larger than 300 MHz, it will be,

$$a(h_r)dB = 3.2(\lg(11.75h_r))^2 - 4.97. \tag{2.15}$$

On the other hand, for the suburban or rural environment, the Hata model path loss will be,

$$P_{L,\text{suburban}}(\text{Hata})dB = P_L(\text{Hata})_{\text{urban}}dB - 2\left[\lg\left(\frac{f}{28}\right)\right]^2 - 5.4 \tag{2.16}$$

and

$$P_{L,\text{rural}}(\text{Hata})dB = P_L(\text{Hata})_{\text{urban}}dB - 4.78[\lg(f)]^2 + 18.33\lg(f) - K, \tag{2.17}$$

where the K varies from 35.94 (countryside) to 40.94 (desert) [1]. One should notice that for a distance from transmitter to receiver larger than 1 m, the Hata propagation model is a good choice to model the transmission period, especially for first-generation wireless communications. Yet as the BS coverage area becomes smaller and smaller, because of BS densification and higher carrier frequencies, eventually it will be of no use. Some new propagation models should be provided, especially for the mmWave mMIMO systems. This will be discussed in the following sections. Furthermore, some other wireless wave propagation models are omitted, like the moving antenna model either in BS or the user side. One can refer to [1, Chapter 2] and Tse and Viswanath [2, Chapter 2].

2.2.5 SHADOWING EFFECTS

Shadowing effects are defined as the effects of received signal power fluctuations due to obstruction between the transmitter and receiver. Therefore, the signal changes as a result of the shadowing mainly come from reflection and scattering during transmittal. The shadowing effects will also result in the wave lights bending, that is, the transmission does not follow a straight line. Apart from *large-scale fading*, these effects are mostly summarized as the *small-scale fading*. To describe them, some models are needed. One of these is the log-normal shadowing. Which is, while denoting the ratio of transmitting power and receiving power as $\gamma = P_t/P_r$, one has [1],

$$P(\gamma) = \frac{\zeta}{\sqrt{2\pi}\sigma_{\gamma_{\text{dB}}}\gamma} \exp\left[-\frac{(10\lg\gamma - \mu_{\gamma_{\text{dB}}})^2}{2\sigma_{\gamma_{\text{dB}}}^2}\right], \tag{2.18}$$

where $\zeta = 10/ln10$, $\sigma_{\gamma_{\text{dB}}}$ yields the mean description of $\gamma_{\text{dB}} = 10\lg\gamma$, mostly obtained by the analytical model or empirical measures, and $\sigma_{\gamma_{\text{dB}}}$ is the standard deviation of γ_{dB}. The shadowing effects to the wireless propagation can be found in Fig. 2.4. While doing the analysis, the shadowing effects can be either incorporated into the path loss lights with LoS or using other propagation models, or can be separately calculated by the shadowing and path loss.

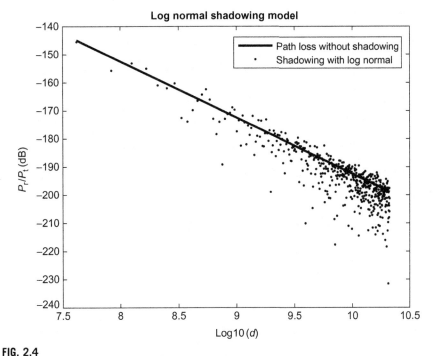

FIG. 2.4

The shadowing model with log normal.

2.3 FROM SISO TO MIMO

In this section, we will give the general evolution of the wireless communications from SISO to MIMO, mostly focus on the cellular coverage area and outage probability analysis, the channel models and therefore capacity and transmission rate analysis based on them.

2.3.1 OUTAGE PROBABILITY AND CELL COVERAGE AREA

As the literature, while combining the shadowing and path loss effects, power ratio of the received to transmitted power reads [1],

$$\eta_{P_r/P_t} dB = 10 \lg K - 10\phi \lg\left(\frac{d}{d_0}\right) - \gamma_{dB}, \tag{2.19}$$

where γ_{dB} yields a Guassian distributed random variable with mean zero and variance $\sigma^2_{\gamma_{dB}}$, and ϕ here is the path loss exponent. Thus while yielding the definition of outage probability that a value falls below a certain threshold, we can get the expression of the power outage probability of the path loss and shadowing by [1],

$$P_{\text{out}}(P_r \leq P_{th}) = 1 - Q\left(\frac{P_{th} - \left(P_t + 10\lg K + 10\phi\lg\left(\frac{d}{d_0}\right)\right)}{\sigma_{\gamma_{dB}}}\right), \tag{2.20}$$

where Q denotes the Q functions [12], and the conversion between the Q function and complementary error function can be given as [12],

$$Q(x) = \frac{1}{2}\text{erfc}\left(\frac{x}{\sqrt{2}}\right). \tag{2.21}$$

Outage probability is defined as the point at which the receiver power value falls below the threshold (where the power value relates to the minimum signal to noise ratio (SNR) within a cellular), one can say that the receiver is out of the range of BS in cellular communications. Hence the cellular coverage area range can be calculated with this expression. Because of the shadowing effects, the cellular coverage area is not always a sphere shape. Yet in theoretical analysis, we adopt the sphere shape or even a circular area for BS coverage area while omitting the height of BS for simplicity (in some cases, the hexagon area assumption is also adopted). Thus, with received power outage probability analysis, the cellular coverage area can be calculated as,

$$
\begin{aligned}
C &= \frac{1}{\pi R^2}\int_0^{2\pi}\int_0^R P(P_r \leq P_{th})r\,dr\,d\theta \\
&= \frac{1}{\pi R^2}\int_0^{2\pi}\int_0^R P(1 - P_{\text{out}}(P_r \leq P_{th}))r\,dr\,d\theta \\
&= \frac{1}{\pi R^2}\int_0^{2\pi}\int_0^R Q\left(\frac{P_{th} - \left(P_t + 10\lg K + 10\phi\lg\left(\frac{d}{d_0}\right)\right)}{\sigma_{\gamma_{dB}}}\right)r\,dr\,d\theta.
\end{aligned}
\tag{2.22}
$$

2.3.2 RAYLEIGH AND RICIAN CHANNEL MODELS

The Rayleigh and Rician channel models are two widely used channel models in wireless communications. The Rician channel assumes that the transmission paths from the transmitter to receiver are comprised of the dominant LoS path and other scattering paths, whereas the Rayleigh channel consists of scattering channels from the transmitter to receiver. While assuming the phase on each path i is $2\pi ft_i$ modulo 2π, where $ft_i = d_i/\lambda$, as in cellular communications, we can always have $d_i \gg \lambda$. In this case, one can say that along the path, the phases uniformly vary between 0 and 2π, whereas different phases are independent. While adding all of the paths together, denoting $\alpha_i(t)$, $t_i(t)$ as the attenuation and propagation delay on path i within time t, the tap gain will be [2],

$$h_1[m] = \alpha_i\left(\frac{m}{W}\right)e^{-j2\pi ft_i(m/W)}\sin\left[l - t_i\left(\frac{m}{W}\right)W\right]. \tag{2.23}$$

This can be modeled with the widely used *circular symmetric complex random variable*, where each tap is the sum of a large number of small independent circular symmetric random variables [2]. Generally, we use the real part of the tap gain to model the zero mean Gaussian random variable. One should be aware that this circular, symmetric, complex, random, variable assumption is used a lot, especially in MIMO system. In the mMIMO system we also adopt this assumption while doing the spectrum efficiency (SE) or energy efficiency (EE) analysis with random matrix theory [13] where the tap gains are assumed to follow the $\mathcal{CN}(0,\sigma_l^2)$ there. This is the so-called Rayleigh fading. The density of the Rayleigh random variable is [2],

$$\frac{x}{\sigma_l^2}\exp\left\{\frac{-x^2}{2\sigma_l^2}\right\}, \tag{2.24}$$

and the density of a squared magnitude is $h_l[m]^2$ [2],

$$\frac{1}{\sigma_l^2}\exp\left\{\frac{-x}{\sigma_l^2}\right\}. \tag{2.25}$$

In contrast, the Rician channel models the channel with a specular path and several scattered paths, where the specular path is large and has a known magnitude [2]. Consider two random Gaussian variables X, Y, where X has a nonzero mean (say m) and Y has a zero mean, both X and Y have equal variance as η^2. Then the transformation,

$$Z = \sqrt{X^2 + Y^2} \tag{2.26}$$

is the Rician distributed. The tap gain of the Rician channel can be given as [2],

$$h_l[m] = \sqrt{\frac{k}{k+1}}\sigma_1 e^{j\theta} + \sqrt{\frac{1}{k+1}}\mathcal{CN}(0,\sigma_l^2), \tag{2.27}$$

where k (K-factor) denotes the ratio of energy in LoS path and scattered paths, which can be given as,

$$k = \frac{m^2}{2\eta^2}. \tag{2.28}$$

Here the first term yields the LoS path with uniform phase θ and the second term the scattering paths, independent of θ.

2.3.3 CAPACITY AND TRANSMISSION RATE ANALYSIS

General analysis of SISO system

Information theory can be rooted back to the Shannon theory invented by Claude Shannon in 1948, to characterize the channel limits in reliable communications. Shannon theory proves that we can approach the maximum transmission rate in a given channel while reducing the error rate with sophisticated encoding methods. The capacity can be derived using the Shannon theory and entropy analysis.

Let us consider a typical communication system, and its signal arriving at the receiver reads,

$$y = h * x + n, \tag{2.29}$$

where h, x, and n yield the channel matrix, transmit signal matrix, and channel noise matrix, respectively. Here * denotes the convolution operation. Typically, the channel noise is assumed to be an additive Gaussian white noise (AWGN) with $\mathcal{CN}(0, \sigma^2)$, where we can always obtain the transmitted signal perfectly, on condition that the noise variance is zero. Or else, some methods should be tackled. In addition, in the case of SISO, the matrix takes the one-dimensional condition. Which results in the capacity of SISO can be derived using the Gaussian channel assumption with power constraint as,

$$C = \max_{p(x):EX^2 \leq P} I(X;Y), \tag{2.30}$$

where $I(X;Y)$ represents the mutual information, which can be obtained by the entropy analysis. In addition, for an X, its entropy is defined as,

$$I(x) = H(x) = \sum_{i=1}^{N} p_i \log_2 \left[\frac{1}{P_i} \right] = \sum_{x \in X} p(x) \log_2 \left[\frac{1}{p(x)} \right] = -\sum_{x \in X} p(x) \log_2[p(x)], \tag{2.31}$$

where $X = [x_1, ..., x_i, ..., x_N]$ is the random variable set, and $p(x) = [p_1, ..., p_i, ..., p_N]$ is the probability mass function set. The mutual information by $I(X;Y)$ can be gained by the following methods,

$$I(X;Y) = H(X) - H(X|Y) = H(Y) - H(hX + N|X) = H(Y) - H(N). \tag{2.32}$$

In general, we use the $I(X;Y) = H(Y) - H(N)$ to calculate the capacity of wireless systems, where the differential entropy can be calculated by $H_d(Y) = \log_2(\pi e \sigma_y^2)$. The received average power is,

$$\sigma_y^2 = E[Y^2] = E[(hX + N)(hX + N)*] = \sigma_x^2 |h|^2 + \sigma_n^2. \tag{2.33}$$

Thus the capacity will be,

$$\begin{aligned} C &= \max_{p(x):EX^2 \leq P} I(X;Y) = H_d(Y) - H_d(N) = \log_2(\pi e \sigma_y^2) - \log_2(\pi e \sigma_n^2) \\ &= \log_2(\pi e \sigma_x^2 |h|^2 + \sigma_n^2) - \log_2(\pi e \sigma_n^2) = \log_2\left(1 + \frac{\sigma_x^2}{\sigma_n^2}|h|^2\right) \\ &= \log_2\left(1 + \frac{P_t}{\sigma_n^2}|h|^2\right). \end{aligned} \tag{2.34}$$

While denoting the SNR as $\rho = \dfrac{P_t}{\sigma_n^2 |h|^2}$. Finally, the capacity of the SISO system will be,

$$C = \max_{p(x):EX^2 \leq P} I(X;Y) = \log_2(1 + \rho) = \log_2\left(1 + \frac{P_t}{\sigma_n^2}|h|^2\right). \tag{2.35}$$

In contrast, in the MIMO system, this expression will yield the transmission rate of each user under the equal power allocation method.

The MIMO system

In the MIMO system, it is known that system capacity can be boosted without extra bandwidth or transmit power by simply increasing the antenna number at the transmitter and receiver sides. With transmitter antenna number N and receiver antenna number M, the received signal reads,

$$Y = Hx + n. \tag{2.36}$$

As we can see here, within all of the channel, transmitted signal x and noise n are tackled with a matrix. Taking the average power allocation and equal Rayleigh channel matrix assumption, the capacity of the MIMO system can be described with the function as,

$$C = \sum_{i=1}^{M} B_i \log_2 \left(\det \left[I_N + \frac{P_{t_i}}{M\sigma_n^2} HH^H \right] \right) = \sum_{i=1}^{M} B_i \log_2 \left(\det \left[I_N + \frac{\rho}{M} HH^H \right] \right), \tag{2.37}$$

where B_i, $\det$, I_N yield the bandwidth for each user, the determinant of a matrix, and the $M \times N$ identify the matrix, respectively, and ρ here is the average SNR in each channel, H^H denotes the conjugate transpose of a matrix H. In addition, the channel matrix can be decomposed as,

$$H = \begin{bmatrix} h_{1,1} & h_{1,2} & \cdots & h_{1,N} \\ h_{2,1} & h_{2,2} & \cdots & h_{2,N} \\ \cdots & \cdots & \ddots & \cdots \\ h_{M,1} & h_{M,2} & \cdots & h_{M,N} \end{bmatrix}. \tag{2.38}$$

Each of the entry in Rayleigh channel matrix, as aforementioned, can be expressed as,

$$h_{i,j} = \alpha + j\beta = \sqrt{\alpha^2 + \beta^2} e^{j \arctan \frac{\beta}{\alpha}} = |h_{i,j}| e^{j\phi_{i,j}}. \tag{2.39}$$

where α and β are random distributed variables and $|h_{i,j}|$ is a Rayleigh distributed random variable. While further assuming the Rayleigh channels are i.i.d. complex zero mean and unit entries, expression of $h_{i,j}$ can be written as,

$$h_{i,j} = \text{Normal}\left(0, \frac{1}{\sqrt{2}}\right) + j \, \text{Normal}\left(0, \frac{1}{\sqrt{2}}\right), \tag{2.40}$$

where Normal yields the normal distribution. Thus we can conclude that $h_{i,j}$ obeys the χ_2^2 distribution with 2 degrees of freedom and $E[h_{i,j}] = 1$.

Since the determination expression of $I_N + \frac{\rho}{M} HH^H$ yields the eigenvalue, thus one can mainly focus on the eigenvalue of HH^H while calculating the capacity. If one recalls that for two matrix $\{A_{p \times q}, B_{q \times p}, p \leq q\}$, eigenvalue of $AB_{p \times p}$ and $BA_{q \times q}$ are equal, thus we conclude that the eigenvalue of HH^H and $H^H H$ are the same and that $\lambda = [\lambda_1, \lambda_2, ..., \lambda_{\min(N,M)}]$ yields the nonzero eigenvalues (as zero eigenvalue has no contribute to the capacity effect) of the B with,

$$B = \begin{cases} HH^H, M < N, \\ H^H H, M > N. \end{cases} \tag{2.41}$$

In this case, the MIMO capacity can be written as,

$$C = \sum_{i=1}^{M} B_i \log_2 \left(\det \left[I_N + \frac{\rho}{M} HH^H \right] \right) = \sum_{i=1}^{M} B_i \log_2 \left(\prod_{i=1}^{\min(N,M)} \left[1 + \frac{\rho}{M} \lambda_i^2 \right] \right). \qquad (2.42)$$

On the other hand, singular value decomposition (SVD) is used a lot in MIMO capacity analysis. In SVD, the channel matrix is decomposed by,

$$H = UDV^H, \qquad (2.43)$$

where D is $M \times N$ matrix, whose nonzero norm by the diagonal entries comprise the singular values of H, and U and V are $M \times M$ and $N \times N$ unitary matrix, respectively. Then Eq. (2.36) can be further written as,

$$Y = UDV^H x + N. \qquad (2.44)$$

While premultiplied by U^H, and with $\widetilde{Y} = U^H Y$, $\widetilde{x} = xV^H$, $\widetilde{N} = U^H N$, we have,

$$\widetilde{Y} = D\widetilde{x} + \widetilde{N}. \qquad (2.45)$$

Adopting λ_i as the eigenvalue of ith channel matrix, the ith receiver signal will be,

$$\widetilde{Y}_i = \lambda_i \widetilde{x}_i + \widetilde{N}_i, \qquad (2.46)$$

where the covariance and trace are,

$$R_{\widetilde{y}\widetilde{y}} = U^H R_{yy} U, R_{\widetilde{x}\widetilde{x}} = V^H R_{yy} V, \qquad (2.47)$$

$$tr(R_{\widetilde{y}\widetilde{y}}) = tr(R_{yy}), tr(R_{\widetilde{x}\widetilde{x}}) = tr(R_{xx}), tr(R_{\widetilde{n}\widetilde{n}}) = tr(R_{nn}). \qquad (2.48)$$

With perfect channel state information (CSI) and equality power allocation mechanism, we can also have the expression as Eq. (2.42) [14, 15].

2.4 FROM MIMO TO mMIMO

The transmission rate balloon is on the way of 5G, where massive MIMO together with other proposals (small cell, HetNets, mmWave, etc.) are believed to be the potential solutions to boost this trend. In massive MIMO, it is proved that with perfect CSI and completely eliminated pilot contamination, one can always achieve better SNR while simply increasing the transmitter antenna number [16, 17]. Yet another side effect brought about by massive MIMO is that of cellular densification; similar to Cooper's law, the capacity increase will mostly be due to a denser and denser cell deployment [18]. The emerging topics in massive MIMO are EE and low latency backhauls to sustain the even faster transmission speed, etc. With more complex user technology joining the Internet service in the 5G era, maybe a total redesign of the network architecture will be needed, such as the proposed information centric networking (ICN) [19], software defined radio (SDR) [20] or named network functions virtualization (NFV). In addition, devices to devices (D2D) communications [21], full duplex communication mechanism [22, 23], and collaboration between different users and cellulars [16] are also believed to be indispensable elements of 5G.

2.4.1 EVEN FASTER TRANSMISSION SPEED

Here for the sake of compactness, we take the assumption that the bandwidth number is equal to one, to study about the antenna number effects to the transmission speed. Furthermore, the study is limited to a one cellular coverage area. We will discuss the study of mMIMO under multicells in the following parts. As demonstrated in the study by Marzetta in [17], one has,

$$\sum_{i=1}^{\min(M,N)} \lambda_i^2 = \begin{cases} tr(\boldsymbol{HH}^H), M < N, \\ tr(\boldsymbol{H}^H\boldsymbol{H}), M > N. \end{cases} \tag{2.49}$$

Here tr denotes the trace of a matrix, where the worse case is that all but one of the singular value is equal to zero. This means only one channel is suitable for transmission. In contrast, the best case is that all of the singular values are equal but not zero. In this case, the bound of capacity of massive MIMO system can be given as, while antenna is growing large under the condition that transmitter antenna number is much greater than the receiver antenna number [17], can be given as,

$$\log_2\left(1 + \frac{\rho tr(\boldsymbol{HH}^H)}{N}\right) \leq C \leq \min(N,M) \times \log_2\left(1 + \frac{\rho tr(\boldsymbol{HH}^H)}{N \min(N,M)}\right). \tag{2.50}$$

While further assuming the coefficient of the magnitude of propagation is normalized to one, we have $tr(N, M) = NM$ [17]. Thus the above functions can be simplified as,

$$\log_2(1 + \rho M) \leq C \leq \min(N,M) \times \log_2\left(1 + \frac{\rho \max(N,M)}{N}\right). \tag{2.51}$$

In addition, as shown in the study by Marzetta [17], under low SNR conditions, simply increasing the transmitter antenna number has no effect on the capacity performance with the function,

$$C_{\rho \to 0} \approx \frac{\rho tr(\boldsymbol{HH}^H)}{N \ln 2} \approx \frac{\rho M}{\ln 2}. \tag{2.52}$$

This mostly happens at the edge of a cellular coverage area. Yet with better SNR conditions, as the study demonstrates, while keeping the transmitter antenna number growing and the receiver antenna constant, as a consequence, we have [24],

$$\left(\frac{\boldsymbol{HH}^H}{N}\right)_{N \gg M} \approx \boldsymbol{I}_M, \tag{2.53}$$

where the achievable rate is [17],

$$C_{N \gg M} \approx \log_2 \det(\boldsymbol{I}_M + \rho \boldsymbol{I}_M) = M \log_2(1 + \rho). \tag{2.54}$$

On the other hand, while keeping the transmitter antenna number constant and increasing the receiver antenna number, under the propagation matrix are asymptotically orthogonal assumptions, so we have [17],

$$\left(\frac{\boldsymbol{HH}^H}{N}\right)_{M \gg N} \approx \boldsymbol{I}_N,$$

$$\det(\boldsymbol{I} + \mathrm{HH}^H) = \det(\boldsymbol{I} + \boldsymbol{H}^H \boldsymbol{H}). \tag{2.55}$$

Thus the achievable rate is [17],

$$C_{M \gg N} \approx \log_2 \det\left(\boldsymbol{I}_N + \frac{\rho}{N}\boldsymbol{H}^H \boldsymbol{H}\right) \approx N \log_2\left(1 + \frac{\rho M}{N}\right). \tag{2.56}$$

In addition, in multicell conditions, with the antenna number growing large (with $\lim \sup_N K/N < \infty$) and all of the impacts added in, such as noise, imperfect CSI, interference, pilot contamination, as the study shows in [25], no matter what kind of beamforming technique, the achievable capacity will approach,

$$R_\infty = \log_2(1 + \gamma_\infty) = \log_2\left(1 + \frac{1}{\alpha(\bar{L} - 1)}\right), \tag{2.57}$$

where γ_∞ yields the SINR with infinite antenna number, $\alpha \in (0, 1]$ is the intercell interference factor, and $\bar{L} = 1 + (\alpha - 1)$ with L representing the cell number.

2.4.2 ENERGY EFFICIENCY

As higher transmission speed needs even higher transmitter power to boost up its transmission, under 5G background, EE becomes another issue that attracts lots of attention worldwide. Nowadays, the focus on EE is confined to the component selection, network coding, and also new network architecture design.

In a component selection study, the cellular zooming mechanism based on mMIMO was proposed [26], where the cellular could be zoomed in to cover more area or zoomed out to save energy, on condition that there were less active users in that area. And the transmission in low user number areas would be handed over to the neighboring cells with coordinated multipoint transmission/reception (CoMP) technology. Under this system, the antenna selection mechanism [27], radio frequency (RF) chain selection mechanism [28], and also a combination of these [29, 30], and other components are proposed in mMIMO EE (with respect to the EE definition) with an achievable transmission rate divided by the consumed energy to sustain the rate:

$$\eta_{\mathrm{EE}} = \frac{C}{P}, \tag{2.58}$$

where C and P denote the achievable capacity (or transmission rate) and consumed power, respectively.

In network coding for EE, most of the literature is based on the two way or multi-way ray network. For instance, the analog network coding (ANC) and physical-layer network coding (PNC) is studied using the two-way wireless relay system for EE by Choi and To [31]; a two-stage maximum energy efficiency (MEE) method is proposed by Zhao et al. [32]. In addition to these, the joint design of network coding and MIMO package scheduling is studied by Zhao and Yang [33]. One can find other similar work in network coding.

Most of the new network architecture for EE under the 5G background is based on the cloud radio access network (C-RAN), whereas the study from China Mobile (CMCC) by their C-RAN white paper, with C-RAN's shared machine room, shows that more energy can be saved with less energy consumption from the air conditioner, grids, etc. In this regard, literature can be found such as Chen [34], where a simple but efficient precoding scheme is proposed to reduce the computation complexity of cooperative transmission, thus lowing the associated power consumption. And an energy consumption model for C-RAN is further studied for capturing the energy consumption of centralized BBU resources, plus an algorithm that optimizes the BBU resource allocation is proposed by Khan et al. [35]. Other literature concerning network architecture is based on the heterogeneous network (HetNet), where not only the C-RAN, but also D2D and other deployments are taken into comprehensive consideration.

2.5 EMERGING TOPICS IN mmWave mMIMO

Other than the aforementioned discussions, mmWave mMIMO is being used to explore new unlicensed frequency resources to boost up the 5G transmission rate requirements. Typically, mmWave is defined as the band ranging from 30 to 300 GHz, both from IEEE and ITU [36]. Most of the recent studies of mmWave mMIMO for 5G are focused on the 28 GHz, 38 GHz, and the 60 GHz band, and also the E-band (71–76 GHz and 81–86 GHz) [36].

However, as the mmWave mMIMO work at even higher frequencies, this brings in different challenges to the research for both indoor and outdoor environments. For instance, even higher frequency in turns means much shorter wavelengths, so that mmWave cannot transmit over a long distance as was the case for prior licensed frequencies. Also as the conditions that mmWave mMIMO is experiencing for this higher frequency are more severe, a new channel model to describe the mmWave transmission will be needed. In addition, due to the short working distance, the small cell with less coverage area and the D2D proposals are addressed a lot, while combining with the mmWave for 5G.

The New York University together with University of Texas-Austin have done some preliminary studies in mmWave mMIMO channel estimation and modeling work. The stochastic geometry and random geometric graph theory are mostly used in mmWave mMIMO topics for signal to interference ratio (SIR), connectivity, coverage, as well as outage probability and throughput analysis. For mmWave mMIMO, one can find in the literature studies by, but not limited to, Sun et al., Bai and Heath, and Alkhateeb [37–39].

In a nutshell, with all of the discussed technologies, in the arena of 5G, some fundamentals and even redesigns of the whole network will be needed to cater to the new era. With mmWave added in, the new challenges waiting to be tackled are not only the new channel modeling, but also the beamforming techniques, the new system design, channel coding method (such as the nonorthogonal multiple access (NOMA) [40–42]) and full duplex communications [43].

REFERENCES

[1] A. Goldsmith, Wireless Communications, Cambridge University Press, Cambridge, UK, 2005.

[2] D. Tse, P. Viswanath, Fundamentals of Wireless Communication, Cambridge University Press, Cambridge, UK, 2004.

[3] L. Hanzo, H. Haas, S. Imre, D. O'Brien, M. Rupp, L. Gyongyosi, Wireless myths, realities, and futures: from 3G/4G to optical and quantum wireless, Proc. IEEE 100 (Special Centennial Issue) (2012) 1853–1888.

[4] F. Pérez Fontán, P. Mariño Espiñeira, Modeling the Wireless Propagation Channel A Simulation Approach With MATLAB, Wiley, New York, 2008.

[5] S.F. Mahmoud, Y.M.M. Antar, High frequency ground wave propagation, IEEE Trans. Antennas Propag. 62 (11) (2014) 5841–5846.

[6] Y.P. Zhang, D. Liu, Antenna-on-chip and antenna-in-package solutions to highly integrated millimeter-wave devices for wireless communications, IEEE Trans. Antennas Propag. 57 (10) (2009) 2830–2841.

[7] H. Mokhtari, P. Lazaridis, Comparative study of lateral profile knife-edge diffraction and ray tracing technique using GTD in urban environment, IEEE Trans. Veh. Technol. 48 (1) (1999) 255–261.

[8] P. Begovic, N. Behlilovic, E. Avdic, Applicability evaluation of Okumura, Ericsson 9999 and winner propagation models for coverage planning in 3.5 GHZ WiMAX systems, in: Proceedings of IWSSIP, 2012, pp. 256–260.

[9] D. Zhang, K. Yu, Z. Zhou, T. Sato, Energy efficiency scheme with cellular partition zooming for massive MIMO systems, in: Proceedings of IEEE ISADS, 2015, pp. 266–271.

[10] P. Keawbunsong, P. Supanakoon, S. Promwong, Hata's path loss model calibration for prediction DTTV propagation in urban area of Southern Thailand, IOP Conf. Ser. Mater. Sci. Eng. 83 (1) (2015) 012013.

[11] M.V.S.N. Prasad, K. Ratnamala, M. Chaitanya, P.K. Dalela, Terrestrial communication experiments over various regions of Indian subcontinent and tuning of Hata's model, Ann. Telecommun. 63 (3–4) (2008) 223–235.

[12] MathWorks, 2015, Available at: http://www.mathworks.com/help/comm/ref/qfuncinv.html.

[13] H.Q. Ngo, E.G. Larsson, T.L. Marzetta, Energy and spectral efficiency of very large multiuser MIMO systems, IEEE Trans. Commun. 61 (4) (2013) 1436–1449.

[14] Y. Liu, Y. Yu, W.-J. Lu, H.B. Zhu, Stochastic multiple-input multiple-output channel model based on singular value decomposition, IET Commun. 9 (15) (2015) 1852–1856.

[15] G. Lebrun, J. Gao, M. Faulkner, MIMO transmission over a time-varying channel using SVD, IEEE Trans. Wirel. Commun. 4 (2) (2005) 757–764.

[16] F. Rusek, D. Persson, B.K. Lau, E.G. Larsson, T.L. Marzetta, O. Edfors, F. Tufvesson, Scaling up MIMO: opportunities and challenges with very large arrays, IEEE Signal Process. Mag. 30 (1) (2013) 40–60.

[17] T.L. Marzetta, Noncooperative cellular wireless with unlimited numbers of base station antennas, IEEE Trans. Wirel. Commun. 9 (11) (2010) 3590–3600.

[18] J. Andrews, Will densification be the death of 5G? IEEE ComSoc CTN, 2015. http://www.comsoc.org/ctn/will-densification-be-death-5g.

[19] C. Fang, F.R. Yu, T. Huang, J. Liu, Y. Liu, A survey of green information-centric networking: research issues and challenges, IEEE Commun. Surv. Tut. 17 (3) (2015) 1455–1472.

[20] A. Luiz Garcia Reis, A.F. Barros, K. Gusso Lenzi, L.G. Pedroso Meloni, S.E. Barbin, Introduction to the software-defined radio approach, IEEE (Revista IEEE America Latina) Latin Am. Trans. 10 (1) (2012) 1156–1161.

[21] D. Feng, L. Lu, Y. Yuan-Wu, G. Li, S. Li, G. Feng, Device-to-device communications in cellular networks, IEEE Commun. Mag. 52 (4) (2014) 49–55.

[22] S. Hong, J. Brand, J. Choi, M. Jain, J. Mehlman, S. Katti, P. Levis, Applications of self-interference cancellation in 5G and beyond, IEEE Commun. Mag. 52 (2) (2014) 114–121.

[23] X. Zhang, W. Cheng, H. Zhang, Full-duplex transmission in phy and mac layers for 5G mobile wireless networks, IEEE Wirel. Commun. 22 (5) (2015) 112–121.

[24] M. Matthaiou, M.R. McKay, P.J. Smith, J.A. Nossek, On the condition number distribution of complex wishart matrices, IEEE Trans. Commun. 58 (6) (2010) 1705–1717.

[25] J. Hoydis, S. Ten Brink, M. Debbah, Massive MIMO in the UL/DL of cellular networks: how many antennas do we need? IEEE J. Sel. Areas Commun. 31 (2) (2013) 160–171.

[26] Z. Niu, Y. Wu, J. Gong, Z. Yang, Cell zooming for cost-efficient green cellular networks, IEEE Commun. Mag. 48 (11) (2010) 74–79.

[27] S. Jin, X. Zhang, Optimal energy efficient scheme for MIMO-based cognitive radio networks with antenna selection, in: Proceedings of CISS, 2015, pp. 1–6.

[28] X. Zhang, S. Zhou, Z. Niu, X. Lin, An energy-efficient user scheduling scheme for multiuser MIMO systems with RF chain sleeping, in: Proceedings of the 2013 IEEE Wireless Communications and Networking Conference (WCNC), 2013, pp. 169–174.

[29] D. Zhang, K. Yu, Z. Zhou, T. Sato, Energy efficiency scheme with cellular partition zooming for massive MIMO systems, in: Proceedings of IEEE ISADS, 2015, pp. 266–271.

[30] Z. Zhou, S. Zhou, J. Gong, Z. Niu, Energy-efficient antenna selection and power allocation for large-scale multiple antenna systems with hybrid energy supply, in: Proceedings of GLOBECOM, 2014, pp. 2574–2579.

[31] J. Choi, D. To, Energy efficiency of HARQ-IR for two-way relay systems with network coding, in: Proceedings of the 18th European Wireless Conference, 2012, pp. 1–5.

[32] M. Zhao, Z. Zhang, W. Zhou, J. Zhu, Maximizing energy efficiency in analog network coding based two-way relay-assisted system, in: Proceedings of WCSP, 2014, pp. 1–6.

[33] M. Zhao, Y. Yang, Packet scheduling with joint design of MIMO and network coding, in: Proceedings of IEEE MASS, 2009, pp. 227–236.

[34] L. Chen, H. Jin, H. Li, J.-B. Seo, Q. Guo, V. Leung, An energy efficient implementation of C-RAN in HetNet, in: Proceedings of IEEE VTC Fall, 2014, pp. 1–5.

[35] M. Khan, R.S. Alhumaima, H.S. Al-Raweshidy, Reducing energy consumption by dynamic resource allocation in C-RAN, in: Proceedings of EuCNC, 2015, pp. 169–174.

[36] Y. Niu, Y. Li, D. Jin, L. Su, A.V. Vasilakos, A survey of millimeter wave communications (mmWave) for 5G: opportunities and challenges, Wirel. Netw. 21 (8) (2015) 2657–2676.

[37] S. Sun, T.S. Rappaport, R.W. Heath, A. Nix, S. Rangan, MIMO for millimeter-wave wireless communications: beamforming, spatial multiplexing, or both? IEEE Commun. Mag. 52 (12) (2014) 110–121.

[38] T. Bai, R.W. Heath, Coverage and rate analysis for millimeter-wave cellular networks, IEEE Trans. Wirel. Commun. 14 (2) (2015) 1100–1114.

[39] A. Alkhateeb, O. El Ayach, G. Leus, R.W. Heath, Channel estimation and hybrid precoding for millimeter wave cellular systems, IEEE J. Sel. Topics Signal Process. 8 (5) (2014) 831–846.

[40] L. Dai, B. Wang, Y. Yuan, S. Han, C.-L. I, Z. Wang, Non-orthogonal multiple access for 5G: solutions, challenges, opportunities, and future research trends, IEEE Commun. Mag. 53 (9) (2015) 74–81.

[41] Y. Saito, Y. Kishiyama, A. Benjebbour, T. Nakamura, A. Li, K. Higuchi, Non-orthogonal multiple access (NOMA) for cellular future radio access, in: Proceedings of the 2013 IEEE 77th Vehicular Technology Conference (VTC Spring), 2013, pp. 1–5.

[42] T.S. Rappaport, S. Sun, R. Mayzus, H. Zhao, Y. Azar, K. Wang, G.N. Wong, J.K. Schulz, M. Samimi, F. Gutierrez, Millimeter wave mobile communications for 5G cellular: it will work! IEEE Access 1 (2013) 335–349.

[43] T. Riihonen, S. Werner, R. Wichman, Hybrid full-duplex/half-duplex relaying with transmit power adaptation, IEEE Trans. Wirel. Commun. 10 (9) (2011) 3074–3085.

Hybrid antenna array for mmWave massive MIMO

3

J.A. Zhang*, X. Huang*, V. Dyadyuk[†], Y. Jay Guo*

University of Technology Sydney, Sydney, NSW, Australia* CSIRO, Sydney, NSW, Australia[†]

CHAPTER OUTLINE

3.1 INTRODUCTION

Massive multiple-input multiple-output (MIMO) is practical in millimeter-wave (mmWave) systems. Thanks to the small wavelength of mmWave, it is feasible to accommodate a large number of antenna elements in a physically limited space. Massive MIMO is also necessary for mmWave cellular communications. The radiation power is low due to the small antenna size, and the propagation attenuation of mmWave is very large. Hence it is necessary to use high-directivity antennas to ensure that sufficiently high signal power can be received for successful signal detection. Furthermore, in order to support mobile users and multiple users at different locations, mmWave radio needs to use steerable directional antennas or configurable

mmWave Massive MIMO. http://dx.doi.org/10.1016/B978-0-12-804418-6.00003-0

antenna arrays instead of high gain dish antennas. Hence, for mmWave cellular communications, massive antenna array is becoming a promising proposition.

Unfortunately, a fully digital implementation of massive mmWave array, that is, using a radio frequency (RF) front end and digital baseband for each antenna, is unrealistic for most commercial applications, although it provides full capacity and flexibility. For example, to achieve an antenna gain of 30 dB, 1000 antenna elements are required. The cost of such a full digital array will be prohibitively high. Full digital mmWave massive array is also impractical due to the tight space constraints. Antenna elements in an array must be placed closely to prevent grating lobes. The analog components, such as the low noise amplifier (LNA) (or power amplifier), frequency converter, and local oscillator (LO) should be tightly packed behind the antenna elements to reduce signal attenuation and distortion. The frontal area of each antenna element is just a few square millimeters, and in a large phased array all connections on the back of the array are constrained to the same area per antenna element. This space constraint poses a major engineering challenge for antenna arrays at mmWave frequencies.

A hybrid array, which consists of multiple analog subarrays with their own respective digital chains, turns out to be a more feasible solution [1–3]. In the hybrid antenna array, antenna elements are grouped into analog subarrays. In its simplest form, only a phase shifter is dedicated to a single antenna element and all the rest of the components are shared by all antenna elements in a subarray. Each analog subarray outputs/accepts only one digital signal, and signals for all the analog subarrays are processed jointly in a digital processor. Such a structure can largely reduce the cost and complexity, and overcome the space constraint problem with a significantly reduced number of hardware components.

Such a hybrid array provides strong capability in achieving both array diversity and multiplexing gain, particularly when we consider the multipath sparsity feature of mmWave propagation channels. It has been shown in propagation measurements [4] that mmWave multipath signal propagation is sparse in both temporal and spatial domains. Only a few multipath signals arrive in several concentrated directions, and the nonline-of-sight (NLOS) component has much lower power compared to the line-of-sight (LOS) component. Hence the analog subarray can achieve large array gain by forming analog beamforming, and using individual digital processing chains for each subarray adds capabilities for multiplexing and multiuser interference (MUI) cancellation.

Undoubtedly, massive hybrid array for mmWave communications faces many challenging design problems in both hardware and signal processing. In this chapter, we will offer a comprehensive overview for mmWave massive hybrid array, covering both hardware design and signal processing techniques. We will review the structure of hybrid arrays in Section 3.2, discuss hardware design problems including antenna and RF chains in Section 3.3, and then investigate the applications of signal processing techniques to hybrid arrays in Section 3.4, covering topics on pure beamforming, single-user MIMO and spatial division multiple access (SDMA) techniques.

3.2 MASSIVE HYBRID ARRAY ARCHITECTURES

Fig. 3.1A shows the architecture of a hybrid array, where the whole array is divided into many analog subarrays. Each subarray includes N antennas and a unit containing RF and intermediate frequency (IF) components. These components can be shared by different antenna elements in different ways, depending on actual implementations. Each subarray is connected to a baseband processor via a digital-to-analog convertor (DAC) in the transmitter or an analog-to-digital convertor (ADC) in the receiver. The signals from all the subarrays are interconnected and can be processed centrally in the baseband processor. For convenience, we denote an array with M subarrays and N antenna elements in each subarray as an $N \times M$ hybrid array. Typically, N is larger than M such that high antenna gain can be achieved at lower cost. The distance between corresponding elements in adjacent subarrays is called *subarray spacing*.

Signals in analog subarray and the digital processor are processed in different domains and different ways. In each subarray, the signal can be simply weighted in the analog domain, for the purpose of achieving mainly array gain. The signal

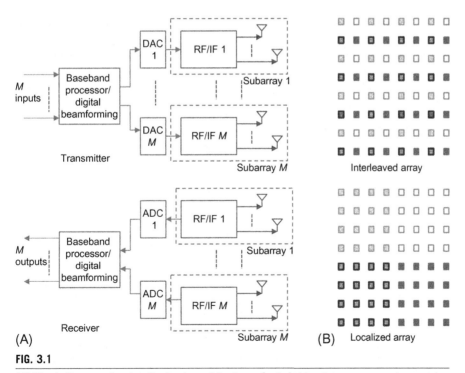

FIG. 3.1

(A) Block diagram showing the basic architecture of a hybrid antenna array system; and (B) two types of array configurations in hybrid uniform square arrays: Interleaved (*upper*) and localized (*bottom*) configurations. Each square represents an antenna element and squares with the same color represent antenna elements in the same analog subarray.

for each antenna element of a subarray can be varied in both magnitude and phase, typically with limited resolution. In the simplest case, only a phase shifter is applied and the signal is weighted by a discrete phase shifting value from a quantized value set of which the size is typically represented through the number of quantization bits. For example, 3-bit quantization means 8 discrete values uniformly distributed over $[-\pi, \pi]$. In the digital processor, signals from/to all the subarrays are jointly processed, and advanced techniques such as spatial precoding/decoding can be implemented, similar to conventional MIMO systems.

Antenna elements in a hybrid array can be configured in various ways to form different topologies with respective advantages and disadvantages. Such a configuration is typically fixed at the fabrication stage. The typical two types of regular configurations are interleaved and localized arrays, as illustrated in Fig. 3.1B for a 16×4 uniform square hybrid array. In an interleaved array, antenna elements in each subarray scatter uniformly over the whole array, while in a localized array, they are adjacent to each other. It is interesting to note that single localized and interleaved subarrays have different beam patterns but the hybrid arrays can have the same pattern when the same digital beamforming vector is used, as demonstrated in Fig. 3.2. The figure clearly shows the difference between the beam patterns of the two individual subarrays, and the similarity of the two array patterns. The larger grating lobe in the interleaved subarray is suppressed in the array pattern, and the localized subarray does not have grating lobe but has wider beamwidth, which is also reduced significantly in the hybrid array. The beam pattern suggests that interleaved arrays have narrower beamwidth and are more suitable for generating multibeam for SDMA applications, while localized arrays can better support systems with relatively

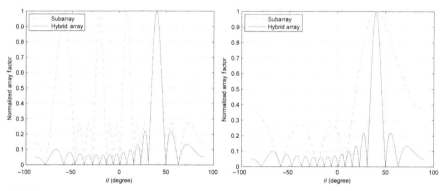

FIG. 3.2

Normalized array factor of a uniform 16×16 square hybrid array with square subarrays in the interleaved (*left*) and localized configuration (*right*). The beam direction is at elevation angle of 40 degrees and azimuth angle of 0 degrees. X-axis is for the elevation angle.

Reproduced with permission from J.A. Zhang, X. Huang, Y. Jay Guo, Adaptive searching and tracking algorithm for AoA estimation in localized hybrid array, 2015 IEEE International Conference on Communication Workshop (ICCW), 8–12 June, 2015, pp. 1095–1100, © 2015 IEEE.

larger angles of arrival (AoAs). Generally speaking, interleaved array is harder to implement due to the space constraint. It offers advantages in beamwidth, but overall is inferior to localized array, as summarized in Table 1 in Ref. [2].

3.3 HARDWARE DESIGN FOR ANALOG SUBARRAY
3.3.1 ANTENNA ARRAYS

A larger antenna array, particularly phased arrays, can be built by assembling together smaller arrays. The assembly can be classified as the "brick" and "tile" approaches [5]. For mmWave arrays where the element spacing is limited and the depth is unconstrained, the brick approach shows a lot of advantages, as Monolithic Microwave Integrated Circuits (MMICs) can be placed sequentially behind the antenna elements. For brick construction the array modules are perpendicular to the aperture plane, with typical radiating elements being dipoles or quasi-Yagi antennas. Microstrip phased arrays are well developed, but are less amenable to brick construction, particularly when the frequency approaches the mmWave region.

Next we provide two examples of array design by the Commonwealth Scientific and Industrial Research Organisation (CSIRO), both suitable for the brick construction. These arrays have built-in fixed phase shifters, and hence have beamforming pointing at preset directions. They can be used as antenna elements in a hybrid array or an analog subarray (with fixed mainbeam direction).

Quasi-Yagi antenna arrays

As reported by Deal et al. [6], a quasi-Yagi antenna is a compact and simple planar antenna that can operate over an extremely wide frequency bandwidth (of the order of 50%) with good radiation characteristics. The compact size of the single element and low mutual coupling between the elements make it ideal for use in a massive array. The antenna is compatible for integration with microstrip-based MMICs. Four-element quasi-Yagi antenna arrays, with phase shift between adjacent array elements fixed at 0, 57, 90, and 125 degrees, were designed and fabricated in CSIRO [7]. The different interelement phase difference is achieved by different microstrip feed networks.

Fig. 3.3 shows an example of the single quasi-Yagi antenna element in the array fabricated on a dielectric alumina substrate with metallization on both sides. The top metallization consists of a microstrip feed, a broadband microstrip-to-coplanar stripline (CPS) balun and two dipoles. One dipole is the driver element fed directly by the CPS and the second dipole (the director) is parasitically fed. The metallization on the bottom plane forms the microstrip ground, and is truncated to create the reflector element for the antenna. The driver on the top plane simultaneously directs the antenna propagation toward the endfire direction, and acts as an impedance-matching parasitic element. The driver element may also be implemented using a folded dipole to give greater flexibility in the design of the driver impedance value and to disable use on a liquid crystal polymer substrate [8]. The alumina substrate has

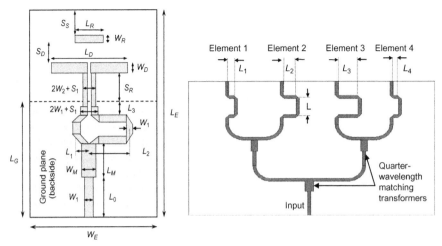

FIG. 3.3

Left: Schematic of the quasi-Yagi antenna array element. $L_E=3$, $W_E=2$, $W_1=0.12$, $L_0=0.45$, $L_G=1.54$, $L_M=0.54$, $W_M=0.205$, $L_1=0.22$, $L_2=0.7$, $L_3=0.1$, $S_1=0.06$, $W_2=0.06$, $L_D=1.29$, $L_R=0.488$, $W_D=0.12$, $W_R=0.12$, $S_R=0.516$, $S_D=0.323$, $S_S=0.383$ (all dimensions in mm), substrate 127 μm alumina ($\varepsilon r=9.9$, tan $\delta=0.0003$). *Right*: Plan view of the microstrip feed network with equal amplitude and phase shift between the outputs.

Reproduced with permission from V. Dyadyuk, X. Huang, L. Stokes, J. Pathikulangara, A.R. Weily, N. Nikolic, J.D. Bunton, Y. Jay Guo, Adaptive antenna arrays for ad-hoc millimetre-wave wireless communications, in: M. Khatib (Ed.), Advanced Trends in Wireless Communications, InTech, 2011, ISBN: 978-953-307-183-1, under license CC BY-NC-SA 3.0. Available from: http://dx.doi.org/10.5772/16000 © 2011.

the following specifications: dielectric thickness 127 μm, metallization thickness 3 μm, dielectric permittivity 9.9, and loss tangent 0.0003. The total area of the substrate is ~2.5 × 3 mm.

The microstrip feed networks were designed using simple T-junction power dividers and quarter-wavelength matching sections. Fig. 3.3 also shows the layout of the network that achieves the required phase shift through varying the length of microstrip lines. The spacing between the elements is 0.48 wavelength, which was selected to minimize the appearance of grating lobes.

Fig. 3.4 shows the measured normalized radiation patterns of the four arrays in the *xz*-plane. The side lobe levels may be improved by using a tapered excitation of the elements instead of the simple equal-amplitude excitation.

Stacked patch antenna with perpendicular feed substrate

An 8-element linear array is developed and reported in Ref. [9], demonstrating the feasibility of using brick array construction for microstrip phased arrays. The antenna uses a radiating element with a new transition, the circularly polarized

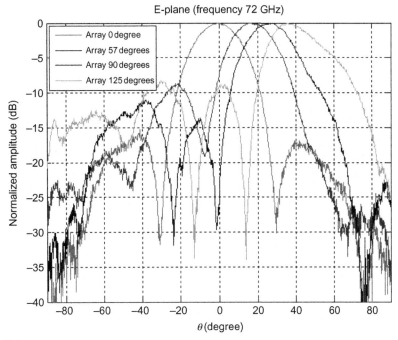

FIG. 3.4

Measured copolar E-plane radiation patterns for all arrays at 72 GHz.

stacked patch antenna with perpendicular feed substrate. Coupling from the patch to the perpendicular microstrip feed line is through a slot in the ground plane of both the patch and the perpendicular substrate.

The individual layer of the antenna is shown in Fig. 3.5. The structure consists of three identical 0.508 mm Rogers RT 6002 substrates, two bonding layers and one ground plane. The first two substrates contain the stacked patch antennas that have truncated corners to realize circular polarization. These two substrates are bonded together by a layer of 35 μm CuClad 6700. A ground plane with a coupling slot is attached to the second substrate. A second bonding layer, used at the back of the ground plane, attaches the edge of the third substrate, which is oriented perpendicular to these layers and contains a microstrip feed line and second coupling slot. The bonding layers are very important as they remove potential air gaps between the perpendicular substrate and stacked patches and reduce the sensitivity of the structure to the gap. This is particularly important for millimeter wave antennas.

An 8-element linear array was built using this stacked patch antenna with perpendicular feed substrate. Five feed networks based on fixed microstrip line delays were developed to test the beamforming performance of this array, with mainbeam pointing to the angles of 0, −10, −20, −30, and −40 degrees. These prototypes are demonstrated to operate at 20 GHz, with measured gains from 9.58 (at −40 degrees) to 12.94 dBic (at 0 degrees) [9].

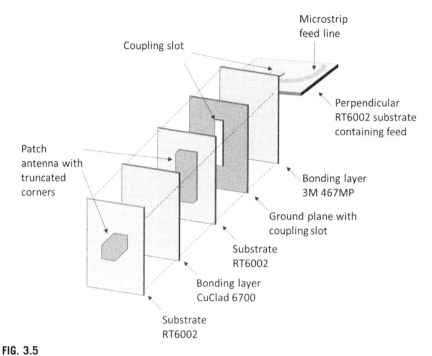

FIG. 3.5

Exploded view showing individual layers of the circular polarized stacked patch antenna with perpendicular feed substrate.

3.3.2 RF CHAIN ARCHITECTURES

The simplest way to implement an analog subarray and realize analog beamforming is through phased array, where the delay/phase of the signal at each antenna is carefully controlled. Ideally, a high-gain pencil beam is generated by a true time delay at each element that compensates exactly for the free-space propagation delay. Developing a low-loss delay-adjustable linear delay line directly at mmWave is very challenging. Some of the promising newer technologies for implementing broadband true time delay include switched-length transmission lines using RF microelectromechanical switches and variable velocity transmission lines based on ferroelectric materials. However, many problems such as reliability and frequency-dependent loss need to be overcome. Equivalent delay can also be implemented by phase shifters in the RF, IF, or LO channels. For a narrowband system operating at the frequencies below 40 GHz, implementing the delay as an equivalent phase shift at the center frequency is a simple option; however it remains challenging at upper mmWave bands, where high-resolution phase shifters with low insertion loss, phase and magnitude errors are not available at present. Hence, phase shifting at the IF or LO frequencies is preferable.

Several possible receiver RF chain architectures are shown in Fig. 3.6. Architectures of the transmitter RF chain are very similar, with a reverse signal flow and the LNA replaced with a power amplifier. All the illustrated architectures use an optional variable attenuator at IF, which can be used for compensating the conversion gain variations and loss in the combining network and calibration of the analog subarray. This also adds finite magnitude adjustment capability to analog subarrays, which has hardly been considered in the literature of signal processing for mmWave hybrid array. The optional band-pass filters (BPFs) may be required for band limiting of

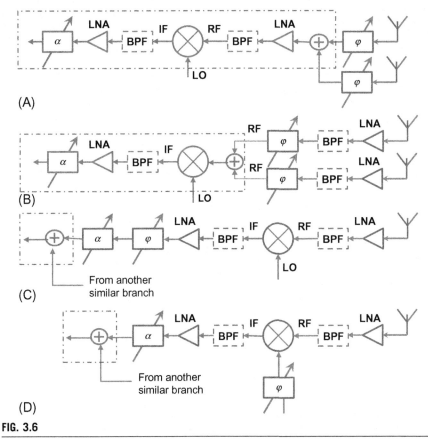

FIG. 3.6

Options for implementing phase shifting in the analog subarray, using an example of two branches/antennas in each subarray. The blocks φ and α denote variable phase shifter and magnitude attenuator, respectively. Blocks in green and in dash-dot boxes represent those able to be shared by antenna elements in a subarray. The circle with a cross denotes where signals from individual antenna elements are combined to input to the shared components. Subfigures show options: (A) Phase shifter at the RF before an LNA, (B) Phase shifter at the RF after an LNA, (C) Phase shifter at the IF, and (D) Phase shifter at the LO.

the IF signal after frequency conversion, and also for band limiting and image rejection (in most applications) at RF frequency.

Fig. 3.6A shows the simplest architecture where only phase shifter and antenna are independent and all the rest of the components are shared by all elements in an analog subarray. This passive power-combining architecture incurs power-combining loss, which increases with the number of antenna elements and operating frequency. Such a power loss can make large passive arrays impractical. A modification of this architecture is shown in Fig. 3.6B, where an individual LNA is applied to each antenna element before the phase shifter. This modification can reduce the noise significantly and provide increased receiver sensitivity. Option Fig. 3.6B can be implemented using either a shared frequency converter (with individual RF chains combined at the input to the mixer) or individual frequency conversion and combining at the IF. Although research prototypes of 2- to 4-bit phase shifters have been reported [10,11], suitable phase shifters for the architectures of Fig. 3.6A and B are not commercially available yet in the upper mmWave frequency band.

Fig. 3.6C and D shows more practical configurations, with phase shift implemented at the IF and LO circuits, respectively. Commercial 6-bit digital phase shifter MMICs are available (e.g., from Hittite Microwave and Triquint) for a range of the LO and IF frequencies suitable for the mmWave arrays. These devices provide 360 degrees of phase change, with a least-significant bit (LSB) of 5.625 degrees that allows analog beamforming with an accuracy of the scan angle to a fraction of a degree. The option in Fig. 3.6D is particularly attractive because the devices in the LO path are typically operated in saturation, and variable loss with phase shift is not a problem.

RF chains shown in Fig. 3.6A–D can be combined in various ways into the antenna array. However, integration of an individual RF chain behind the antenna element is not feasible due to the physical size of the component in existing semiconductor technologies, particularly for the mainstream gallium arsenide (GaAs) chips. Instead, it is possible to integrate all the RF chains for a subarray behind it. For example, for a subarray with the maximum scan angle greater than ±30 degrees, the element size would be 2–2.5 mm. A subarray integration would provide a tight but feasible accommodation for each of the RF, IF, LO, power, and control circuits. A practical implementation for an active 1×4 hybrid array is reported in Ref. [7]. The area of this array is about 100 mm^2 using GaAs MMICs.

The advance of CMOS and silicon germanium (SiGe) BiCMOS technologies will also make the subarray-based integration easier. CMOS and SiGe BiCMOS implementations can have smaller profiles and have very similar physical dimensions of the subarray. Until the early 2000s, operation in mmWave frequency bands was the exclusive domain of III–V compound semiconductors, such as GaAs and indium phosphide (InP). CMOS and SiGe BiCMOS technologies have evolved quickly with aggressive scaling (currently 45 nm CMOS and 0.8 μm SiGe BiCMOS). Integrated circuits employing these technologies have been successfully designed for E-band [7,10,11]. These latest achievements indicate that Si technology is suitable for a massive, short-to-medium-range mmWave array development. However, GaAs circuits still hold superior performance as power amplifier MMICs, as the output power of Si devices is typically only a few milliwatts, while GaAs can achieve larger than 1 W.

A common problem for all phase shifting-based architecture discussed previously is the fractional bandwidth constraint. When an array is scanned with a phase shifter instead of true time delay, the position of the mainbeam varies with frequency, and this effect becomes more pronounced when the beam scanning angle is wider. For a large uniform array the 3-dB fractional bandwidth can be approximated as $0.866\lambda\sin(\theta)/D$, where D is the array diameter, λ is the wavelength, and θ is the maximum scan angle. Thus a fractional bandwidth above 1% is feasible for a large array ($D/\lambda = 120$) at the maximum scan angle not exceeding 45 degrees.

3.3.3 HYBRID ARRAY PROTOTYPES

Prototypes for mmWave hybrid arrays have been built and tested, for example, the CSIRO prototype [7] and the Samsung prototype [12]. These prototypes demonstrate the effectiveness of hybrid arrays in generating beamforming and multiplexing gains.

The CSIRO prototype [3] is a four-channel receiver RF module integrated with a linear endfire quasi-Yagi antenna array with regular dipole driver, structurally similar to that shown in Fig. 3.6C. It operates over the frequency range of 71–76 GHz at the subharmonic LO of 38–39 GHz and IF 1–7 GHz. Typical conversion gain was 6 ± 1 dB over the operating RF and IF frequency range of 71.5–72.5 GHz and 3.5–4.5 GHz, respectively. The maximum magnitude imbalance between each of four channels was below ±1.5 dB.

A simplified schematic, together with a photograph of the assembled four-channel integrated receiver array prototype, is shown in Fig. 3.7. The IF preamplifiers, interconnect, matching, and group delay equalization circuits are developed using a standard commercial thin-film process on ceramic substrate. Phase and magnitude controls for each channel are implemented at IF using 6-bit digital phase shifters HMC649LP6 and attenuators HMC4214LP3, which are available from Hittite Microwave Corporation.

A test bed was set up using this unit and measurement was conducted in an anechoic chamber. Measured antenna array patterns were very close to those predicted by electromagnetic simulations for steering angles within ±40 degrees. Beam steering accuracy of 1 degree has been achieved with 6-bit discrete phase shift and magnitude control at IF. Measured array gain was 9.5 dBi for steering angles below 22 degrees and reduced to ~7.5 dBi at the maximum steering angle of ±42 degrees. A summary of the E-plane measurements is provided in Fig. 3.8, where measured steering angle, −3 dB beamwidth, measured and simulated gain loss due to steering, measured cross-polar ratio, measured magnitudes of the highest side lobe and a grating lobe are plotted as a function of a phase shift between adjacent array elements. Grating lobes were observed only at the steering angles beyond ±43 degrees, confirming simulated predictions.

The Samsung prototype is a 2×2 MIMO system with separated transmitter and receiver arrays in each end, operating at 28 GHz with a bandwidth of 500 MHz. Each

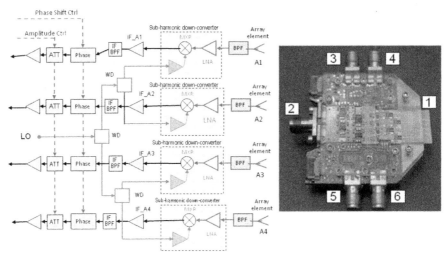

FIG. 3.7

CSIRO prototype, receiver only. *Left*: Simplified schematic configured for analog beamforming; *Right*: Assembled module where 1 is antenna array, 2 is LO input, and 3–6 are IF outputs. The abbreviations BPF, LNA, MxR, WD, phase, ATT, and LO denote a band-pass filter, low noise amplifier, subharmonically pumped mixer, Wilkinson divider, phase shifter, attenuator, and local oscillator, respectively.

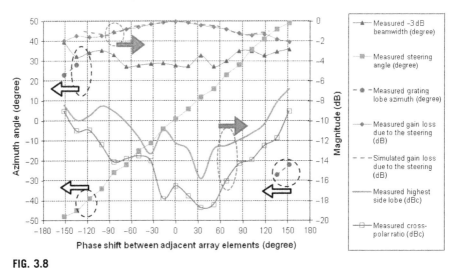

FIG. 3.8

Summary of test results for the CSIRO hybrid array test bed.

array is a uniform planar array with eight horizontal and four vertical elements over an area of 60×30 mm. These elements are grouped into two subarrays. Each subarray has 4 RF units and hence each RF unit connects with eight elements. It is not disclosed in Ref. [12] whether these eight elements are connected to generate fixed beamforming pointing at different directions or not. The total array gain is 18 dBi with 3 dB beamwidth of 10 degrees (horizontal) and 20 degrees (vertical). The beam scanning range is $[-30, 30]$ degrees. Adaptive beamforming techniques are tested in a field trial using the array, demonstrating the support of mobile users up to 8 km/h even in NLOS environments.

3.4 SMART ANTENNA TECHNIQUES

A hybrid antenna array enables various smart-antenna techniques, such as pure beamforming (spatial diversity), MIMO (spatial multiplexing), and SDMA. The LOS or near-LOS channel property, together with the large number of antennas and the hybrid array architecture, cause many new challenges and motivate new signal processing and optimization techniques. For example, channels without rich scattering drive new design for MIMO, such as LOS-MIMO; SDMA becomes more preferable because it can exploit the channel independence between different users, while multiplexing gain for single-user MIMO requires larger arrays. It is also possible to apply rigid mathematical tools based on beamforming to design MIMO and SDMA systems, instead of relying on channel statistics.

A hybrid array is not a simple put-together of multiple analog arrays. Conventionally, an analog phased array uses integer multiples of a fixed value for its phase shifters, based on the signal direction. This is effective when signals concentrate in one direction. A hybrid array provides the capability for optimizing phase shifting values across subarrays. That is, each subarray may form multiple simultaneous beams instead of the traditional single beam in the phased array applications, and the overall beamforming for one targeted user can have contributions from more than one subarray.

3.4.1 ARRAY GEOMETRY

We now use the receiver as an example to describe the signal model in a hybrid array mathematically. The analog subarray is assumed to have a phase shifter applied to the signal at each antenna element.

We consider a square array with $M = M_x \times M_y$ subarrays, and each subarray has $N = N_x \times N_y$ elements, where M_x (or N_x) and M_y (or N_y) are the numbers of subarrays (or elements) placed along the x-axis and y-axis, respectively. The location of the ith element in the mth, $m = m_y M_x + m_x$, subarray is $(X_{i,m}, Y_{i,m})$, where $X_{i,m} = X_{i,0} + m_x d_x^s$, $Y_{i,m} = Y_{i,0} + m_y d_y^s$, $m_x = 0, 1, \ldots, M_x - 1$, and $m_y = 0, 1, \ldots, M_y - 1$, d_x^s and d_y^s are the subarray spacing along the x-axis and y-axis, respectively. The location $(X_{i,0}, Y_{i,0})$ of the ith element, $i = i_y N_x + i_x$, is given by $X_{i,0} = X_{0,0} + i_x d_x^e$, $i_x = 0, 1, \ldots, N_x - 1$, and $Y_{i,0} = Y_{0,0} + i_y d_y^e$, $i_y = 0, 1, ..., N_y - 1$, where d_x^e and d_y^e are the subarray element spacing along the x-axis and y-axis, respectively, and $(X_{0,0}, Y_{0,0})$ is the location of the element numbered $i = 0$. For simplicity, we will let $(X_{0,0}, Y_{0,0}) = (0, 0)$, and drop the subscript x and y when the discussions apply to subarrays and elements in both axes.

3.4.2 PURE BEAMFORMING AND AOA ESTIMATION

Here, pure beamforming is referred to as the generation of single or multiple beams to achieve spatial diversity and mitigate MUI for a user of interest. It typically involves estimating AoA of the incident signals and generating beamforming vectors based on the estimates.

AoA estimation in an mmWave hybrid array is a very important problem and forms the basis for many advanced techniques, such as beamforming, single-user MIMO and spatial division multiuser access. AoA estimation in a hybrid array is quite different from those well-studied ones in either a full analog or digital array. A full analog array generally uses beam scanning to search the AoA [13], while a full digital array can estimate it in one step using, for example, spectrum analysis techniques and maximal likelihood estimator. For a hybrid array, existing algorithms need to be adapted to the special architecture and to different subarray configurations. The AoA estimation algorithms for a hybrid array typically need to be implemented recursively between digital and analog parts. This is because a low-accuracy AoA estimate leads to a low-analog beamforming gain and signal to noise ratio (SNR) at the digital branches, which results in inaccurate AoA estimation. Hence AoA estimation can only be improved recursively by updating analog beamforming weights with the latest estimated AoA values.

Assume that a planar waveform $\tilde{s}(t) = hx(t)$ arrives at the hybrid array from zenith angle θ and azimuth angle ϕ, where $x(t)$ is the transmitted signal, and h is the pathloss. Let $s_m(t)$ denote the received signal output to the digital baseband from the mth subarray. Without considering any distortions such as mutual coupling between antenna elements, $s_m(t)$ can be expressed as

$$s_m(t) = \tilde{s}(t) \sum_{i=0}^{N-1} P_{i,m}(\theta, \phi) e^{j \left[\frac{2\pi}{\lambda_c} \left(X_{i,m} \sin\theta\cos\phi + Y_{i,m} \sin\theta\sin\phi \right) + \alpha_{i,m} \right]} + z_m(t), \quad m = 0, 1, ..., M-1$$

(3.1)

where $P_{i,m}(\theta, \phi)$ is the radiation pattern of the ith element located at $(X_{i,m}, Y_{i,m})$ in the mth subarray, λ_c is the wavelength of the carrier signal, $\alpha_{i,m}$ is the phase shift, and $z_m(t)$ is the additive white Gaussian noise presented at the output of the mth subarray. The noise $z_m(t)$ can be from a single noise source or a sum of multiple noise sources, depending on where the noise is introduced.

Assume that elements in all subarrays have the same omnidirectional radiation pattern, that is, $P_{i,m}(\theta, \phi) = 1$. Then, Eq. (3.1) can be simplified as

$$s_m(t) = \tilde{s}(t) P_m(\theta, \phi) e^{j \frac{2\pi}{\lambda_c} \left(m_x d_x^s \sin\theta\cos\phi + m_y d_y^s \sin\theta\sin\phi \right)} + z_m(t)$$

(3.2)

where $P_m(\theta,\phi) = \sum_{i=0}^{N-1} e^{j\left[\frac{2\pi}{\lambda_c}\left(i_x d_x^e \sin\theta\cos\phi + i_y d_y^e \sin\theta\sin\phi\right) + \alpha_{i,m}\right]}$ is the overall radiation pattern of the mth subarray.

For the estimation of a single AoA value, one technique is to exploit the constant phase difference between corresponding elements in two neighboring subarrays, as proposed in Ref. [1], where a differential beam tracking algorithm and a differential beam search (DBS) algorithm are proposed for interleaved and localized arrays, respectively. The need for two different algorithms is mainly due to the phase ambiguity problem in the localized configuration. DBS can remove the phase ambiguity, at the cost of increased complexity and reduced convergence speed. Exploiting phase difference not only removes the necessity of a known reference signal or signal synchronization, but also leads to a Doppler resilient solution.

The basic principle in the two algorithms is similar, and is based on setting the phase shifter $\alpha_{i,m}$ to be the same for antennas with the same indexes across subarrays, that is, $\alpha_{i,m} = \alpha_i$, and calculating the cross-correlation between neighboring subarrays. For interleaved arrays, the cross-correlation between neighboring subarrays is given by

$$R_x = E\left\{s^*_{m_y M_x + m_x}(t)s_{m_y M_x + m_x + 1}(t)\right\} = E\left\{|\tilde{s}(t)|^2\right\}|P_m(\theta,\phi)|^2 e^{jN_x u_x} \qquad (3.3)$$

and

$$R_y = E\left\{s^*_{m_y M_x + m_x}(t)s_{(m_y+1)M_x + m_x}(t)\right\} = E\left\{|\tilde{s}(t)|^2\right\}|P_m(\theta,\phi)|^2 e^{jN_y u_y} \qquad (3.4)$$

where $u_x = \dfrac{2\pi}{\lambda_c}d_x^s \sin\theta\cos\phi/N_x$, $u_y = \dfrac{2\pi}{\lambda_c}d_y^s \sin\theta\sin\phi/N_y$.

When d_x^s and d_y^s are chosen to be smaller than $\dfrac{\lambda_c}{2}$, u_x and u_y can be estimated from the phase of R_x and R_y without phase ambiguity. Once one set of estimates is obtained, they are applied to generate the phase shifting values $\alpha_{i,m}$, and the process is repeated to achieve improved SNRs. Signals from different subarrays can also be combined constructively using the estimated AoA values to generate better estimates. This process typically converges quickly.

For a localized array, it is impossible to make the subarray space smaller than a half wavelength. When antenna elements are placed at a distance of a half wavelength, there will be N_x (or N_y) possible values from the angle of the cross-correlation values, which is known as the phase ambiguity problem. These possible values are given by

$$\begin{aligned}u_x(q_x) &= \mathrm{mod}\left((\angle\{R_x\} + 2\pi q_x)/N_x, 2\pi\right) - \pi \\ u_y(q_y) &= \mathrm{mod}\left((\angle\{R_y\} + 2\pi q_y)/N_y, 2\pi\right) - \pi\end{aligned} \qquad (3.5)$$

where mod(.) is the modulo operator, and $q_x \in [0, N_x - 1]$, $q_y \in [0, N_y - 1]$. The DBS algorithm independently tests every combination of (q_x, q_y) in each iteration and chooses the one generating the maximum power as the input for the next iteration. The DBS algorithm suffers from convergence problems, primarily due to the estimation error in $\angle\{R\}$ and the phase discontinuity problem. These problems cause the "zigzag" effect, where the estimates jump between correct and wrong values.

The estimation error in $\angle\{R\}$ can have a significant impact on the value of $|P_m(\theta,\phi)|^2$ and hence the choice of the right (q_x, q_y) in the next iteration. Let the phase shifting value used in the (i_x, i_y)th subarray, determined from the estimates, be $\alpha(i_x, i_y) = -(i_x u_x + i_y u_y)$. Let the estimation errors in $\angle\{R\}$ be δ, and $\widehat{q}$ be the actual integer that generates the estimate u. The overall radiation pattern of a subarray can be obtained as

$$P_m(\theta,\phi) = \sum_{i_y=0}^{N_y-1} \sum_{i_x=0}^{N_x-1} e^{j\left[\left(\delta_x + 2\pi\left(q_x - \widehat{q}_x\right)\right)i_x/N_x + \left(\delta_y + 2\pi\left(q_y - \widehat{q}_y\right)\right)i_y/N_y\right]}$$

In Fig. 3.9, we demonstrate how $|P_m(\theta,\phi)|$ changes with different $\Delta q = q - \widehat{q}$ at various estimation errors $\delta_x = \delta_y$, when $N_x = N_y$. The gap between $|P|$ for $\Delta q_x = \Delta q_y = 0$ and the other cases decreases quickly with δ increasing. For small δ such as $\delta < 0.5$, the gap is larger than 20 dB and the right q can be found even at very low SNRs. However, with δ increasing, the gap reduces quickly, and a wrong q may be chosen at a low SNR. In such a situation, larger δ can lead the iteration toward the wrong direction. Therefore, it is critical to have an algorithm that can obtain small δ, particularly at a low SNR.

Phase discontinuity is a special but neglected problem in AoA estimation in a localized hybrid array. It arises when the actual value of $\angle\{R\}$ is close to π and noise causes the estimate to vary around π. For example, an actual $\angle\{R\} = 0.99\pi$ may be

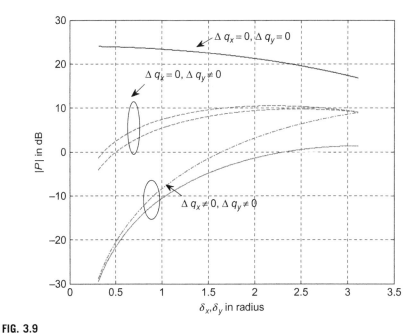

FIG. 3.9

Variation of $|P|$ (in dB) with the estimation error δ and different values of Δq.

estimated as -0.99π. The difference between them is small in terms of angle; however, they lead to significantly different values of $\exp(j\angle\{R\}/N)$ and then u for a given q. Hence the estimate of u for a given q can jump between largely different values, which lead to nonconvergence and the zigzag effect.

To solve these two problems and speed up the convergence, an adaptive searching and tracking (AST) algorithm is proposed in Ref. [14]. The basic idea is to get a good enough initial estimate using scanning/searching, and then use it as an anchor point to determine u from $\angle\{R\}$ in the following iterations. Once an anchor point is obtained, the algorithm intends to end searching and enter to the searching mode to minimize the estimation period. Hence the AST algorithm includes two interchangeable searching and tracking stages, and a mechanism is designed to enable the switching between the two stages. In the searching stage, AST tries to find a good initial estimate for $\angle\{R\}$ through scanning a set of angles uniformly distributed on a unit until the energy of the received signal satisfies a condition, named Condition A. If Condition A is not met, a new set of interpolated angles will be scanned. Once Condition A is met and an initial estimate is obtained, AST will move to the tracking stage, where estimated value will be directly used as the input for the next iteration without searching. The energy of the received signal will also be monitored in the tracking stage. If it violates another condition, called Condition B, the algorithm switches back to the searching stage. There could be several choices for Conditions A and B. In Ref. [14], a threshold on the peak-to-average power ratio of the signals in each iteration is used for Condition A, and the degradation of total power below a threshold for Condition B. The AST algorithm shows great improvement on convergence rate and estimation error performance compared to the DBS algorithm, as illustrated in Ref. [14]. However, it was also observed in the simulation that it does not always converge, which is largely due to the use of nonoptimal switching Conditions A and B.

To avoid the searching process in DBS, a frequency-domain AoA algorithm is proposed in Ref. [15], where the frequency-dependent property of a wideband array and the mutual coupling effect are also considered and mitigated. Unfortunately, this approach relies on the product of two cross-correlation values and has significantly higher noise than the time domain approach.

3.4.3 SINGLE-USER MIMO

In this case, more than one spatial stream is exchanged between two arrays. Due to the multipath sparsity, the channel propagation matrix can be near-singular and conventional MIMO capacity will degrade significantly. Single-user MIMO capacity for a full digital mmWave array has been studied in, for example, Ref. [19] for LOS path only, and Ref. [16] for a two-path channel. Throughput optimization has also been studied for a hybrid array in a sparse but non-LOS channel in, for example, Ref. [17]. When LOS-path is dominating, multiplexing gain is largely limited to the gain achievable by LOS-MIMO, which relies on careful placement of transmitter and receiver antennas.

For a full digital array, the LOS-MIMO capacity, which can be achieved at the Rayleigh distance, depends on the orientation of transmitter and receiver arrays, their distance R, the element spacing and the number of antenna elements, as presented in Ref. [18]. The Rayleigh distance leads to a full-rank and orthogonal MIMO channel matrix, but generally requires impractically large antenna space and array size. For example, in a setup with a carrier frequency 38 GHz, two parallel uniform linear arrays (ULAs) of 16 elements, and $R = 500$ meters, achieving system capacity requires an element spacing of about 0.5 m ($\sim$63 wavelengths). In Ref. [19], system throughput is examined for arrays with closer element space. It is shown that the maximum distance to support multiplexing communications over LOS-MIMO channel is mainly determined by the product of the aperture sizes of the transmitter and receiver antenna arrays, instead of the numbers of antennas at both ends. For communication distance in the order of kilometers, the multiplexing gain is limited to 4, even for a large array size of 5 m.

The capacity results for a full digital array indicate that it is insufficient to use a hybrid array for achieving single-user MIMO capacity, as multiplexing gain is limited. The capacity of a hybrid array is closely related to subarray configurations. Phase shifting values become one part of the channel, and are the major parameters that affect the capacity for a fixed array. Capacity optimization for hybrid arrays, however, remains a challenging open problem, and analytical results are not available yet. In Fig. 3.10, a numerical example of the throughput versus element spacing is shown for two parallel 8×4 and 4×4 hybrid ULAs, assuming equal distance between antenna elements. The throughput is computed using the capacity equation for deterministic channels without applying water-filling power optimization. We use the term throughput instead of capacity here because the curves for both interleaved and localized arrays are obtained when the phase shifting values are chosen to be corresponding to the AoAs of the signal. The upper bound corresponds to the eigen-beamforming, and can only be approached but not be achieved when analog subarrays can only choose discrete phase shifting values. Interesting observations from the figure include the following: (1) the upper bound, as well as the throughput of localized arrays, are convex functions of the element spacing; they reach the maximum at about 15.75 and 31.5 wavelengths for 8×4 and 4×4 arrays, which are $\frac{1}{4}$ and $\frac{1}{2}$ of the element spacing (61 wavelengths) when the capacity of a full digital array is maximized; (2) for a practical array size, the hybrid array achieves throughput very close to the upper bound and the capacity of full digital arrays; and (3) interleaved arrays achieve much lower throughput compared to localized arrays. It is also demonstrated in Ref. [2] that capacity can be increased significantly by using larger subarray spacing than intra-subarray element spacing, when the total array size is fixed.

Overall, single-user MIMO is a less attractive option in mmWave cellular systems due to the limited multiplexing gain and the dependency on distance relationship.

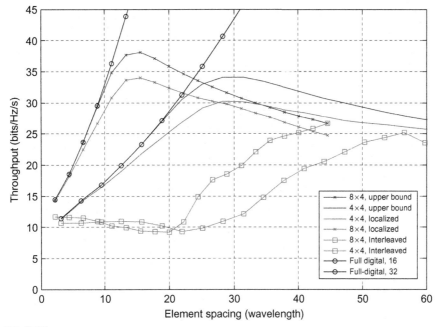

FIG. 3.10

Throughput of LOS-MIMO systems versus uniform element spacing for two parallel 8×4 and 4×4 ULAs at an averaged received SNR of 20 dB. The two curves with legend "full digital" are for the throughput of full digital arrays with 16 and 32 antennas. The upper bounds are obtained by assuming that analog subarrays can implement eigen-beamforming to convert the channel matrix into a diagonal matrix with nonzero elements corresponding to the four largest eigenvalues.

Reproduced with permission from J.A. Zhang, X. Huang, Y. Jay Guo, Adaptive searching and tracking algorithm for AoA estimation in localized hybrid array, 2015 IEEE International Conference on Communication Workshop (ICCW), 8–12 June, 2015, pp. 1095–1100. © 2015 IEEE.

3.4.4 SDMA

In hybrid arrays, SDMA techniques can be applied to support up to M users. Due to the special structure of hybrid arrays, however, user scheduling and beamforming generation are challenging and quite different from those in conventional full digital arrays.

Consider a simple one-dimensional hybrid linear array, based on the square array model as described in Section 3.4.1. Assume that $K \leq M$ users with AoAs θ_k will be connected by the hybrid array simultaneously using SDMA techniques. For LOS channels, the received signal seen at the digital baseband can be represented as

$$\mathbf{y} = \mathbf{A} \begin{pmatrix} \mathbf{W}_0 \\ \vdots \\ \mathbf{W}_{M-1} \end{pmatrix} \mathbf{Hx} + \mathbf{z} \tag{3.6}$$

In Eq. (3.6), $\mathbf{y}$ is an $M \times 1$ vector; $\mathbf{x}$ is a $K \times 1$ vector denoting the transmitted signal from K users; $\mathbf{z}$ is the $M \times 1$ noise vector; $\mathbf{H}$ is a $K \times K$ diagonal matrix with diagonal element h_k denoting the path loss for user k; $\mathbf{A}$ is an $M \times NM$ analog phase shift matrix

$$\mathbf{A} = \begin{pmatrix} \mathbf{a}_0 & \mathbf{0} & \cdots & \mathbf{0} \\ \mathbf{0} & \mathbf{a}_1 & \cdots & \mathbf{0} \\ \vdots & \vdots & \ddots & \vdots \\ \mathbf{0} & \mathbf{0} & \cdots & \mathbf{a}_{M-1} \end{pmatrix}$$

where $\mathbf{a}_m = \left(e^{j\alpha_{0,m}}, \cdots, e^{j\alpha_{N-1,m}} \right)$ is the phase shifting vector in the mth subarray; and $\mathbf{W}_m$ is the $N \times K$ array pattern matrix for the mth subarray with its (i,k)th element being

$$w_{m,i,k} = e^{j\frac{2\pi}{\lambda_c}(id^e + md^s)\sin(\theta_k)}$$

From Eq. (3.6), we can see that the received signal at each subarray is a mix of signals from K users, and hence the phase shifting values will impact all the signals. Optimal user scheduling will typically require joint optimization of phase shifting values and user association, which is very complex, particularly when phase shifting values are discrete. One simpler and suboptimal approach is to separate user allocation and phase shifting value generation. In particular, each subarray only communicates to one user, one user can be served by multiple subarrays, and different users are largely separated in directions and can be served by different subarrays. Such a scheme will be effective when the number of users is sufficiently large, although a qualitative analysis is not available yet.

The design of such a simpler SDMA scheme can be divided into three steps: (1) associate subarrays with users, (2) determine the phase shifting values, and (3) determine precoding and SDMA equalization coefficients to mitigate MUI.

The first step can be realized by simply choosing users with largely separated AoAs, together with similar path loss values. Once users are associated with subarrays, phase shifting values can be determined using one of many possible metrics, such as maximizing analog beamforming gain, and maximizing the ratio of signal to interference and noise [20]. Because the phase shifting values are discrete, codebook construction and searching techniques [21], which are similar to those studied for MIMO channels, can be applied to speed up this process. The second and third steps can also be processed jointly to improve system performance. For example, in Ref. [20], the two steps are combined and the beamforming design is formulated as a signal to interference and noise ratio (SINR) constrained power minimization problem. It is further simplified as a semidefinite programming problem by assuming large K-factor Rician channels based on channel sparsity and dominating LOS propagation.

More advanced and complex beamformer design has been investigated by considering more general cases where cross-subarray modulation is applied. That is, different users' signals are precoded and mapped to multiple subarrays, and each subarray will form multiple beams pointing at multiple users. These techniques largely exploit channel sparsity to simplify beamformer design. For example, the optimization metric, such as the capacity or mutual information of a hybrid array system, is generally a nonconvex function under the constraints of analog phase shifters. To make the optimization problem tractable, approximation is applied to simplify the metric by exploiting a large number of antennas, the channel sparsity and the high correlation of the channel matrix. One example can be seen from Ref. [22], where joint spatial division and multiplexing (JSDM) is proposed for mmWave hybrid arrays. The JSDM scheme first partitions users with similar covariance channel matrixes into the same group, and then determines the spatial division prebeamformer and the multiuser MIMO (MU-MIMO) precoder for each group. The prebeamformer is determined according to the covariance matrix and hence does not require real-time channel feedback. The MU-MIMO precoder is determined by using instantaneous channel values, which are not difficult to obtain thanks to the considerable array dimension reduction after the prebeamforming.

3.5 CONCLUSIONS

In this chapter, we introduced a hybrid array architecture for mmWave cellular communications, where antenna elements are grouped into multiple analog subarrays, and a single digital signal is received from or sent to each subarray. RF components can be shared by multiple antenna units in each analog subarray. In the simplest case, each subarray is a phased array where only an adjustable phase shifter is private to each antenna. Such a hybrid array provides an affordable and spatially feasible solution for mmWave massive arrays, and can achieve comparable performance with a full digital array thanks to the temporal and spatial sparsity of mmWave propagation channels.

We first presented the architecture, highlighting two typical configurations of interleaved and localized arrays, and made a brief comparison of their performance. We then discussed four optional hardware implementations of this architecture, which provide different trade-offs between complexity and performance. Building massive antenna arrays through promising brick construction is demonstrated with quasi-Yagi antenna arrays and stacked patch antennas with perpendicular feed substrate. Two practical prototypes of hybrid arrays, developed by CSIRO and Samsung, are also introduced.

After reviewing the hardware work, this chapter further discussed signal processing techniques for hybrid arrays, with focus on AoA estimation, single-user LOS-MIMO and its capacity, and SDMA techniques. We highlighted the difference for the AoA estimation problem between hybrid arrays and conventional analog and digital arrays, and reviewed several solutions for the problem. We also show that the capacity of single-user LOS-MIMO is very limited by the practical size of the array, and hence in general the single-user multiplexing gain is small. SDMA is a better way of exploiting multiplexing gain for the mmWave hybrid array. SDMA techniques, including the approaches of using one or more subarrays for a single user and cross-subarray modulations, are discussed. The former has a lower complexity and is easier to implement, while the latter can achieve better performance with significantly increased complexity.

The hybrid array is an emerging architecture with many challenging problems to be addressed. Nevertheless, it is a practical solution for massive array and is very promising for 5G mmWave cellular communications.

REFERENCES

[1] X. Huang, Y.J. Guo, J. Bunton, A hybrid adaptive antenna array, IEEE Trans. Wirel. Commun. 9 (5) (2010) 1770–1779.

[2] J.A. Zhang, X. Huang, V. Dyadyuk, Y.J. Guo, Massive hybrid antenna array for millimeter-wave cellular communications, IEEE Wirel. Commun. 22 (1) (2015) 79–87.

[3] Y.J. Guo, X. Huang, V. Dyadyuk, A hybrid adaptive antenna array for long rang mm-Wave communications, IEEE Antennas Propag. Mag. 54 (2) (2012) 271–282.

[4] T.S. Rappaport, S. Sun, R. Mayzus, H. Zhao, Y. Azar, K. Wang, G.N. Wong, J.K. Schulz, M. Samimi, F. Gutierrez, Millimeter wave mobile communications for 5G cellular: it will work! IEEE Access 1 (2013) 335–349.

[5] R.J. Mailloux, Phased Array Antenna Handbook, Artech House, Norwood, MA, 2005.

[6] W.R. Deal, N. Kaneda, J. Sor, Q.Y. Qian, T. Itoh, A new quasi-Yagi antenna for planar active antenna arrays, IEEE Trans. Microw. Theory Tech. 0018-9480, 48 (6) (2000) 910–918.

[7] V. Dyadyuk, X. Huang, L. Stokes, J. Pathikulangara, A.R. Weily, N. Nikolic, J. D. Bunton, Y. Jay Guo, Adaptive antenna arrays for ad-hoc millimetre-wave wireless communications, in: M. Khatib (Ed.), Advanced Trends in Wireless Communications, InTech, ISBN: 978-953-307-183-1, 2011. Available from: http://www.intechopen.com/books/advanced-trends-in-wireless-communications/adaptive-antenna-arrays-for-ad-hoc-millimetre-wave-wireless-communications.

[8] N. Nikolic, A.R. Weily, Compact E-band planar quasi-Yagi antenna with folded dipole driver, IET Microw. Antennas Propag. 4 (11) (2010) 1728–1734.

[9] A.R. Weily, N. Nikolic, Circularly polarized stacked patch antenna with perpendicular feed substrate, IEEE Trans. Antennas Propag. 61 (10) (2013) 5274–5278.

[10] A. Natarajan, S.K. Reynolds, M. Tsai, et al., A fully-integrated 16-element phased-array receiver in SiGe BiCMOS for 60-GHz communications, IEEE J. Solid State Circuits 46 (5) (2011) 1059–1075.

[11] S. Drago, M.C.A. van Schie, A.J.M. de Graauw, et al., A 60 GHz wideband low noise eight-element phased array RX front-end for beam steering communication applications in 45 nm CMOS, in: IEEE RFIC Symposium, 2012.

[12] W. Roh, J.-Y. Seol, J. Park, B. Lee, J. Lee, Y. Kim, J. Cho, K. Cheun, F. Aryanfar, Millimeter-wave beamforming as an enabling technology for 5G cellular communications: theoretical feasibility and prototype results, IEEE Commun. Mag. 52 (2) (2014) 106–113.

[13] P. Liao, R. York, A new phase-shifterless beam-scanning technique using arrays of coupled oscillators, IEEE Trans. Microw. Theory Tech. 41 (10) (1993) 1810–1815.

[14] J.A. Zhang, X. Huang, Y. Jay Guo, Adaptive searching and tracking algorithm for AoA estimation in localized hybrid array, in: 2015 IEEE International Conference on Communication Workshop (ICCW), 8–12 June, 2015, pp. 1095–1100.

[15] X. Huang, Y.J. Guo, Frequency-domain AoA estimation and beamforming with wideband hybrid arrays, IEEE Trans. Wirel. Commun. 10 (8) (2011) 2543–2553.

[16] W. Cai, P. Wang, Y. Li, Y. Zhang, B. Vucetic, Deployment optimization of uniform linear antenna arrays for a two-path millimeter wave communication system, IEEE Commun. Lett. 19 (4) (2015) 669–672.

[17] O. El Ayach, S. Abu-Surra, S. Rajagopal, Z. Pi, R.W. Heath Jr., Spatially sparse precoding in millimeter wave MIMO systems, IEEE Trans. Wirel. Commun. 13 (3) (2013) 1499–1513.

[18] I. Sarris, A.R. Nix, Design and performance assessment of high-capacity MIMO architectures in the presence of a line-of-sight component, IEEE Trans. Veh. Technol. 56 (4) (2007) 2194–2202.

[19] P. Wang, Y. Li, X. Yuan, L. Song, B. Vucetic, Tens of gigabits wireless communications over E-band LoS MIMO channels with uniform linear antenna arrays, IEEE Trans. Wirel. Commun. 13 (7) (2014) 3791–3805.

[20] S. Wu, L. Chiu, K. Lin, T. Chang, Robust hybrid beamforming with phased antenna arrays for downlink SDMA in indoor 60 GHz channels, IEEE Trans. Wirel. Commun. 12 (9) (2013). 4542, 4557.

[21] V. Raghavan, R.W. Heath, A.M. Sayeed, Systematic codebook designs for quantized beamforming in correlated MIMO channels, IEEE J. Sel. Areas Commun. 25 (7) (2007) 1298–1310.

[22] A. Alkhateeb, O. El Ayach, G. Leus, R.W. Heath Jr., Hybrid precoding for millimeter wave cellular systems with partial channel knowledge, in: Proceedings of Information Theory and Applications (ITA) Workshop, 2013.

Encoding and detection in mmWave massive MIMO

4

S.A.R. Naqvi, S.A. Hassan, Z. Mulk

National University of Sciences & Technology (NUST), Islamabad, Pakistan

CHAPTER OUTLINE

4.1 INTRODUCTION

In conventional ultra high frequency (UHF) wireless cellular systems, multiple antennas are used at the base station (BS), which provide high throughput as well as improved quality of service to its users. Considering a multiple cell geometry, where each cell is equipped with tens of antenna making a massive multiple input multiple output (MIMO) system, it is obvious that the knowledge of channel state information (CSI) at the BS plays an important role in achieving high system performance. The most efficient way of obtaining CSI is through reciprocity that uses uplink training of pilots [1, 2]. The major constraint while considering a multiple cell scenario, using a UHF network, is the allocation of pilot signals to the users. This affects the system performance in a great way as the CSI is further dependent upon the pilot allocation scheme. The frequent mobility of users shortens the channel coherence period and as a result the length of pilot sequence is limited. Therefore, considering the scarcity of bandwidth it is not feasible to allocate distinct orthogonal pilot signals to users in each cell. The major issue of reusing nonorthogonal pilot

mmWave Massive MIMO. http://dx.doi.org/10.1016/B978-0-12-804418-6.00004-2

signals in different cells is commonly known as pilot contamination, which limits the achievable throughput [3]. It is caused when the CSI at BS is corrupted because of pilot signals from neighboring cells using the same frequency.

As mentioned earlier, the primary bottleneck in assigning unique, orthogonal pilot signals to clients in each cell is the limited bandwidth offered by the UHF band. Greater bandwidth is also required to provide enhanced data rates: a salient feature of the upcoming fifth generation (5G) technology. With a channel bandwidth of approximately 500 MHz and the possibility of using larger antenna arrays due to small wavelength [4], investigation into the use of millimeter wave (mmWave) technology in future cellular networks is already underway [5, 6]. The subsequent discussion will aim to compare the performance of networks using UHF and mmWave technologies, considering not only the different bandwidths on offer, but also the distinct path loss models applied to transmissions in each system.

In this chapter, a regular hexagonal geometry with random deployment of users is considered in order to study the impact of pilot contamination for a more general scenario. Specifically, we study a conventional frequency reuse hex geometry and derive the throughput of the network when multiple users are reusing the same pilots leading to pilot contamination problems. We restrict the scope to only the first tier of co-channel interferers. Finally, with the help of numerical simulations, relations between cell throughput, number of antennas, number of users per cell, pilot reuse factors, and coherence period have been studied. It is shown that the reuse pattern has a strong impact on the system achievable throughput for UHF bands, whereas pilot reuse in mmWave has virtually no effect on interferences arising from users in the neighboring cells.

4.2 BACKGROUND

Marzetta [7] has shown in his work that as the number of antennas increases to infinity, the throughput eventually saturates. Gopalakrishnan et al. [8] and Ngo et al. [9] derived the asymptotic throughput bound, which shows that the throughput is limited by pilot contamination. Different techniques have been studied in order to reduce the effect of pilot contamination. Tadilo et al. [10] proposed a novel pilot optimization and channel estimation algorithm to reduce the weighted sum mean square error (WSMSE). Papadopoulos et al. [11] and Huh et al. [12] improved the throughput with the help of cooperative communication in which they divided the mobile terminals in different groups. Appaiah et al. [13] used asynchronous transmission of pilots to reduce the correlation error for channel estimation and Jose et al. [3] worked on regularized zero force precoding. Yang et al. [14] derived the throughput for a line geometry of cells where users are co-located in each cell.

mmWave-related communication is an emerging field in wireless technologies. In this respect Rappaport et al. [6] provide motivation for these new systems by proposing methodology and hardware for measurements. In recent publications [15–19], the authors have analyzed coverage and rate trends in mmWave cellular networks.

Moreover, Bai et al. [4] summarize the distinguishing features of mmWave networks, apart from introducing models for blockage and beamforming in such systems.

The rest of the chapter is organized thus; we first develop the system model, before proceeding to derive the lower bound for achievable cell throughput and the mathematical expressions different interference terms. Subsequently, the numerical results for average cell throughput and other parameters are discussed. Finally, we draw conclusions at the end of the chapter.

4.3 SYSTEM MODEL

Consider a tier-1 hexagonal geometry of a cellular system, where each cell has a radius R. By tier-1 we refer to the area of interfering cells that causes co-channel interference (CCI) and is considered the main source of pilot contamination. The reuse factor of the hexagonal geometry is given by Q, where $Q = \{1, 3, 4, 7, \ldots\}$. For instance, in Fig. 4.1, six interfering cells for the center cell constitute tier-1 geometry with $Q = 1$. Each cell contains an M-antenna BS that serves K randomly deployed single antenna mobiles. This chapter will include scenarios with Q being equal to 1, 3, or 7. Next, we proceed to determining the "path loss factor" for both

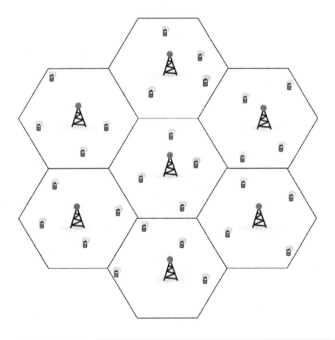

FIG. 4.1

Random deployment of users in a hexagonal geometry with $Q = 1$.

UHF and mmWave networks. To this end, we need to define the path loss models for either system. The path loss for mmWave link $L_{\text{mm}}(r)$, in dB, is modeled as,

$$L_{\text{mm}}(r) = \rho + 10\alpha \log(r) + \chi_{\text{mm}}, \tag{4.1}$$

whereas the path loss for the UHF link, $L_{\text{UHF}}(r)$, in dB, is given by,

$$L_{\text{UHF}}(r) = 20\log\left(\frac{4\pi}{\lambda_{\text{c}}}\right) + 10\alpha \log(r) + \chi_{\text{UHF}}. \tag{4.2}$$

In Eq. (4.1), χ_{mm} is the zero mean log normal random variable for the mmWave link, modeling the effects of shadowing. In addition, the fixed path loss in Eq. (4.1) is given by $\rho = 32.4 + 20\log(f_{\text{c}})$, where f_c is the carrier frequency. Similarly, in Eq. (4.2), χ_{UHF} represents shadowing in UHF links. The path loss exponents for both UHF and mmWave links are denoted by α.

Now, using these models, we extract the respective path loss factors for either system, which are given as,

$$\zeta_{i,j,k}^{\text{mm}} = 10^{-0.1(L_{\text{mm}}(r))}, \tag{4.3}$$

$$\zeta_{i,j,k}^{\text{UHF}} = 10^{-0.1(L_{\text{UHF}}(r))}, \tag{4.4}$$

where $\zeta_{i,j,k}^{\text{mm}}$ is the path loss factor for the mmWave link and $\zeta_{i,j,k}^{\text{UHF}}$ is that for the UHF link. For the sake of generalization, all subsequent mathematical manipulations will use ζ to represent the path loss factor in both networks.

A Rayleigh block fading channel has been assumed where all the channel coefficients remain constant for a block of T symbols; T being the channel coherence period. Considering the geometry, the channel matrix between the antenna i and mobiles in cell j is $\mathbf{C}_{i,j} \in \mathcal{C}^{M \times K}$ where all the entries of $\mathbf{C}_{i,j}$ are independent and identically distributed (i.i.d.) complex Gaussian with zero mean and unit variance. Fig. 4.2 further illustrates the mode of pilot reuse in the system under investigation,

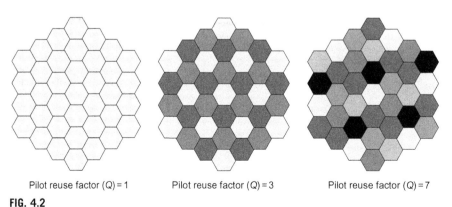

Pilot reuse factor $(Q) = 1$ Pilot reuse factor $(Q) = 3$ Pilot reuse factor $(Q) = 7$

FIG. 4.2

Pilot reuse under consideration.

wherein each color represents a set of orthogonal pilots that are also mutually orthogonal from color to color. As a final remark, all the time/frequency resources allocated to payload data transmission are used in all the cells, as in [20].

4.4 MULTICELL UPLINK COMMUNICATION

All the mobiles in this uplink scenario communicate with their respective BS in two stages: uplink training with pilot reuse and actual transmission of the data.

4.4.1 UPLINK TRAINING WITH PILOT REUSE

In each cell, orthogonal pilots are assigned to the users, which are reused by other cells by the factor Q. Assume a predesigned pilot sequence matrix,

$$\Psi = \left[\Psi_0, \ldots, \Psi_{Q-1}\right] \in \mathbb{C}^{QK \times QK}, \tag{4.5}$$

where Ψ is divided among all cells. Assume that Ψ_0 is assigned to the Cell 0 which is considered to be the serving cell (center cell from Fig. 4.1) and the pilot sequence is further reused by Cell lQ where $1 \leq |l| \leq \lfloor \frac{L}{Q} \rfloor$ and L is the total number of cells in geometry. The pilot $\Psi_{q,k}$ is assigned to a single user k of a cell in order to remove intracell interference with the help of the orthogonality of the pilot signals.

We only analyze Cell 0 here for brevity. In a single time-frequency block, all the mobiles send their pilot signals to the BS. The signals received at BS 0 are,

$$\mathbf{P}_0 = \sqrt{\phi_\tau QK} \sum_{j=0}^{L-1} \mathbf{C}_{0,j} \ \Psi_{(j)}^* + \mathbf{W}_0, \ \in \mathbb{C}^{M \times K}, \tag{4.6}$$

where $\mathbf{C}_{0,j} = \left[\zeta_{0,j,1}\mathbf{c}_{0,j,1}, \ldots, \zeta_{0,j,K}\mathbf{c}_{0,j,K}\right]$ is the channel matrix between the mobiles of cell j and BS 0, $\mathbf{c}_{0,j,k}$ is the channel vector between BS 0 and cell j's kth mobile user and $(j) = (j \bmod Q)$. The variable $\mathbf{W}_0$ is i.i.d. $\mathcal{CN}(0,1)$ and ϕ_τ is the average transmission power per mobile. The factor $\sqrt{QK}$ guarantees that average power is ϕ_τ.

The received signal $\mathbf{P}_0$ is projected onto $\Psi_{0,k}$ in order to estimate $\mathbf{c}_{0,0,k}$ at the BS 0. After normalization, the resulting signals are,

$$\bar{\mathbf{p}}_{0,k} = \mathbf{c}_{0,0,k} + \sum_{l \neq 0} \sqrt{\frac{\zeta_{0,lQ,k}}{\zeta_{0,0,k}}} \mathbf{c}_{0,lQ,k} + \frac{\mathbf{w}_{0,k}}{\sqrt{\zeta_{0,0,k}\phi_\tau KQ}} \in \mathbb{C}^{M \times 1}. \tag{4.7}$$

A minimum mean squared error (MMSE) estimator [21] is applied to the received vector of pilots to get,

$$\hat{\mathbf{c}}_{0,0,k} = \mathbf{Y}_{cp}\mathbf{Y}_{pp}^{-1}\bar{\mathbf{P}}_{0,k}, \tag{4.8}$$

where $\mathbf{Y}_{pp}$ and $\mathbf{Y}_{cp}$ are the correlation and cross-correlation matrices, respectively. As all the channel vectors are independent, therefore, the cross-correlation matrix becomes $\mathbf{Y}_{cp} = \mathbf{I}_M$. The correlation matrix $\mathbf{Y}_{pp}$ is given as,

$$\mathbf{Y}_{pp} = \mathbf{I}_M + \underbrace{\sum_{l \neq 0} \frac{\zeta_{0,lQ,k}}{\zeta_{0,0,k}}}_{} \mathbf{I}_M + \underbrace{\frac{1}{\zeta_{0,0,k} \; \phi_\tau KQ}}_{a_2} \mathbf{I}_M. \tag{4.9}$$

From the above equation, we can define $\sigma_\tau^2 = (1 + a_1 + a_2)^{-1}$ and now apply the MMSE decomposition,

$$\mathbf{c}_{0,0,k} = \hat{\mathbf{c}}_{0,0,k} + \tilde{\mathbf{c}}_{0,0,k}, \tag{4.10}$$

where $\tilde{\mathbf{c}}_{0,0,k}$ is the independent uncorrelated estimation error. Both the entries are i.i.d. where $\hat{\mathbf{c}}_{0,0,k}$ is $\mathcal{CN}(0, \sigma_\tau^2)$ and $\tilde{\mathbf{c}}_{0,0,k}$ is $\mathcal{CN}(0, 1 - \sigma_\tau^2)$.

4.4.2 ACTUAL TRANSMISSION OF DATA

In the next $T - KQ$ slots, all the mobiles transmit their data. This takes place in all cells after the uplink training of pilots. The received signal at BS 0 is,

$$r_0 = \sum_{j,k} \sqrt{\phi \zeta_{0,j,k}} \; \mathbf{c}_{0,j,k} \; r_{j,k} + \mathbf{n}_0, \tag{4.11}$$

where ϕ is the average power consumed by a mobile to transmit the data $r_{j,k}$ by a user k in cell j and $\mathbf{n}_o$ is the noise, which is $\mathcal{CN}(0,1)$. At BS 0, maximal ratio combining (MRC) is applied to receive the kth mobile's signal. The unit norm vector is denoted as,

$$\mathbf{u}_k = \frac{\hat{\mathbf{c}}_{0,0,k}}{\|\hat{\mathbf{c}}_{0,0,k}\|} \; \in \; \mathbb{C}^{M \times 1}. \tag{4.12}$$

After applying maximum ratio combining and normalization, we get,

$$\mathbf{u}_k^* \bar{\mathbf{r}}_0 \; = \mathbf{u}_k^* \hat{\mathbf{c}}_{0,0,k} \; r_{0,k} + \sqrt{\frac{1}{\phi_\tau \zeta_{0,0,k}}} w_k \tag{4.13}$$

$$+ \mathbf{u}_k^* \tilde{\mathbf{c}}_{0,0,k} \; r_{0,k} + \mathbf{u}_k^* \sum_{i \neq k} \sqrt{\frac{\zeta_{0,0,i}}{\zeta_{0,0,k}}} \mathbf{c}_{0,0,i} r_{0,i} \tag{13a}$$

$$+ \mathbf{u}_k^* \sum_{j \neq 0} \sum_{i=1}^K \sqrt{\frac{\zeta_{0,j,i}}{\zeta_{0,0,k}}} \mathbf{c}_{0,j,i} r_{j,i}. \tag{13b}$$

In Eq. (4.13), the two terms are desired signal and the noise respectively. In Eq. (13a) the terms refer to intracell interference and in Eq. (13b) the term denotes intercell interference.

4.4.3 **ACHIEVABLE CELL THROUGHPUT**

The average throughput of cell 0 using the abovementioned scheme is achieved by the following derivation. Tier-1 BSs of hexagonal geometry with M-antennas are deployed. Each BS serves K randomly located single antenna mobiles of its own cell with path loss factor $\zeta_{0,0,k}$. The inter cell interference in Eq. (13b) can be re-written as,

$$\mathbf{u}_k^* \sum_{j\neq 0} \sum_{i=1}^K \sqrt{\frac{\zeta_{0,j,i}}{\zeta_{0,0,k}}} \mathbf{c}_{0,j,i} r_{j,i} \quad -\mathbf{u}_k^* \sum_{l\neq 0} \sqrt{\frac{\zeta_{0,lQ,k}}{\zeta_{0,0,k}}} \mathbf{c}_{0,lQ,k} r_{lQ,k} \tag{4.14}$$

$$+\mathbf{u}_k^* \sum_{l\neq 0} \sqrt{\frac{\zeta_{0,lQ,k}}{\zeta_{0,0,k}}} \mathbf{c}_{0,lQ,k} r_{lQ,k}, \tag{14a}$$

where Eq. (14a) represents the pilot contamination. For a single user k, we can calculate the throughput as,

$$\mathcal{R} \geq \log_2 \left(1 + \frac{\left|\mathbf{u}_k^* \hat{\mathbf{c}}_{0,0,k}\right|^2}{\frac{1}{\phi \zeta_{0,0,k}} + I_1 + I_2 + I_3} \right), \tag{4.15}$$

where $I_1, I_2,$ and I_3 are intracell, intercell, and interference due to pilot contamination, respectively, and given as,

$$I_1 = \left|\mathbf{u}_k^* \tilde{\mathbf{c}}_{0,0,k}\right|^2 + \sum_{i\neq k} \left|\mathbf{u}_k^* \sqrt{\frac{\zeta_{0,0,i}}{\zeta_{0,0,k}}} \mathbf{c}_{0,0,i}\right|^2, \tag{4.16}$$

$$I_2 = \sum_{j\neq 0} \sum_{i=1}^K \frac{\zeta_{0,j,i}}{\zeta_{0,0,k}} \left|\mathbf{u}_k^* \mathbf{c}_{0,j,i}\right|^2 - \sum_{l\neq 0} \frac{\zeta_{0,lQ,k}}{\zeta_{0,0,k}} \left|\mathbf{u}_k^* \mathbf{c}_{0,lQ,k}\right|^2, \tag{4.17}$$

$$I_3 = \sum_{l\neq 0} \frac{\zeta_{0,lQ,k}}{\zeta_{0,0,k}} \left|\mathbf{u}_k^* \mathbf{c}_{0,lQ,k}\right|^2. \tag{4.18}$$

Both the terms I_1 and I_2 are independent of $\left|\mathbf{u}_k^* \hat{\mathbf{c}}_{0,0,k}\right|^2$, whereas I_3 is dependent upon $\mathbf{c}_{0,0,k}$ which also shows dependence upon $\left|\mathbf{u}_k^* \hat{\mathbf{c}}_{0,0,k}\right|^2$. In order to analyze I_3, MMSE decomposition is applied on $\mathbf{c}_{0,lQ,k}$. Hence we can write $\hat{\mathbf{c}}_{0,lQ,k}$ as $\hat{\mathbf{c}}_{0,0,k}$ because the MMSE decomposition for both the terms is proportional. Thus,

$$\mathbf{c}_{0,lQ,k} = \sqrt{\frac{\zeta_{0,lQ,k}}{\zeta_{0,0,k}}} \hat{\mathbf{c}}_{0,0,k} + \tilde{\mathbf{c}}_{0,lQ,k}, \tag{4.19}$$

where $\tilde{\mathbf{c}}_{0,lQ,k}$ is i.i.d. $\mathcal{CN}(0, 1 - \frac{\zeta_{0,lQ,k}}{\zeta_{0,0,k}} \sigma_\tau^2)$ and denotes the estimation error and is not dependent upon $\hat{\mathbf{c}}_{0,lQ,k}$. Re-write I_3 as,

$$I_3 = \sum_{l \neq 0} \frac{\zeta_{0,IQ,k}^2}{\zeta_{0,0,k}^2} \left| \mathbf{u}_k^* \hat{\mathbf{c}}_{0,0,k} \right|^2 + \sum_{l \neq 0} \frac{\zeta_{0,IQ,k}}{\zeta_{0,0,k}} \left| \mathbf{u}_k^* \tilde{\mathbf{c}}_{0,IQ,k} \right|^2$$
$$+ \sum_{l \neq 0} \left(\left(\frac{\zeta_{0,IQ,k}}{\zeta_{0,0,k}} \right)^{\frac{3}{2}} \left(\mathbf{u}_k^* \hat{\mathbf{c}}_{0,0,k} \tilde{\mathbf{c}}^*_{0,IQ,k} \mathbf{u}_k + \mathbf{u}_k^* \tilde{\mathbf{c}}_{0,IQ,k} \hat{\mathbf{c}}^*_{0,0,k} \mathbf{u}_k \right) \right),$$

(4.20)

and denote the rate conditioned on $\hat{\mathbf{c}}^*_{0,0,k}$ by $\bar{\mathcal{R}}$. To achieve the lower bound of $\bar{\mathcal{R}}$, convexity of $\log_2 \left(1 + \frac{1}{x} \right)$ is used, therefore,

$$\bar{\mathcal{R}} \geq \log_2 \left(1 + \frac{\left| \mathbf{u}_k^* \hat{\mathbf{c}}_{0,0,k} \right|^2}{\frac{1}{\phi \zeta_{0,0,k}} + \mathbb{E}[I_1] + \mathbb{E}[I_2] + \mathbb{E}[I_3]} \right),$$

(4.21)

where $\mathbb{E}[I_1]$, $\mathbb{E}[I_2]$, and $\mathbb{E}[I_3]$ are given as,

$$\mathbb{E}[I_1] = (1 - \sigma_\tau^2) + \sum_{i \neq k} \frac{\zeta_{0,0,i}}{\zeta_{0,0,k}},$$

(4.22)

$$\mathbb{E}[I_2] = \sum_{j \neq 0} \sum_{i=1}^{K} \frac{\zeta_{0,j,i}}{\zeta_{0,0,k}} - \sum_{l \neq 0} \frac{\zeta_{0,IQ,k}}{\zeta_{0,0,k}},$$

(4.23)

$$\mathbb{E}[I_3] = \sum_{l \neq 0} \frac{\zeta_{0,IQ,k}^2}{\zeta_{0,0,k}^2} \left| \mathbf{u}_k^* \hat{\mathbf{c}}_{0,0,k} \right|^2 + \sum_{l \neq 0} \frac{\zeta_{0,IQ,k}}{\zeta_{0,0,k}} \left(1 - \frac{\zeta_{0,IQ,k}}{\zeta_{0,0,k}} \sigma_\tau^2 \right).$$

(4.24)

The last two terms in Eq. (4.20) have zero mean hence neglected in Eq. (4.24). Note that,

$$\left| \mathbf{u}_k^* \hat{\mathbf{c}}_{0,0,k} \right|^2 = \sum_{q=1}^{M} \hat{c}_{k,q}^2,$$

(4.25)

where the term $\hat{c}_{k,q}$ is i.i.d. $\mathcal{CN}(0, \sigma_\tau^2)$. Hence, $\left| \mathbf{u}_k^* \hat{\mathbf{c}}_{0,0,k} \right|^2$ has Gamma distribution with parameters (M, σ_τ^2). The right hand side of Eq. (4.21) is written as,

$$\log_2 \left(1 + \frac{\Gamma_k}{\frac{1}{\phi \zeta_{0,0,k}} + \Omega_1 + \Omega_2 + \sum_{l \neq 0} \frac{\zeta_{0,IQ,k}^2}{\zeta_{0,0,k}^2} \Gamma_k} \right),$$

(4.26)

where $\Gamma_k = \left| \mathbf{u}_k^* \hat{\mathbf{c}}_{0,0,k} \right|^2$, Ω_1 is the intracell interference and Ω_2 is the intercell interference given as,

$$\Omega_1 = (1 - \sigma_\tau^2) + \sum_{i \neq k} \frac{\zeta_{0,0,i}}{\zeta_{0,0,k}},$$

(4.27)

$$\Omega_2 = \sum_{j \neq 0} \sum_{i=1}^{K} \frac{\zeta_{0,j,i}}{\zeta_{0,0,k}} - \sum_{l \neq 0} \frac{\zeta_{0,IQ,k}^2}{\zeta_{0,0,k}^2} \sigma_\tau^2.$$

(4.28)

The inverse Gamma distribution $(1/\Gamma_k)$ has a mean value of $1/(M-1)\sigma_\tau^2$. Therefore,

$$\mathcal{R}_k \geq \log_2 \left(1 + \frac{1}{(\frac{1}{\phi \zeta_{0,0,k}} + \Omega_1 + \Omega_2)/\Gamma_k + \sum_{l \neq 0} \frac{\zeta_{0,lQ,k}^2}{\zeta_{0,0,k}^2}} \right) \tag{4.29}$$

$$\mathcal{R}_k \geq \log_2 \left(1 + \frac{1}{(\frac{1}{\phi \zeta_{0,0,k}} + \Omega_1 + \Omega_2)/\mathbb{E}\left[\frac{1}{\Gamma_k}\right] + \sum_{l \neq 0} \frac{\zeta_{0,lQ,k}^2}{\zeta_{0,0,k}^2}} \right) \tag{4.30}$$

$$\mathcal{R}_k \geq \log_2 \left(1 + \frac{(M-1)\sigma_\tau^2}{\frac{1}{\phi \zeta_{0,0,k}} + \Omega_1 + \Omega_2 + \Omega_3} \right) \tag{4.31}$$

where $\Omega_3 = (M-1)\sum_{l \neq 0} \frac{\zeta_{0,lQ,k}^2}{\zeta_{0,0,k}^2}\sigma_\tau^2$ is the pilot contamination.

The average achievable throughput for cell 0 is given by,

$$\mathcal{R} \geq K(1 - \frac{QK}{T})\log_2 \left(1 + \frac{(M-1)\sigma_\tau^2}{\frac{1}{\phi \zeta_{0,0,k}} + \Omega_1 + \Omega_2 + \Omega_3} \right), \tag{4.32}$$

where ϕ is the average data power, T is the channel coherence period, and σ_τ^2 is the normalized estimation power,

$$\sigma_\tau^2 = \left(1 + \sum_{l \neq 0} \frac{\zeta_{0,lQ,k}}{\zeta_{0,0,k}} + \frac{1}{\sqrt{\phi_\tau KQ}} \right)^{-1}. \tag{4.33}$$

In Eq. (4.32) the constants Ω_1, Ω_2, and Ω_3 are defined as,

$$\Omega_1 = (1 - \sigma_\tau^2) + \sum_{i \neq k} \frac{\zeta_{0,0,i}}{\zeta_{0,0,k}}, \tag{4.34}$$

$$\Omega_2 = \sum_{j \neq 0} \sum_{i=1}^{K} \frac{\zeta_{0,j,i}}{\zeta_{0,0,k}} - \sum_{l \neq 0} \frac{\zeta_{0,lQ,k}^2}{\zeta_{0,0,k}^2}\sigma_\tau^2, \tag{4.35}$$

$$\Omega_3 = (M-1)\sigma_\tau^2 \sum_{l \neq 0} \frac{\zeta_{0,lQ,k}^2}{\zeta_{0,0,k}^2}. \tag{4.36}$$

4.5 RESULTS

In this section, we evaluate the effect of different interferences, pilot reuse factor, and network topology on the average cell throughput. Different simulations have been carried out to study the parameters upon which maximum throughput is achieved. For all simulations, we assume coherence period $T = 100$, radius $R = 800$ m, and $\phi_t = \phi = 100$ mW, unless noted otherwise. However, it is worth remembering that recent studies regarding the outdoor channel propagation characteristics for the mmWave network, referred to in Roh et al. [22], consider transmission links established for a distance of up to only 200–300 m. The value of R is chosen only for a fair comparison between the two modes of transmission under investigation. All the mobiles in a UHF system operate at 3 GHz frequency and 10 MHz bandwidth. On the other hand, the mmWave network uses a carrier frequency of 94 GHz and a bandwidth of 2 GHz. The path loss exponents are assumed to be 3.8 for UHF (as in Li et al. [14]) and 3.3 for mmWave, unless otherwise stated. In order to obtain ergodic cell throughput, the Monte-Carlo method is applied for a single user having random location in every trial.

Since the mathematical manipulations of the previous section have provided us with expressions for not only the pilot contamination, but also the intra- and intercellular interferences in the medium, we begin by contrasting the interference patterns in the networks under discussion. The first graph, as shown in Fig. 4.3, displays the said information for the UHF link for 5, 10, 15, and 20 users per cell, with the number of antennas, M, being fixed at 100.

Note that the intra- and the intercellular interferences increase steadily with an increase in the number of users for either value of Q. On average, both these interferences have the same impact on the throughput because the location of user is random; for distances closer to the BS, the intracell interference dominates and for distances far away, the intercell interference dominates. Hence, both generally have the same contribution to the cell throughput. It may also be seen that the pilot contamination has separate fixed values for the two reuse factors, independent of the total number of users. This is because, for tier-1, the number of interferers causing pilot contamination will always be six for any reuse factor. However, the figure clearly shows the potency of pilot reuse, as pilot contamination falls from approximately 0 dBm for $Q = 1$, to -40 dBm for $Q = 7$. Simply stated, the greater the value of the pilot reuse factor, the greater the distance of the interferer from the BS, resulting in a greater path loss experienced by the interfering signal and, hence, lesser pilot contamination.

Next, we consider the performance of the mmWave link in terms of interference for the same number of users and antennas, as shown in Fig. 4.4. The gradual increase in inter- and intracellular interferences mirrors the trend seen previously in UHF systems. However, there are two other observations to be noted. Firstly, the intercellular interference in the mmWave network for both pilot reuse factors is much smaller than that seen in the previous graph. The reader may notice that there is a difference of nearly 10 dBm in the intercellular interferences for the two systems at $K = 20$, for

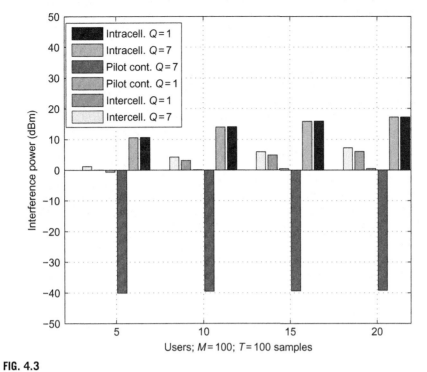

FIG. 4.3

Interference components in UHF systems using $Q = 1$ and 7.

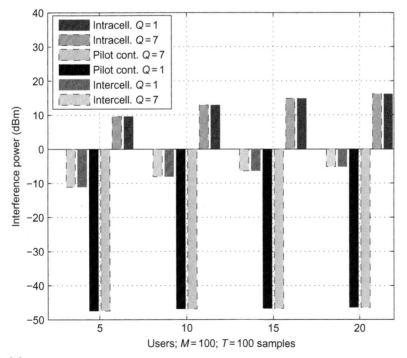

FIG. 4.4

Interference components in mmWave systems using $Q = 1$ and 7.

either value of Q. Secondly, the pilot contamination is not only the same for $Q = 1$ and $Q = 7$, but also remarkably smaller than in Fig. 4.3. As an indicator, the pilot contamination for $Q = 1$ is -48 dBm, down from 0 dBm for the UHF-based communication. These results allude to the advantage afforded by the mmWave link's increased path loss, which drowns out the interference from the neighboring cells to the extent that a higher pilot reuse factor has no further impact on pilot contamination.

Now, we vary the coherence period, T in samples, while keeping Q fixed at 3. Fig. 4.5 plots the average sum-rate against T for 6, 12, and 18 users in both UHF and mmWave networks. When we consider only six users in a cell, the throughput does not change much with the increase in T but when K goes to 18, the throughput starts increasing gradually as T increases. It can be seen that for lower T, $K = 18$ has the least performance. This is because, it is very difficult to generate orthogonal pilots for more users in lesser coherence period and hence the pilot contamination dominates the system. When T increases beyond a specific value, $K = 18$ outperforms the rest of two cases. The mmWave system performs consistently better than that of UHF for the same value of K primarily due to decreased interference in the system.

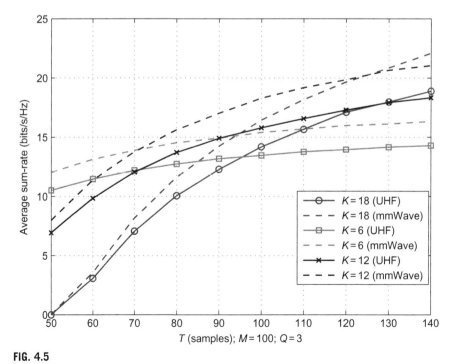

FIG. 4.5

Average throughput versus the coherence period for $Q = 3$.

Fig. 4.6 illustrates the relationship between the average throughput and M for both UHF and mmWave links employing a pilot reuse factor of 3. As a preliminary observation, it may be stated that the *sum-rate* is higher for greater values of K, as M increases. Here again it may be noted that the mmWave network for the same value of K performs better than the UHF network under the assumed conditions, as explained previously. All plots start to approach saturation at higher values of M. Another point worth noting is that, at lower values of M, the performance of the mmWave system with 5 users per cell is nearly equivalent to that of the UHF system with 10 users per cell.

Extending the previous discussion, we now fix K at a low value and then observe the effect of Q, M and the mode of transmission on the system performance. The results are shown in Fig. 4.7. Once again, all plots start to saturate at higher values of M. Moreover, we see that the average sum-rate for $Q = 1$ for the mmWave link nearly equals that for the system operating at UHF with $Q = 7$ at $M = 100$.

Early on, it was mentioned that mmWave is a candidate for being the enabling technology of 5G networks due to its superior bandwidth as compared to UHF. We conclude this Results section by highlighting this very point. The data-sets in Fig. 4.8 were achieved by multiplying the results from Eq. (4.32) by the bandwidth of the respective transmission technique. As such, the average rate is measured simply

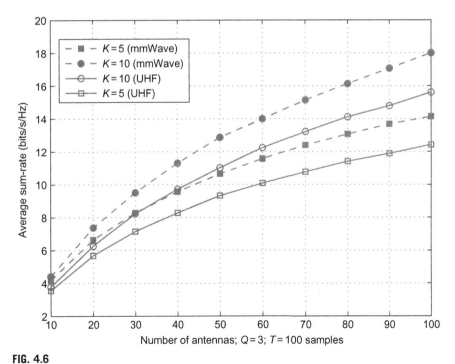

FIG. 4.6

Average throughput versus number of base station antennas at $Q = 3$.

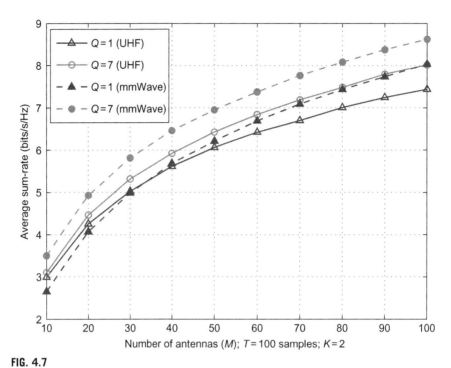

FIG. 4.7

Average throughput versus number of base station antennas at $K = 2$.

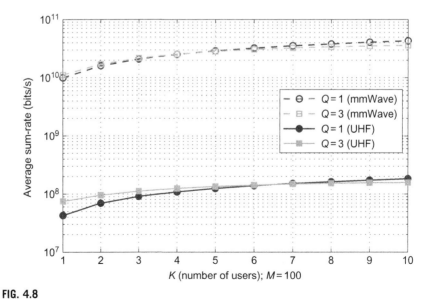

FIG. 4.8

Average throughput versus K.

in *bits per second* in this case. It is obvious from the figure that the mmWave link outperforms the UHF link. Additionally, the plots for $Q = 1$ and $Q = 3$ for the mmWave network provide virtually similar results for all values of K used. This further re-enforces the fact that at such high bandwidth, mmWave links need not incorporate pilot reuse into the network.

4.6 CONCLUSION

In this chapter we evaluated the lower bound for nonasymptotic throughput of an uplink massive MIMO system with hexagonal geometry and random deployment of users, using MRC receivers. This expression was then used to gauge the performance of the mmWave and UHF links under similar conditions. It was observed that, for UHF, the effect of pilot contamination diminishes when we have a reuse factor greater than 1. mmWave, on the other hand, shows very little variation in its already depleted pilot contamination levels for increase values of this factor. Moreover, the relationship between the number of users in a cell, number of BS antennas, duration of coherence period and average cell throughput has also been studied through simulations. Through these studies, we can calculate the lower bound of average throughput for a cell that can be achieved for the different values of parameters like K, T, Q, and M.

REFERENCES

[1] T.L. Marzetta, B.M. Hochwald, Fast transfer of channel state information in wireless systems, IEEE Trans. Signal Process. 54 (2006) 1268–1278.

[2] L. Lu, G.Y. Li, A.L. Swindlehurst, A. Ashikhmin, R. Zhang, An overview of massive MIMO: benefits and challenges, IEEE J. Sel. Top. Signal Process. 8 (5) (2014) 742–758.

[3] J. Jose, A. Ashikhmin, T.L. Marzetta, S. Vishwanath, Pilot contamination problem in multi-cell TDD systems, in: IEEE International Symposium on Information Theory, July, 2009, pp. 2184–2188.

[4] T. Bai, A. Alkhateeb, R. Heath, Coverage and capacity of millimeter-wave cellular networks, IEEE Commun. Mag. 52 (9) (2014) 70–77.

[5] T.S. Rappaport, R.W. Heath, R.C. Daniels, J.N. Murdock, Millimeter Wave Wireless Communication, Prentice Hall, Upper Saddle River, NJ, 2014.

[6] T.S. Rappaport, S. Sun, R. Mayzus, H. Zhao, Y. Azar, K. Wang, G.N. Wong, J.K. Schulz, M. Samimi, F. Gutierrez, Millimeter wave mobile communications for 5G cellular: it will work! IEEE Access 1 (2013) 335–349.

[7] T.L. Marzetta, Noncooperative cellular wireless with unlimited number of BS antenna, IEEE Trans. Wirel. Commun. 9 (11) (2010) 3590–3600.

[8] B. Gopalkrishnan, N. Jindal, An analysis of pilot contamination on multi-user MIMO cellular systems with many antennas, in: Proceedings of IEEE SPAWC, June, 2011, pp. 381–385.

[9] H.Q. Ngo, T.L. Marzetta, E.G. Larsson, Analysis of pilot contamination effect in very large multicell multiuser MIMO systems for physical channel models, in: Proceedings of IEEE ICASSP, May, 2011, pp. 3464–3467.

[10] T.E. Bogale, L.B. Le, Pilot optimization and channel estimation for multiuser massive MIMO systems, in: Proceedings of IEEE CISS, March, 2014.

[11] H.C. Papadopoulos, G. Caire, S.A. Ramprashad, Achieving large spectral efficiencies from MU-MIMO with tens of antennas: location adaptive TDD MU-MIMO design and user scheduling, in: Proceedings of IEEE ASILOMAR, November, 2010, pp. 636–643.

[12] H. Huh, G. Caire, H.C. Papadopoulos, S.A. Ramprashad, Achieving 'massive MIMO' spectral efficiency with a not-so-large number of antennas, IEEE Trans. Wirel. Commun. 11 (9) (2012) 3226–3239.

[13] K. Appaiah, A. Ahsikhmin, T. Marzetta, Pilot contamination reduction in multi-user TDD systems, in: Proceedings of IEEE ICC, May, 2010.

[14] Y. Li, Y.H. Nam, B.L. Ng, J. Zhang, A non-asymptotic throughput for massive MIMO cellular uplink with pilot reuse, in: Proceedings of the Wireless Communication Symposium—Globecom, 2012, pp. 4500–4504.

[15] S. Rangan, T.S. Rappaport, E. Erkip, Millimeter wave cellular wireless networks: potentials and challenges, Proc. IEEE 102 (3) (2014) 366–385.

[16] A. Ghosh, T.A. Thomas, M.C. Cudak, R. Ratasuk, P. Moorut, F.W. Vook, T. S. Rappaport, G.R. MacCartney, S. Sun, S. Nie, Millimeter wave enhanced local area systems: a high data rate approach for future wireless networks, IEEE J. Sel. Areas Commun. 32 (6) (2014) 1152–1163.

[17] T. Bai, R.W. Heath, Coverage and rate analysis for millimeter-wave cellular networks, IEEE Trans. Wirel. Commun. 14 (2) (2015) 1100–1114.

[18] M.N. Kulkarni, S. Singh, J.G. Andrews, Coverage and rate trends in dense urban mmWave cellular networks, in: Proceedings of IEEE Global Communications Conference (GLOBECOM), 8–12 December, 2014, pp. 3809–3814.

[19] S. Singh, M.N. Kulkarni, A. Ghosh, J.G. Andrews, Tractable model for rate in self-backhauled millimeter wave cellular networks, IEEE J. Sel. Areas Commun. 33 (10) (2015) 2196–2211.

[20] E. Bjornson, E.G. Larsson, M. Debbah, Massive MIMO for maximal spectral efficiency: how many users and pilots should be allocated? IEEE Trans. Wirel. Commun. 99 (2015) 1–16.

[21] G.L. Stuber, Principles of Mobile Communication, Springer, New York, NY, 2011.

[22] W. Roh, J. Seol, J. Park, B. Lee, J. Lee, Y. Kim, J. Cho, K. Cheun, F. Aryanfar, Millimeter-wave beamforming as an enabling technology for 5G cellular communications: theoretical feasibility and prototype results, IEEE Commun. Mag. 52 (2) (2014) 106–113.

Precoding for mmWave massive MIMO

X. Gao*, L. Dai*, Z. Gao†, T. Xie*, Z. Wang*

Tsinghua University, Beijing, China *Beijing Institute of Technology, Beijing, China†*

CHAPTER OUTLINE

5.1 INTRODUCTION

While the fundamentals of precoding (beamforming) are the same regardless of carrier frequency, signal processing in millimeter-wave (mmWave) massive multiple-input multiple-output (MIMO) systems is subject to a set of nontrivial practical constraints [1]. The traditional analog beamforming for mmWave communications is usually realized by the analog circuitry such as the phase shifter (PS) network with only a small number of RF chains, which places constant amplitude constraints on the design of the analog beamformer [2]. As a result, the analog beamforming will suffer from serious performance loss, although it is simple to be implemented in the hardware. By contrast, the digital precoding at low frequencies, such as the dirty

mmWave Massive MIMO. http://dx.doi.org/10.1016/B978-0-12-804418-6.00005-4

paper coding (DPC), can control both the signal's phase and amplitude to cancel interferences and achieve the optimal performance [3]. However, the digital precoding requires a dedicated baseband and RF chain for each antenna element, which is costly and energy-intensive at present. If digital precoding is directly applied in mmWave massive MIMO systems with a large number of antennas (e.g., hundreds or even thousands), the associated enormous number of RF chains will bring prohibitively high cost and energy consumption. For example, the energy consumption of one RF chain (including digital-to-analog converter, up converter, etc.) at mmWave is about 250 mW [1], and for a mmWave massive MIMO system with 64 antennas, 16 W will be consumed just by RF chains, not to mention the transmitted energy. To solve this problem, the hybrid analog and digital precoding is proposed [4]. Specifically, it divides the optimal digital precoder into two steps. In the first step, a small-size digital precoder (realized by a small number of RF chains) is employed to cancel interferences, while in the second step, a large-size analog beamformer (realized by a large number of analog phase shifters without RF chains) is used to increase the antenna array gain. Therefore, hybrid precoding can significantly reduce the number of required RF chains without obvious performance loss via careful design, which makes it a promising precoding technology for mmWave massive MIMO systems.

The rest of this chapter is organized as follows. Section 5.2 briefly introduces the channel model of mmWave massive MIMO. After that, traditional schemes, including digital precoding and analog beamforming, are introduced. Finally, hybrid digital and analog precoding is discussed, which is more appropriate for mmWave massive MIMO systems.

5.2 CHANNEL MODEL FOR mmWave MASSIVE MIMO

The high free-space path loss is a characteristic of mmWave propagation, leading to limited spatial selectivity or scattering [5]. On the other hand, the large tightly packed antenna arrays are characteristics of mmWave transceivers, leading to high levels of antenna correlation [6]. This feature of tightly packed arrays in sparse scattering environments makes many of the statistical fading distributions used in traditional MIMO analysis inaccurate for mmWave channel modeling. For this reason, we adopt a narrowband clustered channel representation, based on the extended Saleh-Valenzuela model, which allows us to accurately capture characteristics in mmWave channels [5,6]. Using the clustered channel model, the channel matrix $\mathbf{H}$ is assumed to be a sum of the contributions of L propagation paths. Therefore, the discrete-time narrowband channel $\mathbf{H}$ can be written as [5–7]

$$\mathbf{H} = \sqrt{\frac{N_t N_r}{L}} \sum_{l=1}^{L} \alpha_l \Lambda_r\left(\phi_l^r, \theta_l^r\right) \Lambda_t\left(\phi_l^t, \theta_l^t\right) \mathbf{a}_r\left(\phi_l^r, \theta_l^r\right) \mathbf{a}_t^H\left(\phi_l^t, \theta_l^t\right), \tag{5.1}$$

where α_l is the complex gain of the lth path, whereas ϕ_l^r (θ_l^r) and ϕ_l^t (θ_l^t) are its azimuth (elevation) angles of arrival and departure (AoAs/AoDs), respectively. The functions $\Lambda_r(\phi_l^r, \theta_l^r)$ and $\Lambda_t(\phi_l^t, \theta_l^t)$ represent the receiver and transmitter antenna element gain at the corresponding AoAs/AoDs. For simplicity but without loss of generality, $\Lambda_r(\phi_l^r, \theta_l^r)$ and $\Lambda_t(\phi_l^t, \theta_l^t)$ can be set as one within the range of AoDs/AoAs [4,7]. Finally, the vectors $\mathbf{a}_r(\phi_l^r, \theta_l^r)$ and $\mathbf{a}_t(\phi_l^t, \theta_l^t)$ represent the normalized receiver and transmitter array response vectors at an azimuth (elevation) angle of ϕ_l^r (θ_l^r) and ϕ_l^t (θ_l^t), respectively, which depend on the transmitter and receiver antenna array structure. For the uniform linear array (ULA) with N elements, the array response vector can be presented as [4,7,8]

$$\mathbf{a}_{\mathrm{ULA}}(\phi) = \frac{1}{\sqrt{N}}\left[1, e^{j\frac{2\pi}{\lambda}d\sin(\phi)}, ..., e^{j(N-1)\frac{2\pi}{\lambda}d\sin(\phi)}\right]^T, \tag{5.2}$$

where λ denotes the wavelength of the signal and d is the interelement spacing. Note that we do not include θ in the arguments of $\mathbf{a}_{\mathrm{ULA}}$ as the array's response is invariant in the elevation domain. In the case of a uniform planar array (UPA) with W_1 and W_2 elements ($W_1 W_2 = N$) on horizontal and vertical, respectively, the array response vector can be given by [4,7,8]

$$\mathbf{a}_{\mathrm{UPA}}(\phi, \theta) = \frac{1}{\sqrt{N}}\left[1, ..., e^{j\frac{2\pi}{\lambda}d(x\sin(\phi)\sin(\theta) + y\cos(\theta))}, ..., e^{j\frac{2\pi}{\lambda}d((W_1-1)\sin(\phi)\sin(\theta) + (W_2-1)\cos(\theta))}\right]^T,$$
$$\tag{5.3}$$

where $0 \le x \le W_1 - 1$ and $0 \le y \le W_2 - 1$. Considering UPAs are of interest in mmWave beamforming because they (1) yield smaller antenna array dimensions; (2) facilitate packing more antenna elements in a reasonably sized array; and (3) enable beamforming in the elevation domain (also known as 3D beamforming).

5.3 DIGITAL PRECODING

Digital precoding is a traditional technology widely used in low-frequency MIMO systems [9]. The basic idea of digital precoding is to control both the phases and amplitudes of original signals to cancel interferences in advance. Generally speaking, digital precoding can be divided into two categories: linear precoding, where the transmitted signals are composed by a linear combination of the original signals; and nonlinear precoding, where the transmitted signals are obtained in a nonlinear way. Besides, the digital precoding can be also divided into single-user precoding and multiuser precoding. In this section, we will first focus on single-user systems and discuss some simple linear digital precoding schemes, such as matched filter (MF) and zero-forcing (ZF) precoding. After that, multiuser systems are considered and the classical nonlinear block diagonalization (BD) precoding will be introduced.

5.3.1 SINGLE-USER DIGITAL PRECODING

In this subsection, we mainly focus on simple linear digital precoding. Consider the single-user mmWave massive MIMO system with digital precoding as shown in Fig. 5.1, where the base station (BS) employs N_t antennas to simultaneously transmit

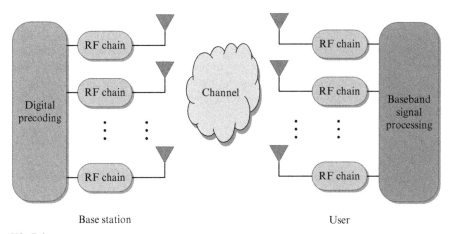

FIG. 5.1

Architecture of digital precoding for single-user mmWave massive MIMO system.

N_r data streams to a user with N_r antennas ($N_r < N_t$). The BS applies an $N_t \times N_r$ digital precoder **D** using its N_t RF chains and the transmitted signal can be presented by

$$\mathbf{x} = \mathbf{Ds}, \tag{5.4}$$

where **s** is the $N_r \times 1$ original signal vector before precoding with normalized power as $\mathbb{E}(\mathbf{ss}^H) = (1/N_r)\mathbf{I}_{N_r}$. Note that to meet the total transmit power constraint, **D** should also satisfy $\|\mathbf{D}\|_F^2 = \mathrm{tr}(\mathbf{DD}^H) = N_r$.

Under a narrowband system [2–4], the received signal vector **y** of size $N_r \times 1$ can be correspondingly presented as

$$\mathbf{y} = \sqrt{\rho}\mathbf{HDs} + \mathbf{n}, \tag{5.5}$$

where **H** is the $N_r \times N_t$ channel matrix with normalized power $\mathbb{E}\left(\|\mathbf{H}\|_F^2\right) = N_t N_r$ as introduced in Section 5.2, ρ denotes the average received power, and **n** is an additive white Gaussian noise (AWGN) vector, whose entries follow the independent and identical distribution (i.i.d.) $\mathcal{CN}(0, \sigma_n^2)$. Moreover, we assume the channel matrix **H** is perfectly known at the BS to enable precoding.

The simplest linear digital precoding is MF precoding [10], which can be presented as

$$\mathbf{D} = \sqrt{\frac{N_r}{\mathrm{tr}(\mathbf{FF}^H)}}\mathbf{F},$$
$$\mathbf{F} = \mathbf{H}^H. \tag{5.6}$$

MF precoding can maximize the received signal-to-noise ratio (SNR) at the user side. However, it usually involves severe interferences among different data streams. To this end, the well-known ZF precoding is proposed [10]. The corresponding digital precoder **D** can be presented as

$$D = \sqrt{\frac{N_r}{\mathrm{tr}\left(\mathbf{FF}^H\right)}}\mathbf{F},$$

$$\mathbf{F} = \mathbf{H}^H\left(\mathbf{HH}^H\right)^{-1}, \tag{5.7}$$

which can entirely eliminate the interferences among different data streams. However, because the digital precoder $\mathbf{D}$ is required to satisfy the total transmitter power constraint, ZF precoding may enhance the power of noise, leading to some performance loss compared with the ideal channel capacity. To make a better trade-off between the received SNR and interferences, Wiener filter (WF) precoding is proposed [10], which can be presented as

$$D = \sqrt{\frac{N_r}{\mathrm{tr}\left(\mathbf{FF}^H\right)}}\mathbf{F},$$

$$\mathbf{F} = \mathbf{H}^H\left(\mathbf{HH}^H + \frac{\sigma_n^2 N_r}{\rho}\mathbf{I}\right)^{-1}. \tag{5.8}$$

5.3.2 MULTIUSER DIGITAL PRECODING

Consider the multiuser mmWave system as shown in Fig. 5.2, where the BS equips N_{BS} antennas and N_{BS} RF chains to simultaneously communicate with U mobile stations (MSs). Each MS equips N_{MS} antennas. The total number of data streams for communication is $N_{MS}U$ ($N_{MS}U \leq N_{BS}$).

On the downlink, the BS employs an $N_{BS} \times N_{MS}U$ digital precoder in the baseband $\mathbf{D} = [\mathbf{D}_1, \mathbf{D}_2, ..., \mathbf{D}_U]$, where $\mathbf{D}_u$ of size $N_{BS} \times N_{MS}$ is the digital precoder for the

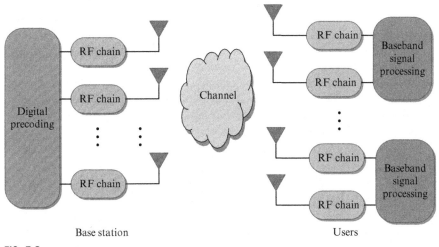

FIG. 5.2

Architecture of digital precoding for multiuser mmWave massive MIMO systems.

uth user. Also, $\mathbf{D}_u$ should satisfy the total transmit power constraint $\|\mathbf{D}_u\|_F = N_{MS}$. Consider the narrowband block-fading channel model as in Refs. [2–4]. The received signal vector $\mathbf{r}_u$ observed by the uth MS can be presented by

$$\mathbf{r}_u = \mathbf{H}_u \sum_{n=1}^{U} \mathbf{D}_n \mathbf{s}_n + \mathbf{n}_u, \tag{5.9}$$

where $\mathbf{s}_n$ of size $N_{MS} \times 1$ is the original signal vector before precoding with normalized power, $\mathbf{H}_u$ of size $N_{MS} \times N_{BS}$ denotes the mmWave massive MIMO channel matrix between the BS and the uth MS, and $\mathbf{n}_u$ is an AWGN vector, whose entries follow the independent and identical distribution (i.i.d.) $\mathcal{CN}\left(0, \sigma_n^2\right)$.

It can be observed from Eq. (5.9) that the terms $\mathbf{H}_u \mathbf{D}_n \mathbf{s}_n$ for $n \neq u$ are interferences to the uth MS. Therefore, a straightforward idea is to design all $\mathbf{D}_n$ to satisfy $\mathbf{H}_u \mathbf{D}_n = \mathbf{0}$, which is called BD precoding [11]. Specifically, we can first define the matrix $\overline{\mathbf{H}}_u$ as

$$\overline{\mathbf{H}}_u = \left[\mathbf{H}_1^H, ..., \mathbf{H}_{u-1}^H, \mathbf{H}_{u+1}^H, ..., \mathbf{H}_U^H\right]^H, \tag{5.10}$$

and $\mathbf{D}_n$ should lie in the null space of $\overline{\mathbf{H}}_u$. To do so, we need to compute the singular value decomposition (SVD) of $\overline{\mathbf{H}}_u$ as

$$\overline{\mathbf{H}}_u = \overline{\mathbf{U}}_u \overline{\mathbf{\Lambda}}_u \overline{\mathbf{V}}_u^H = \overline{\mathbf{U}}_u \overline{\mathbf{\Lambda}}_u \left[\overline{\mathbf{V}}_u^{\text{nonzero}}, \overline{\mathbf{V}}_u^{\text{zero}}\right]^H, \tag{5.11}$$

where $\overline{\mathbf{V}}_u^{\text{nonzero}}$ and $\overline{\mathbf{V}}_u^{\text{zero}}$ are composed of right singular vectors that correspond to nonzero singular values and zero singular values of $\overline{\mathbf{H}}_u$, respectively. Then, the digital precoder $\mathbf{D}_u$ for the uth MS can be designed to include the first N_{MS} columns of $\overline{\mathbf{V}}_u^{\text{zero}}$, that is, $\mathbf{D}_u = \overline{\mathbf{V}}_u^{\text{zero}}(:, 1 : N_{MS})$.

5.3.3 SUMMARY OF DIGITAL PRECODING

Beside the schemes introduced previously, there are also some excellent digital precoding schemes, such as the optimal DPC [3] and the near-optimal Tomlinson-Harashima (TH) precoding [12]. These schemes, however, usually involve high computational complexity. Therefore, we only mention them here without further discussion. In general, digital precoding can usually achieve satisfying performance because it can control both the phases and amplitudes of signals. Unfortunately, digital precoding has high energy consumption and hardware cost due to one dedicated RF chain is required for each antenna, which makes it difficult to be directly extended to mmWave massive MIMO systems with a large number of antennas.

5.4 ANALOG BEAMFORMING

Analog beamforming is developed in point-to-point mmWave systems with large antenna arrays [13]. In such a system, only one RF chain is employed to transmit a single data stream and analog beamforming is utilized to control the phases of original signals to achieve the maximal antenna array gain and effective SNR. In this section, we will first introduce the widely used analog beamforming scheme called beam steering, where the analog beamforming vectors are restricted to the array response vectors such as Eqs. (5.2) and (5.3). Then, we will focus on the more practical system, where the perfect channel state information (CSI) is not available, and introduce some beam training schemes to obtain the best analog beamforming vectors without knowing the channel matrix.

5.4.1 BEAM STEERING

Consider the single-user mmWave massive MIMO system with analog beamforming as shown in Fig. 5.3, where the BS employs N_t antennas but only one RF chain to transmit one data stream to a user with N_r antennas and one RF chain. Define $\mathbf{f}$ of size $N_t \times 1$ as the analog beamforming vector at the BS and $\mathbf{w}$ of size $N_r \times 1$ as the analog combining vector at the user. We aim to design $\mathbf{f}$ and $\mathbf{w}$ to maximize the effective SNR, which can be presented by

$$(\mathbf{w}^{\text{opt}}, \mathbf{f}^{\text{opt}}) = \arg\max \left| \mathbf{w}^H \mathbf{H} \mathbf{f} \right|^2$$
$$\text{s.t. } w_i = \sqrt{N_r^{-1}} e^{j\varphi_i}, \ \forall i, \tag{5.12}$$
$$f_l = \sqrt{N_t^{-1}} e^{j\phi_l}, \ \forall l.$$

Define the SVD of the channel as $\mathbf{H} = \mathbf{U}\mathbf{\Sigma}\mathbf{V}^H$. Then, we know that the optimal unconstraint solutions to Eq. (5.12) should be $\mathbf{w}^{\text{opt}} = \mathbf{U}(:,1)$ and $\mathbf{f}^{\text{opt}} = \mathbf{V}(:,1)$. However, they do not obey the amplitude constraint in Eq. (5.12). To the best of our knowledge, no analytical results are available for Eq. (5.12) in general and beamforming algorithms are limited to search over a fixed number of predefined directions. A possible method to solve this problem is to design the practical solutions $\mathbf{f}$ and $\mathbf{w}$ satisfying the amplitude constraint to be as close to the optimal unconstraint

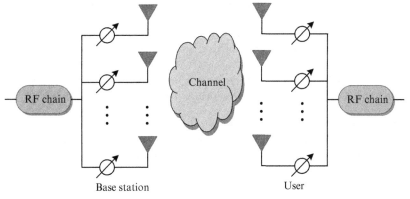

FIG. 5.3

Architecture of analog beamforming for single-user mmWave massive MIMO systems.

solutions $\mathbf{f}^{opt}$ and $\mathbf{w}^{opt}$ as possible. Theorem 5.1 shows that for mmWave systems with large antenna arrays, such design can be significantly simplified.

Theorem 5.1 *Each right singular vector of the matrix channel* $\boldsymbol{H}$ *given by the model Eq. (5.1) with* $L = o(N_t)$ *and* $L = o(N_r)$ *converges in chordal distance to an array response vector* $\boldsymbol{a}_t(\phi^t_1, \theta^t_1)$. *Each left singular vector similarly converges to* $\boldsymbol{a}_r(\phi^r_1, \theta^r_1)$. *The singular values in turn converge to* $N_t N_r |\alpha_1|^2 / L$.

Proof See Ref. [14]. □

Theorem 5.1 states that the channel representation "converges" to its SVD and as a result, the optimal SVD beamforming vector is in fact given in closed form and equal to the array response vector in the strongest direction. Therefore, for the mmWave systems with large N_t and N_r, the beamforming vector $\mathbf{f}$ and combining vector $\mathbf{w}$ can be simply restricted to the array response vectors such as Eqs. (5.2), (5.3), which are sufficiently close to the optimal unconstraint solutions. In other words, we can select $\mathbf{f} = \mathbf{a}_t\left(\phi^t_{k^*}, \theta^t_{k^*}\right)$ and $\mathbf{w} = \mathbf{a}_r\left(\phi^r_{k^*}, \theta^r_{k^*}\right)$, where $k^* = \arg\max_l |\alpha_l|^2$, to steer the beam in the strongest direction, which is expected to achieve the near-optimal performance, especially for large N_t and N_r.

5.4.2 BEAM TRAINING

When we discuss beam steering, we assume that the perfect CSI can be obtained by both the BS and the user, which is impractical in the realistic systems. Actually, for the mmWave systems with only one RF chain, the BS or the user cannot directly observe the channel matrix; rather, it observes a noisy version of the effective channel of a smaller size [2]. In other words, it will suffer from a subspace sampling limitation. Such limitation means the traditional channel estimation algorithms as will be discussed in Chapter 6 cannot be directly employed, also it is not practically possible to estimate all elements of the channel matrix, not to mention the perfect CSI. Without the full CSI, the problem is now converted into a general problem of subspace sampling for beam training [13]. The BS and the user must collaborate to search the best beamformer and combiner pair from predefined codebooks during the beam training.

According to the discussion in Section 5.4.1, we know that the codebook can be designed based on the beam steering scheme, which can be presented as [2]

$$\mathbf{f} \in \mathcal{F} = \left\{ \mathbf{a}_t\left(\overline{\phi}^t_1, \overline{\theta}^t_1\right), \mathbf{a}_t\left(\overline{\phi}^t_2, \overline{\theta}^t_2\right), \dots, \mathbf{a}_t\left(\overline{\phi}^t_{|\mathcal{F}|}, \overline{\theta}^t_{|\mathcal{F}|}\right) \right\}, \tag{5.13}$$

$$\mathbf{w} \in \mathcal{W} = \left\{ \mathbf{a}_r(\overline{\phi}^r_1, \overline{\theta}^r_1), \mathbf{a}_t(\overline{\phi}^r_2, \overline{\theta}^r_2), \dots, \mathbf{a}_t\left(\overline{\phi}^r_{|\mathcal{W}|}, \overline{\theta}^r_{|\mathcal{W}|}\right) \right\}, \tag{5.14}$$

where $\overline{\phi}^t_i$ ($\overline{\theta}^t_i$) and $\overline{\phi}^r_l$ ($\overline{\theta}^r_l$) are the quantified azimuth (elevation) angle of departure and arrival, respectively, which are assumed to uniformly cover the whole range of AoDs/AoAs. The most intuitive yet optimal beam training scheme is to exhaustively search all possible $|\mathcal{F}||\mathcal{W}|$ pairs of beamforming and combining vectors based on the maximization effective SNR criterion Eq. (5.12). However, in mmWave systems, the potentially large number of antennas and substantial beamforming gain requirement will necessitate the codebook sizes of $|\mathcal{F}|$ and $|\mathcal{W}|$ to be very large, which means that the exhaustive search may involve unaffordable overhead.

To solve this problem, Ref. [13] proposes a hierarchical beam training scheme. First, we construct a series of codebooks $\mathcal{F}_1, \mathcal{F}_2, ..., \mathcal{F}_K$ ($\mathcal{W}_1, \mathcal{W}_2, ..., \mathcal{W}_K$) with increasing resolution as shown in Fig. 5.4A. Then, the BS and the user jointly train the beams at the first level (lowest resolution codebook $\mathcal{F}_1$) by sending training data, which consists of three steps: (1) the BS sends training data by one possible beamforming vector from the codebook $\mathcal{F}_1$ to the user, and the user can determine the best combining vector according to Eq. (5.12) (Fig. 5.4B); (2) the user and the BS swap their roles and determine the best beamforming vector in a similar method (Fig. 5.4C); and (3) they feed back the index of the selected beamforming vector to each other (Fig. 5.4D). Such procedure described previously will be repeated with a higher resolution codebook within the chosen beam until the last level (highest resolution codebook $\mathcal{F}_K$) is considered. In this way, the hierarchical beam training can effective reduce the overhead, compared with the exhaustive search. It is also worth pointing out that every time we move to the next level, we can use shorter training sequences due to additional array gain.

5.4.3 SUMMARY OF ANALOG BEAMFORMING

Besides the introduced analog beamforming and beam training schemes, there are also some related works. For example, Ref. [2] designs a new codebook for analog beamforming, which can achieve better performance than the beam steering codebook. In Ref. [15], a single-sided beam training scheme is proposed to find the optimal beam by two steps: the combiner is fixed to exhaustively search the best precoder at first, and then this best precoder is fixed to exhaustively search the best combiner.

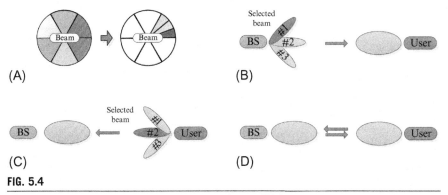

(A) (B) (C) (D)

FIG. 5.4

An example of a hierarchical beam training scheme. (A) Multilevel codebook, (B) beam sweep at BS side, (C) beam sweep at user side, and (D) feedback phase.

This scheme has been adopted by the standard IEEE 802.11ad. In summary, analog beamforming requires only one RF chain, which makes it easier to be implemented. However, analog beamforming usually suffers from serious performance loss, since only the phases of the transmit signals can be controlled. More importantly, analog beamforming is usually employed by single-user single-stream systems and the extension to multiuser multistream systems does not seem straightforward.

5.5 HYBRID PRECODING

As we mentioned previously, both digital precoding and analog beamforming face some challenges when we extend them to mmWave massive MIMO systems. To solve this problem, hybrid analog and digital precoding is proposed. Specifically, it divides the optimal digital precoding into two steps. In the first step, a small-size digital precoder is employed to cancel interferences, while in the second step, a large-size analog beamformer is used to increase the antenna array gain.

On the structural basis, hybrid beamforming can be divided into two categories: the fully connected architecture, where each RF chain is connected to all BS antennas via PSs; and the subconnected architecture, where each RF chain is connected to only a subset of BS antennas. Besides, hybrid precoding can be also divided according to single-user or multiuser. In this section, we will first focus on the single-user system, where two classical hybrid beamforming schemes, that is, the spatially sparse hybrid precoding (fully connected architecture) [4] and the successive interference cancellation (SIC)-based hybrid precoding (subconnected architecture) [16] are introduced. After that, the multiuser system is considered and a two-stage hybrid precoding scheme [17] is investigated. Finally, we make a summary of hybrid precoding and provide some deep insights.

5.5.1 SINGLE-USER HYBRID PRECODING

System model

Consider the single-user mmWave massive MIMO system with hybrid precoding as shown in Fig. 5.5, where the BS employs N_t antennas to simultaneously transmit N_s data streams to a user with N_r antennas. To enable multistream transmission, the BS is equipped with N_t^{RF} RF chains such that $N_s \leq N_t^{RF} \leq N_t$. The BS applies an $N_t^{RF} \times N_s$ digital precoder $\mathbf{D}$ using its N_t^{RF} RF chains, followed by an $N_t \times N_t^{RF}$ analog beamformer $\mathbf{A}$ using analog circuitry such as phase shifters (PSs) [4]. Then, the transmitted signal can be presented by

$$\mathbf{x} = \mathbf{ADs}, \tag{5.15}$$

where $\mathbf{s}$ is the $N_s \times 1$ original signal vector before precoding with normalized power as $\mathbb{E}(\mathbf{ss}^H) = (1/N_s)\mathbf{I}_{N_s}$.

For simplicity, we consider a narrowband system, which can be regarded as a reasonable first step [2–4] because the coherence bandwidth is usually very large

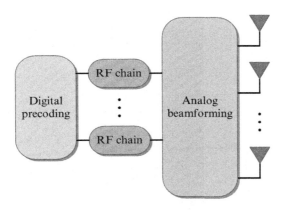

FIG. 5.5

Hardware architecture of hybrid precoding for single-user mmWave massive MIMO systems.

at mmWave (on the order of 100 MHz [2]). The received signal vector $\mathbf{y}$ of size $N_r \times 1$ can be correspondingly presented as

$$\mathbf{y} = \sqrt{\rho}\mathbf{HADs} + \mathbf{n}. \tag{5.16}$$

We assume the channel matrix $\mathbf{H}$ is perfectly known at both the BS and user to enable precoding. In practical systems, CSI at the receiver can be obtained via training and subsequently shared with the transmitter via limited feedback.

Spatially sparse hybrid precoding (fully connected)
Fully connected architecture

For the fully connected architecture, each RF chain is connected to all BS antennas via PSs as shown in Fig. 5.6. Because the analog beamformer $\mathbf{A}$ is realized by analog phase shifters, all elements of $\mathbf{A}$ have the same amplitude $1/\sqrt{N_t}$ but different phases. Besides, the total transmit power constraint is enforced by normalizing $\mathbf{D}$ to satisfy $\|\mathbf{AD}\|_F^2 = N_s$. No further hardware-related constraints are placed on the analog beamformer and digital precoder for the fully connected architecture.

Basic idea

We seek to design $(\mathbf{A}, \mathbf{D})$ to maximize the sum rate $R(\mathbf{A}, \mathbf{D})$ achieved by Gaussian signaling over the mmWave channel

$$R(\mathbf{A}, \mathbf{D}) = \log_2\left(\left|\mathbf{I} + \frac{\rho}{N_s\sigma_n^2}\mathbf{HADD}^H\mathbf{A}^H\mathbf{H}^H\right|\right). \tag{5.17}$$

With the design of $(\mathbf{A}, \mathbf{D})$, the corresponding sum-rate optimization problem can be presented as

$$\begin{aligned}
(\mathbf{A}^{\text{opt}}, \mathbf{D}^{\text{opt}}) &= \arg\max_{\mathbf{A}, \mathbf{D}} R(\mathbf{A}, \mathbf{D}), \\
\text{s.t} \quad &\mathbf{A} \in \mathcal{F}, \\
&\|\mathbf{AD}\|_F^2 = N_s,
\end{aligned} \tag{5.18}$$

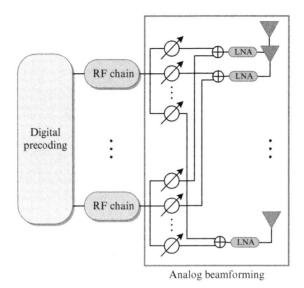

FIG. 5.6

Fully connected architecture of hybrid precoding for single-user mmWave massive MIMO system.

where $\mathcal{F}$ is the set containing all feasible analog beamformers, that is, the set of $N_t \times N_t^{RF}$ matrices with constant-magnitude entries. It is known that there are no general solutions to Eq. (5.18) in the presence of the nonconvex amplitude constraint $\mathbf{A} \in \mathcal{F}$. To find practical solutions that can be implemented in the system of Fig. 5.6, Ref. [4] proposes an approximation of Eq. (5.18) that is easier to be solved. The basic idea of such approximation is to transform the achievable sum rate into the "distance" between the practical hybrid precoder $\mathbf{AD}$ and the optimal unconstrained precoder $\mathbf{P}_{opt}$.

Let us start by defining the ordered SVD of the channel matrix $\mathbf{H}$ as $\mathbf{H} = \mathbf{U}\mathbf{\Sigma}\mathbf{V}^H$, where $\mathbf{U}$ is an $N_r \times \text{rank}(\mathbf{H})$ unitary matrix, $\mathbf{\Sigma}$ is a $\text{rank}(\mathbf{H}) \times \text{rank}(\mathbf{H})$ diagonal matrix including the singular values of $\mathbf{H}$ in a decreasing order, and $\mathbf{V}$ is an $N_t \times \text{rank}(\mathbf{H})$ unitary matrix. By exploiting the SVD of $\mathbf{H}$, we can rewrite Eq. (5.18) as

$$R(\mathbf{A}, \mathbf{D}) = \log_2 \left(\left| \mathbf{I} + \frac{\rho}{N_s \sigma_n^2} \mathbf{\Sigma}^2 \mathbf{V}^H \mathbf{A}\mathbf{D}\mathbf{D}^H \mathbf{A}^H \mathbf{V} \right| \right). \tag{5.19}$$

Further, we can decompose the matrices $\mathbf{\Sigma}$ and $\mathbf{V}$ as

$$\mathbf{\Sigma} = \begin{bmatrix} \mathbf{\Sigma}_1 & \mathbf{0} \\ \mathbf{0} & \mathbf{\Sigma}_2 \end{bmatrix}, \quad \mathbf{V} = [\mathbf{V}_1 \ \mathbf{V}_2], \tag{5.20}$$

where $\mathbf{\Sigma}_1$ is of size $N_s \times N_s$ and $\mathbf{V}_1$ is of size $N_t \times N_s$. We can observe that the optimal unconstrained precoder $\mathbf{P}_{opt}$ is simply given by $\mathbf{P}_{opt} = \mathbf{V}_1$. Unfortunately, the precoder $\mathbf{V}_1$ cannot be expressed as $\mathbf{AD}$ with $\mathbf{A} \in \mathcal{F}$, which means it cannot be realized by the architecture of interest as shown in Fig. 5.6. However, if we can design a practical hybrid precoder **AD** sufficiently "close" to the optimal unconstrained precoder $\mathbf{V}_1$, it can be expected to achieve the near-optimal performance. To verify this idea, Ref. [4] makes the following system assumption:

Approximation 5.1 We assume that the mmWave system parameters $(N_t, N_r, N_t^{\text{RF}}, N_r^{\text{RF}})$, as well as the parameters of the mmWave propagation channel $(L, \alpha_l \dots)$, are such that the hybrid precoders **AD** can be made sufficiently "close" to the optimal unitary precoder $\mathbf{P}_{\text{opt}} = \mathbf{V}_1$. Mathematically, this "closeness" is defined by the following two equivalent approximations:

(1)   The eigenvalues of the matrix $\mathbf{I} - \mathbf{V}_1^H \mathbf{ADD}^H \mathbf{A}^H \mathbf{V}_1$ are small. In the case of mmWave precoding, this can be equivalently stated as $\mathbf{V}_1^H \mathbf{AD} \approx \mathbf{I}$.

(2)   The singular values of the matrix $\mathbf{V}_2^H \mathbf{AD}$ are small; alternatively $\mathbf{V}_2^H \mathbf{AD} \approx \mathbf{0}$.

The detailed explanation for the reasonability of Approximation 5.1 can be found in Ref. [4]. Based on Approximation 5.1, the sum rate $R(\mathbf{A}, \mathbf{D})$ (5.19) can be simplified as

$$R(\mathbf{A}, \mathbf{D}) = \log_2 \left(\left| \mathbf{I} + \frac{\rho}{N_s \sigma_n^2} \boldsymbol{\Sigma}_1^2 \right| \right) - \left(N_s - \left\| \mathbf{V}_1^H \mathbf{AD} \right\|_F^2 \right). \tag{5.21}$$

We observe from Eq. (5.21) that the relationship between the sum rate $R(\mathbf{A}, \mathbf{D})$ and the designed hybrid precoder **AD** is only determined by the term $\left\| \mathbf{V}_1^H \mathbf{AD} \right\|_F^2$. According to Approximation 5.1, the term $\left\| \mathbf{V}_1^H \mathbf{AD} \right\|_F^2$ can be approximately maximized by instead maximizing $\text{tr}(\mathbf{V}_1^H \mathbf{AD})$. Besides, maximizing $\text{tr}(\mathbf{V}_1^H \mathbf{AD})$ is again equivalent to minimizing $\left\| \mathbf{P}_{\text{opt}} - \mathbf{AD} \right\|_F$. Therefore, the sum-rate optimization problem (5.18) can be well approximated by

$$\begin{aligned} (\mathbf{A}^{\text{opt}}, \mathbf{D}^{\text{opt}}) = \underset{\mathbf{A}, \mathbf{D}}{\arg \min} \; & \left\| \mathbf{P}_{\text{opt}} - \mathbf{AD} \right\|_F, \\ \text{s.t} \; & \mathbf{A} \in \mathcal{F}, \\ & \left\| \mathbf{AD} \right\|_F^2 = N_s, \end{aligned} \tag{5.22}$$

which is equivalent to find the projection of $\mathbf{P}_{\text{opt}}$ onto the set of hybrid precoders of the form **AD** with $\mathbf{A} \in \mathcal{F}$. Unfortunately, the nonconvex constraint of the feasible set $\mathcal{F}$ makes finding such a projection both analytically (in closed form) and algorithmically intractable. To obtain the near-optimal and closed-form solution to Eq. (5.22), we need to set more restrictions on $\mathcal{F}$. Specifically, by exploiting the structure of the mmWave massive MIMO channel matrix **H** as introduced in Section 5.2, we notice that near-optimal hybrid precoders can be found by further restricting $\mathcal{F}$ to be the set of vectors of the form $\mathbf{a}_t(\phi_l^t, \theta_l^t)$ and solving

$$\begin{aligned} (\mathbf{A}^{\text{opt}}, \mathbf{D}^{\text{opt}}) = \underset{\mathbf{A}, \mathbf{D}}{\arg \min} \; & \left\| \mathbf{P}_{\text{opt}} - \mathbf{AD} \right\|_F, \\ \text{s.t} \; & \mathbf{A}(:, i) \in \left\{ \mathbf{a}_t(\phi_l^t, \theta_l^t), \; \forall l \right\}, \\ & \left\| \mathbf{AD} \right\|_F^2 = N_s, \end{aligned} \tag{5.23}$$

where $\mathbf{A}(:,i)$ presents the ith column of $\mathbf{A}$. Eq. (5.23) aims to find the best low-dimensional representation of $\mathbf{P}_{\mathrm{opt}}$ using the basis vectors $\mathbf{a}_t(\phi^t_l, \theta^t_l)$. In other words, Eq. (5.23) consists of selecting the "best" N_t^{RF} array response vectors and finding their optimal baseband combination. Note that the constraint of $\mathbf{A}(:,i)$ can be directly embedded into the optimization objective function to obtain the following equivalent problem:

$$\widetilde{\mathbf{D}}^{\mathrm{opt}} = \arg\min_{\widetilde{\mathbf{D}}} \left\| \mathbf{P}_{\mathrm{opt}} - \mathbf{A}_t \widetilde{\mathbf{D}} \right\|_F,$$

$$\text{s.t } \left\| \mathrm{diag}\left(\widetilde{\mathbf{D}} \widetilde{\mathbf{D}}^H \right) \right\|_0 = N_t^{\mathrm{RF}},$$

$$\left\| \mathbf{A}_t \widetilde{\mathbf{D}} \right\|_F^2 = N_s, \tag{5.24}$$

where $\mathbf{A}_t = \left[\mathbf{a}_t\left(\phi^t_1, \theta^t_1\right), \ldots, \mathbf{a}_t\left(\phi^t_L, \theta^t_L\right) \right]$ is an $N_t \times L$ matrix containing all array response vectors and $\widetilde{\mathbf{D}}$ is an $L \times N_s$ matrix. $\mathbf{A}_t$ and $\widetilde{\mathbf{D}}$ are defined as auxiliary matrices, from which we obtain $\mathbf{A}^{\mathrm{opt}}$ and $\mathbf{D}^{\mathrm{opt}}$, respectively. Specifically, according to the sparsity constraint $\left\| \mathrm{diag}\left(\widetilde{\mathbf{D}} \widetilde{\mathbf{D}}^H \right) \right\|_0 = N_t^{\mathrm{RF}}$, it is implied that $\widetilde{\mathbf{D}}$ cannot have more than N_t^{RF} nonzero rows. Therefore, when only N_t^{RF} rows of $\widetilde{\mathbf{D}}$ are nonzero, only N_t^{RF} columns of the matrix $\mathbf{A}_t$ are effectively "selected." Consequently, the baseband precoder $\mathbf{D}^{\mathrm{opt}}$ can be obtained by the N_t^{RF} nonzero rows of $\widetilde{\mathbf{D}}^{\mathrm{opt}}$ and the RF precoder $\mathbf{A}^{\mathrm{opt}}$ can be selected by the corresponding N_t^{RF} columns of $\mathbf{A}_t$, respectively.

So far, the problem of jointly designing $\mathbf{A}$ and $\mathbf{D}$ has been transformed into a sparsity constrained matrix reconstruction problem with one variable. That means extensive classical algorithms on sparse reconstruction [18] now can be used to obtain the solution of Eq. (5.24) and design a hybrid precoder with near-optimal performance. To make this more clear, we can first consider the simplest case of single-stream transmission, that is, $N_s = 1$. Then, Eq. (5.24) is simplified to

$$\widetilde{\mathbf{d}}^{\mathrm{opt}} = \arg\min_{\widetilde{\mathbf{d}}} \left\| \mathbf{p}_{\mathrm{opt}} - \mathbf{A}_t \widetilde{\mathbf{d}} \right\|_F,$$

$$\text{s.t } \left\| \widetilde{\mathbf{d}} \right\|_0 = N_t^{\mathrm{RF}}, \quad \left\| \mathbf{A}_t \widetilde{\mathbf{d}} \right\|_F^2 = N_s. \tag{5.25}$$

Eq. (5.25) is equivalent to the typical problem in sparse signal recovery, which can be solved, for example, by relaxing the sparsity constraint and using convex optimization to solve its $l_2 - l_1$ relaxation.

In the more general case of multiple-stream transmission, that is, $N_s \geq 2$, the problem in Eq. (5.24) is equivalent to the typical problem of sparse signal recovery with multiple measurement vectors, also known as the simultaneously sparse approximation problem [19]. Such a problem can be solved by the well-known concept of orthogonal matching pursuit (OMP) [18,19]. Algorithm 1 provides the pseudocode for the precoder solution obtained by OMP [4].

ALGORITHM 1 SPATIALLY SPARSE PRECODING

Input: $\mathbf{P}_{opt}$

  1: $\mathbf{A} = $ Empty Matrix

  2: $\mathbf{P}_{res} = \mathbf{P}_{opt}$

  3: **for** $i \leq N_t^{RF}$ **do**

  4:    $\mathbf{\Psi} = \mathbf{A}_t^H \mathbf{P}_{res}$

  5:    $k = \arg\max_{l=1,\ldots,L} \left(\mathbf{\Psi}\mathbf{\Psi}^H\right)_{l,l}$

  6:    $\mathbf{A} = \left[\mathbf{A}\,\middle|\,\mathbf{A}_t^{(k)}\right]$

  7:    $\mathbf{D} = \left(\mathbf{A}^H\mathbf{A}\right)^{-1}\mathbf{A}^H\mathbf{P}_{opt}$

  8:    $\mathbf{P}_{res} = \dfrac{\mathbf{P}_{opt} - \mathbf{A}\mathbf{D}}{\left\|\mathbf{P}_{opt} - \mathbf{A}\mathbf{D}\right\|_F}$

  9: **end for**

10: $\mathbf{D} = \sqrt{N_s}\,\dfrac{\mathbf{D}}{\left\|\mathbf{A}\mathbf{D}\right\|_F}$

11: **return A,D**

In summary, after the initialization (Steps 1 and 2), Algorithm 1 starts by finding the vector $\mathbf{a}_t(\phi_l^t, \theta_l^t)$ along which the optimal precoder has the maximum projection in Step 5. Then, in Step 6, it appends the selected column vector $\mathbf{a}_t(\phi_l^t, \theta_l^t)$ to the analog beamformer $\mathbf{A}$. After the dominant vector is found, and the least-squares solution to $\mathbf{D}$ is calculated in Step 7, the contribution of the selected vector is removed in Step 8 and the algorithm proceeds to find the column along which the "residual precoding matrix" $\mathbf{P}_{res}$ has the largest projection. The process continues until all N_t^{RF} precoding vectors have been selected and output the designed $N_t \times N_t^{RF}$ analog beamformer $\mathbf{A}$ and $N_t^{RF} \times N_s$ baseband precoder $\mathbf{D}$, which minimizes $\left\|\mathbf{P}_{opt} - \mathbf{A}\mathbf{D}\right\|_F$. Step 10 ensures that the transmit power constraint $\left\|\mathbf{A}\mathbf{D}\right\|_F^2 = N_s$ is exactly satisfied.

Performance evaluation

In this section, the simulation results are provided to evaluate the performance of the spatially sparse hybrid precoding, compared with the optimal unconstrained precoding $\mathbf{P}_{opt} = \mathbf{V}_1$ (also known as fully digital precoding) and the beam steering precoding whose columns are strict to the array steering vectors (also known as fully analog beamforming). For the channel introduced in Section 5.2, the parameters are set as follows: We assume the propagation environment has $L = 3$ paths with uniformly distributed azimuth and elevation AoA/AoDs [4]. The BS sector angle is assumed to be 60 degrees wide in the azimuth domain and 20 degrees wide in elevation for outdoor deployments. In contrast, the user has relatively smaller antenna arrays of omnidirectional elements because the user must be able to steer beams in any direction as their location and orientation in real systems is random. The interelement spacing d is assumed to be a half wavelength. The SNR is defined as ρ/σ_n^2.

Fig. 5.7 shows the achievable sum-rate comparison in a 64×16 mmWave massive MIMO system with square planar arrays at both the BS and the user. The BS employs $N_t^{RF} = 4$ RF chains to simultaneously transmit $N_s = 1$ or 2 streams. We observe from Fig. 5.7 that the spatially sparse hybrid precoding can achieve the sum rate quite close to that of the optimal unconstrained solution in the case $N_s = 1$, while in the case of $N_s = 2$

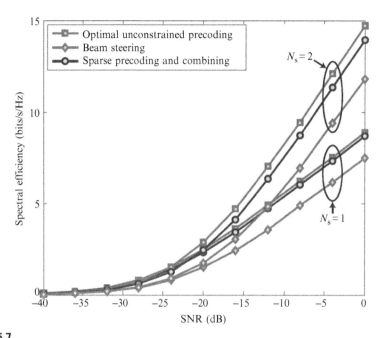

FIG. 5.7

Sum-rate comparison in a 64×16 mmWave massive MIMO system with $N_t^{RF} = 4$.

its performance is within a small gap from the optimal unconstrained solution. This implies that the spatially sparse hybrid precoding can very accurately approximate the channel's dominant singular vectors and achieve near-optimal performance. When compared with traditional beam steering [14], we can observe that the spatially sparse hybrid precoding enjoys obvious improvement in the achievable sum-rate.

Fig. 5.8 shows the achievable sum-rate comparison in a 256×64 mmWave massive MIMO system, where the BS employs $N_t^{RF} = 6$ RF chains and both the BS and the user are equipped with square planar arrays. We can observe from Fig. 5.8 that the spatially sparse hybrid precoding can achieve almost the perfect performance in both $N_s = 1$ and $N_s = 2$ cases. Furthermore, we note that although beam steering is expected to achieve a satisfying performance for the mmWave massive MIMO system with large arrays as discussed in Section 5.4.1, the spatially sparse hybrid precoding still outperforms beam steering by $\sim$5 dB.

Summary of the spatially sparse hybrid precoding

The basic idea of the spatially sparse precoding is to formulate the sum-rate optimization problem as a sparse approximation problem. Then, classical algorithms in sparse signal recovery, such as OMP algorithm, can be utilized. This algorithm takes an optimal unconstrained precoder as input, and approximates it as a linear combination of beam steering vectors that can be applied by analog circuitry together with a digital precoder at baseband to obtain the near-optimal practical hybrid precoder. Simulation results show

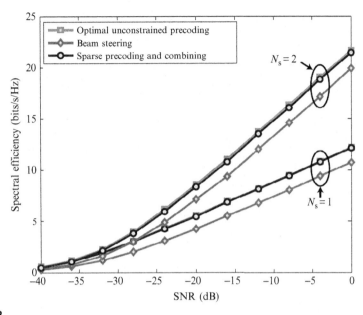

FIG. 5.8

Sum-rate comparison in a 256×64 mmWave massive MIMO system with $N_t^{RF} = 6$.

that for typical mmWave massive MIMO systems, the spatially sparse hybrid precoding can achieve a performance quite close to the optimal unconstrained precoder, while the number of required RF chains is significantly reduced.

Based on the spatially sparse hybrid precoding, several evolved hybrid precoding schemes have been proposed. In Ref. [20], a low-complexity version of the spatially sparse hybrid precoding is proposed. The main contributions of this work include: (1) derivation and integration of a matrix-inversion-bypass OMP algorithm to eliminate the matrix inversion operations; (2) development of a specific precoding reconstruction algorithm for the hardware implementation by considering the mmWave channel properties; and (3) design and implementation of a precoding reconstruction processor in an application-specific integrated circuit (ASIC) chip. In Ref. [21], the mmWave massive MIMO system with partial channel knowledge is considered, where the BS and the user only know their own local angles of arrival (AoAs). In such a scenario, Ref. [21] proposes a modified spatially sparse hybrid precoding accordingly to achieve the near-optimal performance. In Ref. [7], the spatially sparse hybrid precoding is combined with channel estimation, and a multiresolution codebook is designed to estimate the AoA/AoD of each path.

SIC-based hybrid precoding (subconnected)
Subconnected architecture

For the subconnected architecture, each RF chain is connected to only a subset of BS antennas, as shown in Fig. 5.9. Compared with the fully connected architecture, the subconnected architecture can significantly reduce the number of required PSs from

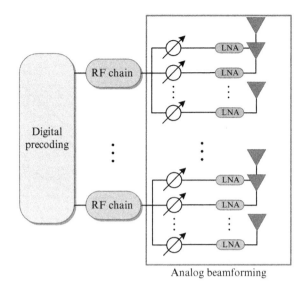

Analog beamforming

FIG. 5.9

Subconnected architecture of hybrid precoding for single-user mmWave massive MIMO system.

$N_t \times N_t^{RF}$ to N_t, which will bring three advantages: (1) it can save the energy to excite PSs; (2) it can save the energy to compensate for the insertion loss of PSs; and (3) it involves lower computational complexity due to the simplified hardware architecture.

Consider the system where the BS has N_t antennas but only N_t^{RF} RF chains. Each RF chain is connected to one subantenna array with a fixed and small number of antennas M, that is, $N_t = N_t^{RF}M$. To fully achieve the spatial multiplexing gain, the BS usually transmits $N_s = N_t^{RF}$ independent data streams to users employing N_r receiver antennas [22]. For the subconnected architecture, the digital precoder $\mathbf{D}$ of size $N_t^{RF} \times N_s = N_s \times N_s$ can be specialized to be a diagonal matrix as $\mathbf{D} = \mathrm{diag}[d_1, d_2, ..., d_{N_s}]$, where $d_n \in \mathbb{R}$ for $n = 1, 2, ..., N_s$ [22]. Then the role of $\mathbf{D}$ essentially performs some power allocation. In addition, unlike the fully connected architecture, the analog beamformer $\mathbf{A}$ of size $N_t \times N_t^{RF} = N_t \times N_s$ in the sub connected architecture will enjoy a special block diagonal structure as

$$\mathbf{A} = \begin{bmatrix} \bar{\mathbf{a}}_1 & \mathbf{0} & ... & \mathbf{0} \\ \mathbf{0} & \bar{\mathbf{a}}_2 & & \mathbf{0} \\ \vdots & & \ddots & \vdots \\ \mathbf{0} & \mathbf{0} & ... & \bar{\mathbf{a}}_{N_s} \end{bmatrix}_{N_t \times N_s}, \tag{5.26}$$

where $\bar{\mathbf{a}}_n \in \mathbb{C}^{M \times 1}$ is the analog weighting vector for the nth subantenna array with M antennas, whose elements have the same amplitude $1/\sqrt{M}$ but different phases.

Basic idea

We aim to maximize the total achievable rate $R(\mathbf{P})$ of mmWave MIMO systems by designing the hybrid precoder $\mathbf{P} = \mathbf{AD}$, which can be expressed as

$$R(\mathbf{P}) = \log_2\left(\left|\mathbf{I} + \frac{\rho}{N_s\sigma_n^2}\mathbf{H}\mathbf{P}\mathbf{P}^H\mathbf{H}^H\right|\right). \tag{5.27}$$

Based on the previous description, the hybrid precoding matrix $\mathbf{P}$ can be presented as $\mathbf{P} = \mathbf{A}\mathbf{D} = \text{diag}\{\bar{\mathbf{a}}_1, \ldots, \bar{\mathbf{a}}_{N_s}\} \cdot \text{diag}\{d_1, \ldots, d_{N_s}\}$. Therefore, there will exist three constraints for the design of $\mathbf{P}$:

Constraint 1: $\mathbf{P}$ should be a block diagonal matrix similar to the form of $\mathbf{A}$ as shown in Eq. (5.26); that is, $\mathbf{P} = \text{diag}\{\bar{\mathbf{p}}_1, \ldots, \bar{\mathbf{p}}_{N_s}\}$, where $\bar{\mathbf{p}}_n = d_n\bar{\mathbf{a}}_n$ is the $M \times 1$ nonzero vector of the nth column $\mathbf{p}_n$ of $\mathbf{P}$, that is,
$\mathbf{p}_n = \left[\mathbf{0}_{1\times M(n-1)}, \bar{\mathbf{p}}_n^T, \mathbf{0}_{1\times M(N_s-n)}\right]^T$.
Constraint 2: The nonzero elements of each column of $\mathbf{P}$ should have the same amplitude because the digital precoding matrix $\mathbf{D}$ is a diagonal matrix, and the amplitude of nonzero elements of the analog beamforming matrix $\mathbf{A}$ is fixed to $1/\sqrt{M}$.
Constraint 3: The Frobenius norm of $\mathbf{P}$ should satisfy $\|\mathbf{P}\|_F \leq N_s$ to meet the total transmit power constraint, where N_s is the number of RF chains equal to the number of transmitted data streams.

Unfortunately, the nonconvex constraints 1 and 2 on $\mathbf{P}$ make the maximization of the total achievable rate (5.27) very difficult to be solved. However, based on the special block diagonal structure of the hybrid precoding matrix $\mathbf{P}$, we can observe that the precoding on different subantenna arrays is independent. This inspires us to decompose the total achievable rate (5.27) into a series of subrate optimization problems, each of which only considers one subantenna array.

In particular, we can divide the hybrid precoding matrix $\mathbf{P}$ as $\mathbf{P} = \left[\mathbf{P}_{N_s-1} \ \mathbf{p}_{N_s}\right]$, where $\mathbf{p}_{N_s}$ is the N_sth column of $\mathbf{P}$, and $\mathbf{P}_{N_s-1}$ is an $N_sM \times (N_s-1)$ matrix containing the first (N_s-1) columns of $\mathbf{P}$. Then, the total achievable rate $R(\mathbf{P})$ in Eq. (5.27) can be rewritten as [16]

$$\begin{aligned}
R(\mathbf{P}) &\overset{(a)}{=} \log_2(|\mathbf{T}_{N_s-1}|) + \log_2\left(\left|\mathbf{I} + \frac{\rho}{N_s\sigma_n^2}\mathbf{T}_{N_s-1}^{-1}\mathbf{H}\mathbf{p}_{N_s}\mathbf{p}_{N_s}^H\mathbf{H}^H\right|\right) \\
&\overset{(b)}{=} \log_2(|\mathbf{T}_{N_s-1}|) + \log_2\left(1 + \frac{\rho}{N_s\sigma_n^2}\mathbf{p}_{N_s}^H\mathbf{H}^H\mathbf{T}_{N_s-1}^{-1}\mathbf{H}\mathbf{p}_{N_s}\right),
\end{aligned} \tag{5.28}$$

where (a) is obtained by defining the auxiliary matrix $\mathbf{T}_{N_s-1} = \mathbf{I} + \frac{\rho}{N_s\sigma_n^2}\mathbf{H}\mathbf{P}_{N_s-1}\mathbf{P}_{N_s-1}^H\mathbf{H}^H$, and (b) is true due to the fact that $|\mathbf{I} + \mathbf{X}\mathbf{Y}| = |\mathbf{I} + \mathbf{Y}\mathbf{X}|$ by defining $\mathbf{X} = \mathbf{T}_{N_s-1}^{-1}\mathbf{H}\mathbf{p}_{N_s}$ and $\mathbf{Y} = \mathbf{p}_{N_s}^H\mathbf{H}^H$. Note that the second term $\log_2\left(1 + \frac{\rho}{N_s\sigma_n^2}\mathbf{p}_{N_s}^H\mathbf{H}^H\mathbf{T}_{N_s-1}^{-1}\mathbf{H}\mathbf{p}_{N_s}\right)$ on the right side of Eq. (5.28) is the achievable subrate of the N_sth subantenna array, while the first term $\log_2(|\mathbf{T}_{N_s-1}|)$ shares the same form as Eq. (5.27). This observation implies that we can further decompose $\log_2(|\mathbf{T}_{N_s-1}|)$ using a similar method in Eq. (5.28) as

$$\log_2(|\mathbf{T}_{N_s-2}|) + \log_2\left(1 + \frac{\rho}{N_s\sigma_n^2}\mathbf{p}_{N_s-1}^H\mathbf{H}^H\mathbf{T}_{N_s-2}^{-1}\mathbf{H}\mathbf{p}_{N_s-1}\right)$$

Then, after N such decompositions, the total achievable rate R in Eq. (5.27) can be presented as

$$R = \sum_{n=1}^{N_s} \log_2 \left(1 + \frac{\rho}{N_s \sigma_n^2} \mathbf{p}_n^H \mathbf{H}^H \mathbf{T}_{n-1}^{-1} \mathbf{H} \mathbf{p}_n \right), \tag{5.29}$$

where we have $\mathbf{T}_n = \mathbf{I} + \frac{\rho}{N_s \sigma_n^2} \mathbf{H} \mathbf{p}_n \mathbf{p}_n^H \mathbf{H}^H$ and $\mathbf{T}_0 = \mathbf{I}_{N_s}$. From Eq. (5.29), we observe that the total achievable rate optimization problem can be transformed into a series of subrate optimization problems of subantenna arrays, which can be optimized one by one. After that, inspired by the idea of SIC for multiuser signal detection [23], we can optimize the achievable subrate of the first subantenna array and update the matrix $\mathbf{T}_1$. Then, a similar method can be utilized to optimize the achievable subrate of the second subantenna array. Such procedure will be executed until the last subantenna array is considered. Fig. 5.10 shows the diagram of the proposed SIC-based hybrid precoding. Next, we will discuss how to optimize the achievable subrate of each subantenna array.

Next, we focus on the subrate optimization problem of the nth subantenna array, which can be directly applied to other subantenna arrays. According to Eq. (5.29), the subrate optimization problem of the nth subantenna array by designing the nth precoding vector $\mathbf{p}_n$ can be stated as

$$\mathbf{p}_n^{\mathrm{opt}} = \arg \max_{\mathbf{p}_n \in \mathcal{F}} \log_2 \left(1 + \frac{\rho}{N_s \sigma_n^2} \mathbf{p}_n^H \mathbf{G}_{n-1} \mathbf{p}_n \right), \tag{5.30}$$

where $\mathbf{G}_{n-1}$ is defined as $\mathbf{G}_{n-1} = \mathbf{H}^H \mathbf{T}_{n-1}^{-1} \mathbf{H}$, $\mathcal{F}$ is the set of all feasible vectors satisfying the three constraints described previously. Note that the nth precoding vector $\mathbf{p}_n$ only has M nonzero elements from the $(M(n-1)+1)$th one to the (Mn)th one. Therefore, the subrate optimization problem (5.30) can be equivalently written as

$$\overline{\mathbf{p}}_n^{\mathrm{opt}} = \arg \max_{\overline{\mathbf{p}}_n \in \overline{\mathcal{F}}} \log_2 \left(1 + \frac{\rho}{N_s \sigma_n^2} \overline{\mathbf{p}}_n^H \overline{\mathbf{G}}_{n-1} \overline{\mathbf{p}}_n \right), \tag{5.31}$$

where $\overline{\mathcal{F}}$ includes all possible $M \times 1$ vectors satisfying *Constraint 2* and *Constraint 3*, $\overline{\mathbf{G}}_{n-1}$ of size $M \times M$ is the corresponding submatrix of $\mathbf{G}_{n-1}$ by only keeping the

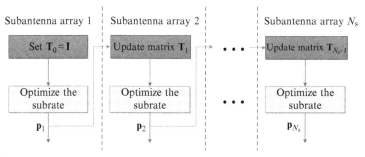

FIG. 5.10

Diagram of the SIC-based hybrid precoding.

rows and columns of $\mathbf{G}_{n-1}$ from the $(M(n-1)+1)$th one to the (Mn)th one, which can be presented as

$$\overline{\mathbf{G}}_{n-1} = \mathbf{R}\mathbf{G}_{n-1}\mathbf{R}^H = \mathbf{R}\mathbf{H}^H\mathbf{T}_{n-1}^{-1}\mathbf{H}\mathbf{R}^H, \tag{5.32}$$

where $\mathbf{R} = \begin{bmatrix} \mathbf{0}_{M \times M(n-1)} & \mathbf{I}_M & \mathbf{0}_{M \times M(N_s - n)} \end{bmatrix}$ is the corresponding selection matrix.

Define the SVD of the Hermitian matrix $\overline{\mathbf{G}}_{n-1}$ as $\overline{\mathbf{G}}_{n-1} = \mathbf{V}\boldsymbol{\Sigma}\mathbf{V}^H$, where $\boldsymbol{\Sigma}$ is an $M \times M$ diagonal matrix containing the singular values of $\overline{\mathbf{G}}_{n-1}$ in a decreasing order, and $\mathbf{V}$ is an $M \times M$ unitary matrix. It is known that the optimal unconstrained precoding vector of Eq. (5.31) is the first column $\mathbf{v}_1$ of $\mathbf{V}$, that is, the first right singular vector of $\overline{\mathbf{G}}_{n-1}$ [16]. However, according to the constraints mentioned above, we cannot directly choose $\overline{\mathbf{p}}_n^{\text{opt}}$ as $\mathbf{v}_1$ because the elements of $\mathbf{v}_1$ do not obey the constraint of the same amplitude (i.e., *Constraint 2*). To find a feasible solution to the subrate optimization problem (5.31), we need to further convert Eq. (5.31) into another form, which is given by Proposition 5.1.

Proposition 5.1 *The optimization problem (5.31)*

$$\overline{\mathbf{p}}_n^{\text{opt}} = \arg \max_{\overline{\mathbf{p}}_n \in \mathcal{F}} \log_2 \left(1 + \frac{\rho}{N_s \sigma_n^2} \overline{\mathbf{p}}_n^H \overline{\mathbf{G}}_{n-1} \overline{\mathbf{p}}_n \right)$$

is equivalent to the following problem:

$$\overline{\mathbf{p}}_n^{\text{opt}} = \arg \min_{\overline{\mathbf{p}}_n \in \mathcal{F}} \|\mathbf{v}_1 - \overline{\mathbf{p}}_n\|_2^2, \tag{5.33}$$

where v_1 is the first right singular vector of $\overline{\mathbf{G}}_{n-1}$.

Proof See Ref. [16].   □

Proposition 5.1 indicates that we can find a feasible precoding vector $\overline{\mathbf{p}}_n$, which is sufficiently close (in terms of Euclidean distance) to the optimal but impractical precoding vector $\mathbf{v}_1$, to maximize the achievable subrate of the nth subantenna array. Because $\overline{\mathbf{p}}_n = d_n \overline{\mathbf{a}}_n$, the target $\|\mathbf{v}_1 - \overline{\mathbf{p}}_n\|_2^2$ in Eq. (5.33) can be rewritten as

$$\|\mathbf{v}_1 - \overline{\mathbf{p}}_n\|_2^2 \overset{(a)}{=} 1 + d_n^2 - 2d_n \text{Re}\left(\mathbf{v}_1^H \overline{\mathbf{a}}_n\right) = \left(d_n - \text{Re}\left(\mathbf{v}_1^H \overline{\mathbf{a}}_n\right)\right)^2 + \left(1 - \left[\text{Re}\left(\mathbf{v}_1^H \overline{\mathbf{a}}_n\right)\right]^2\right), \tag{5.34}$$

where (a) is obtained based on the fact that $\mathbf{v}_1^H \mathbf{v}_1 = 1$ and $\overline{\mathbf{a}}_n^H \overline{\mathbf{a}}_n = 1$ because $\mathbf{v}_1$ is the first column of the unitary matrix $\mathbf{V}$ and each element of $\overline{\mathbf{a}}_n$ has the same amplitude $1/\sqrt{M}$.

From Eq. (5.34), we observe that the distance between $\overline{\mathbf{p}}_n$ and $\mathbf{v}_1$ consists of two parts. The first one is $\left(d_n - \text{Re}\left(\mathbf{v}_1^H \overline{\mathbf{a}}_n\right)\right)^2$, which can be minimized to zero by choosing $d_n = \text{Re}\left(\mathbf{v}_1^H \overline{\mathbf{a}}_n\right)$. The second one is $\left(1 - \left[\text{Re}\left(\mathbf{v}_1^H \overline{\mathbf{a}}_n\right)\right]^2\right)$, which can be minimized by maximizing $|\text{Re}(\mathbf{v}_1^H \overline{\mathbf{a}}_n)|$. Note that both $\overline{\mathbf{a}}_n$ and $\mathbf{v}_1$ have a fixed power of one, that is, $\mathbf{v}_1^H \mathbf{v}_1 = 1$ and $\overline{\mathbf{a}}_n^H \overline{\mathbf{a}}_n = 1$. Therefore, the optimal $\overline{\mathbf{a}}_n^{\text{opt}}$ to maximize $|\text{Re}(\mathbf{v}_1^H \overline{\mathbf{a}}_n)|$ is

$$\overline{\mathbf{a}}_n^{\text{opt}} = \frac{1}{\sqrt{M}} e^{j\text{angle}(\mathbf{v}_1)}, \tag{5.35}$$

where angle($\mathbf{v}_1$) denotes the phase vector of $\mathbf{v}_1$, that is, each element of $\bar{\mathbf{a}}_n^{\text{opt}}$ shares the same phase as the corresponding element of $\mathbf{v}_1$. Accordingly, the optimal choice of d_n^{opt} is

$$d_n^{\text{opt}} = \text{Re}\left(\mathbf{v}_1^H \bar{\mathbf{a}}_n\right) = \frac{1}{\sqrt{M}}\text{Re}\left(\mathbf{v}_1^H e^{j\text{angle}(\mathbf{v}_1)}\right) = \frac{\|\mathbf{v}_1\|_1}{\sqrt{M}}. \tag{5.36}$$

Based on Eqs. (5.35), (5.36), the optimal solution $\bar{\mathbf{p}}_n^{\text{opt}}$ to the optimization problem (5.33) (or equivalently (5.31)) can be obtained by

$$\bar{\mathbf{p}}_n^{\text{opt}} = d_n^{\text{opt}} \bar{\mathbf{a}}_n^{\text{opt}} = \frac{1}{M}\|\mathbf{v}_1\|_1 e^{j\text{angle}(\mathbf{v}_1)}. \tag{5.37}$$

It is worth pointing out that $\mathbf{v}_1$ is the first column of the unitary matrix $\mathbf{V}$, each element v_i of $\mathbf{v}_1$ (for $i = 1,\ldots,M$) has the amplitude less than one. Therefore, we have $\left\|\bar{\mathbf{p}}_n^{\text{opt}}\right\|_2^2 \leq 1$. Note that for all subantenna arrays, the optimal solution $\bar{\mathbf{p}}_n^{\text{opt}}$ for $n = 1,2,\ldots,N_s$ have a similar form. Thus, we can conclude that

$$\left\|\mathbf{P}^{\text{opt}}\right\|_F^2 = \left\|\text{diag}\left\{\bar{\mathbf{p}}_1^{\text{opt}}, \ldots, \bar{\mathbf{p}}_N^{\text{opt}}\right\}\right\|_F^2 \leq N_s, \tag{5.38}$$

which demonstrates that the total transmit power constraint (*Constraint 3*) is satisfied.

After we have acquired $\bar{\mathbf{p}}_n^{\text{opt}}$ for the nth subantenna array, the matrices $\mathbf{T}_n = \mathbf{I} + \frac{\rho}{N_s \sigma_n^2}\mathbf{H}\mathbf{P}_n\mathbf{P}_n^H\mathbf{H}^H$ (5.29) and $\overline{\mathbf{G}}_n = \mathbf{R}\mathbf{H}^H\mathbf{T}_n^{-1}\mathbf{H}\mathbf{R}^H$ (5.32) can be updated. Then, the method described previously for the nth subantenna array can be reused again to optimize the achievable subrate of the $(n+1)$th subantenna array. To sum up, solving the subrate optimization problem of the nth subantenna array consists of the following three steps:

Step 1: Execute the SVD of $\overline{\mathbf{G}}_{n-1}$ to obtain $\mathbf{v}_1$.

Step 2: Let $\bar{\mathbf{p}}_n^{\text{opt}} = \frac{1}{M}\|\mathbf{v}_1\|_1 e^{j\text{angle}(\mathbf{v}_1)}$ be the optimal solution to the current nth subantenna array.

Step 3: Update matrices $\mathbf{T}_n = \mathbf{I} + \frac{\rho}{N_s \sigma_n^2}\mathbf{H}\mathbf{P}_n\mathbf{P}_n^H\mathbf{H}^H$ and $\overline{\mathbf{G}}_n = \mathbf{R}\mathbf{H}^H\mathbf{T}_n^{-1}\mathbf{H}\mathbf{R}^H$ for the next $(n+1)$th subantenna array.

Performance evaluation

In this section, we provide the simulation results of the achievable rate to evaluate the performance of the proposed SIC-based hybrid precoding. We compare the performance of SIC-based hybrid precoding with the spatially sparse precoding [4] and the optimal unconstrained precoding based on the SVD of the channel matrix, which are both with fully connected architecture. Additionally, we also include the conventional analog precoding [24] and the optimal unconstrained precoding (i.e., $\bar{\mathbf{p}}_n^{\text{opt}} = \mathbf{v}_1$), which are both with subconnected architecture as benchmarks for comparison.

The simulation parameters are described as follows. We generate the channel matrix according to the channel model described in Section 5.2. The number of effective channel paths is $L = 3$ [4]. The carrier frequency is set as 28 GHz. Both the transmitter and receiver antenna arrays are ULAs with antenna spacing $d = \lambda/2$. Because the BS usually employs the directional antennas to eliminate interference and increase antenna gain, the AoDs are assumed to follow the uniform distribution within $[-\pi/6, \pi/6]$ [4]. Meanwhile, due to the random position of users, we assume that the AoAs follow the uniform distribution within $[-\pi, \pi]$, which means the omni-directional antennas are adopted by users [4]. Finally, the SNR is defined as ρ/σ_n^2.

Fig. 5.11 shows the achievable rate comparison in an mmWave massive MIMO system, where $N_t \times N_r = 64 \times 16$ and the number of RF chains is $N_t^{RF} = N_s = 8$. We can observe from Fig. 5.11 that the SIC-based hybrid precoding outperforms the conventional analog precoding with subconnected architecture in the whole simulated SNR range. Meanwhile, Fig. 5.11 also verifies the near-optimal performance of SIC-based hybrid precoding because it can achieve about 99% of the rate achieved by the optimal unconstrained precoding with subconnected architecture. More importantly, Fig. 5.11 shows that the performance of SIC-based hybrid precoding is close to the spatially sparse precoding and the optimal unconstrained precoding with fully connected architecture. For example, when SNR = 0 dB, SIC-based hybrid precoding can achieve more than 90% of the rate achieved by the near-optimal spatially sparse precoding in both simulated mmWave MIMO configurations.

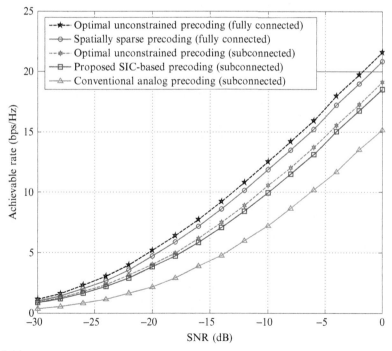

FIG. 5.11

Achievable rate comparison for an $N_t \times N_r = 64 \times 16$ ($N_t^{RF} = N_s = 8$) mmWave massive MIMO system.

Next, we evaluate the energy efficiency of the SIC-based hybrid precoding. Based on the energy consumption model in Ref. [25], the energy efficiency η can be defined as

$$\eta = \frac{R}{P_{total}} = \frac{R}{P_t + N_{RF}P_{RF} + N_{PS}P_{PS}} \, (\text{bps/Hz/W}) \tag{5.39}$$

where $P_{total} \triangleq P_t + N_{RF}P_{RF} + N_{PS}P_{PS}$ is the total energy consumption, P_t is the transmitted energy, P_{RF} is the energy consumed by RF chain, P_{PS} is the energy consumed by PS (including the energy for the excitation and the energy for the compensation of insertion loss [8]), and N_{RF} and N_{PS} are the numbers of required RF chains and PSs, respectively.

Here, we use the practical values $P_{RF} = 250\,\text{mW}$ [26], $P_{PS} = 1\,\text{mW}$ [27], and $P_t = 1\,\text{W}$ (about 30 dBm) in a small-cell transmission scenario [27] because mmWave is more likely to be applied in small cells. Fig. 5.12 shows the energy efficiency comparison against the number of RF chains N_t^{RF}, where SNR $= 0$ dB, $N_t \times N_r = 64 \times 64$ ($N_t^{RF} = 1, 2, 4, \ldots, 64$ to ensure that M is an integer). We can observe that both the spatially sparse precoding and the SIC-based precoding can achieve higher energy efficiency than the optimal unconstrained precoding (also known as the fully digital

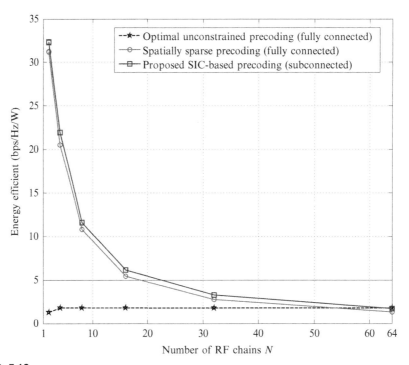

FIG. 5.12

Energy efficiency comparison against the number of RF chains N_t^{RF}, where $N_t \times N_r = 64 \times 64$, SNR $= 0$ dB.

precoding), especially when the number of RF chains N_t^{RF} is limited (e.g., $N_t^{RF} \leq 30$). Besides, we can also observe that the SIC-based precoding is more energy efficient than the spatially sparse precoding.

Summary of the SIC-based hybrid precoding

The basic idea of SIC-based hybrid precoding can be summarized as follows: (1) it decomposes the total achievable rate optimization problem with nonconvex constraints into a series of simple subrate optimization problems, each of which only considers one subantenna array; (2) it proves that maximizing the achievable subrate of each subantenna array is equivalent to simply seeking a precoding vector sufficiently close (in terms of Euclidean distance) to the unconstrained optimal solution; and (3) it maximizes the achievable subrate of each subantenna array one by one until the last subantenna array is considered. Simulation results verify that SIC-based hybrid precoding is near optimal and enjoys higher energy efficiency than the spatially sparse precoding and the fully digital precoding.

The evolved version of SIC-based hybrid precoding can be also found in Ref. [16], where a low-complexity algorithm is proposed to realize SIC-based hybrid precoding, which can avoid the need for the SVD and matrix inversion.

5.5.2 MULTIUSER HYBRID PRECODING

System model

Consider the multiuser mmWave system as shown in Fig. 5.13, where the base station equips N_{BS} antennas and N_{RF} RF chains ($N_{RF} \leq N_{BS}$) to simultaneously communicate with U MSs. Each MS equips N_{MS} antennas but only one RF chain. That means we focus on the case where the BS communicates with every MS via only one stream. Therefore the total number of data streams for communication is $N_S = U \leq N_{RF}$. To fully achieve the spatial multiplexing gain, we assume that the BS will use U out of the N_{RF} available RF chains to serve such U MSs.

We focus on the fully connected architecture as shown in Fig. 5.6. On the downlink, the BS employs a $U \times U$ digital precoder in the baseband $\mathbf{D} = [\mathbf{d}_1, \mathbf{d}_2, ..., \mathbf{d}_U]$

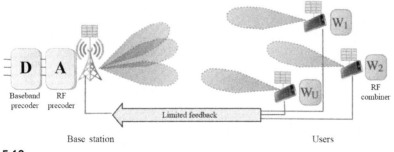

FIG. 5.13

Hardware architecture of hybrid precoding for multiuser mmWave massive MIMO systems.

followed by an $N_{BS} \times U$ analog precoder $\mathbf{A} = [\mathbf{a}_1, \mathbf{a}_2, ..., \mathbf{a}_U]$. Then, the transmitted signal can be presented as

$$\mathbf{x} = \mathbf{ADs}, \quad (5.40)$$

where $\mathbf{s} = [s_1, s_2, ..., s_U]^T$ of size $U \times 1$ is the original signal vector before precoding with normalized power $\mathbb{E}(\mathbf{ss}^H) = (\rho/U)\mathbf{I}_U$, and ρ is the average transmitted power. We assume equal power allocation among different MSs' streams. Because the analog precoder $\mathbf{A}$ is implemented using analog PSs, all the elements of $\mathbf{A}$ have the same constant amplitude N_{BS}^{-1}. Besides, the total transmitted power constraint is enforced by normalizing $\mathbf{D}$ to satisfy $\|\mathbf{AD}\|_F^2 = U$.

Also consider the narrowband block-fading channel model. The received signal vector $\mathbf{r}_u$ observed by the uth MS can be presented by

$$\mathbf{r}_u = \mathbf{H}_u \sum_{n=1}^{U} \mathbf{Ad}_n s_n + \mathbf{n}_u, \quad (5.41)$$

where $\mathbf{H}_u$ of size $N_{MS} \times N_{BS}$ denotes the mmWave massive MIMO channel matrix between the BS and the uth MS, and $\mathbf{n}_u$ is an AWGN vector, whose entries follow the independent and identical distribution (i.i.d.) $\mathcal{CN}(0, \sigma_n^2)$.

At the uth MS, an analog combiner $\mathbf{w}_u$ is used to combine the received signal $\mathbf{r}_u$ as

$$y_u = \mathbf{w}_u^H \mathbf{r}_u = \mathbf{w}_u^H \mathbf{H}_u \sum_{n=1}^{U} \mathbf{Ad}_n s_n + \mathbf{w}_u^H \mathbf{n}_u, \quad (5.42)$$

where $\mathbf{w}_u$ has similar constraints as the analog precoder $\mathbf{A}$, that is, all the elements of $\mathbf{w}_u$ have the same amplitude N_{MS}^{-1} but different phases. Here, we assume that only the analog combining is used at the MS side because MS is more likely to implement cheaper and simpler hardware with lower power consumption.

Two-stage hybrid precoding
Basic idea
Our goal is to design the analog precoder $\mathbf{A}$, the digital precoder $\mathbf{D}$ at the BS, and the analog combiners $\{\mathbf{w}_u\}_{u=1}^{U}$ at MSs to maximize the sum rate $R = \sum_{u=1}^{U} R_u$, where R_u is the rate achieved by the uth MS as expressed by

$$R_u = \log_2 \left(1 + \frac{\frac{P}{U}|\mathbf{w}_u^H \mathbf{H}_u \mathbf{Ad}_u|^2}{\frac{P}{U}\sum_{n \neq u}|\mathbf{w}_u^H \mathbf{H}_u \mathbf{Ad}_n|^2 + \sigma_n^2} \right). \quad (5.43)$$

Then, the precoding design problem can be formulated to find $\mathbf{A}^{opt}$, $\mathbf{D}^{opt}$, and $\{\mathbf{w}_u^{opt}\}_{u=1}^{U}$ that solve

$$\left\{ \mathbf{A}^{opt}, \mathbf{D}^{opt}, \{\mathbf{w}_u^{opt}\}_{u=1}^{U} \right\} = \arg\max \sum_{u=1}^{U} \log_2 \left(1 + \frac{\frac{P}{U}|\mathbf{w}_u^H \mathbf{H}_u \mathbf{Ad}_u|^2}{\frac{P}{U}\sum_{n \neq u}|\mathbf{w}_u^H \mathbf{H}_u \mathbf{Ad}_n|^2 + \sigma_n^2} \right)$$

$$\text{s.t. } \mathbf{a}_u \in \mathcal{F}, \quad u = 1, 2, ..., U,$$
$$\mathbf{w}_u \in \mathcal{W}, \quad u = 1, 2, ..., U,$$
$$\|\mathbf{AD}\|_F^2 = U. \quad (5.44)$$

The problem in Eq. (5.44) is a mixed integer programming problem. Its solution requires a search over the entire $\mathcal{F}^U \times \mathcal{W}^U$ space of all possible $\{\mathbf{a}_u\}_{u=1}^{U}$ and $\{\mathbf{w}_u\}_{u=1}^{U}$ combinations. Further, the digital precoder $\mathbf{D}$ needs to be jointly designed with the analog precoding/combining vectors. Unfortunately, the optimal solution to Eq. (5.44) is not known in general even without the RF constraints, and only iterative solutions exist. To this end, Ref. [17] proposes a two-stage multiuser hybrid precoding. The main idea of the proposed algorithm is to divide the design of the hybrid precoder into two stages as shown in Algorithm 2. In the first stage, the BS analog precoder and the MS analog combiners are jointly designed to maximize the desired signal power of each user, neglecting the resulting interference among users. In the second stage, the BS digital precoder is designed to manage the multiuser interference.

ALGORITHM 2 TWO-STAGE HYBRID PRECODING

Inputs: $\mathcal{F}$BS analog precoding codebook
$\quad\quad\quad\quad$ $\mathcal{W}$MS analog combining codebook
First stage: Single-user analog precoding/combining design
For each MS u, $u = 1, 2, ..., U$
$\quad$ The BS and MS u select $\mathbf{v}_u^{\text{opt}}$ and $\mathbf{g}_u^{\text{opt}}$ respectively that solve
$$\left\{\mathbf{g}^{\text{opt}}, \mathbf{v}_u^{\text{opt}}\right\} = \arg\max_{\substack{\forall \mathbf{g}_u \in \mathcal{W} \\ \forall \mathbf{v}_u \in \mathcal{F}}} \left\| \mathbf{g}_u^H \mathbf{H}_u \mathbf{v}_u \right\|$$
$\quad$ MS u sets $\mathbf{w}_u = \mathbf{g}_u^{\text{opt}}$
BS sets $\mathbf{A} = \left[\mathbf{v}_1^{\text{opt}}, \mathbf{v}_2^{\text{opt}}, ..., \mathbf{v}_U^{\text{opt}}\right]$
Second stage: Multiuser digital precoding design
For each MS u, $u = 1, 2, ..., U$
$\quad$ MS u estimates its effective channel $\overline{\mathbf{h}}_u = \mathbf{w}_u^H \mathbf{H}_u \mathbf{A}$
$\quad$ MS u feeds back $\overline{\mathbf{h}}_u$ to the BS
BS employs ZF digital precoder $\mathbf{D} = \overline{\mathbf{H}}^H \left(\overline{\mathbf{H}} \overline{\mathbf{H}}^H\right)^{-1}$, where $\overline{\mathbf{H}} = \left[\overline{\mathbf{h}}_1^T, \overline{\mathbf{h}}_2^T, ..., \overline{\mathbf{h}}_U^T\right]^T$
$$\mathbf{d}_u = \frac{\mathbf{d}_u}{\|\mathbf{A}\mathbf{d}_u\|_F}, u = 1, 2, ..., U$$

Specifically, Algorithm 2 can be summarized as follows. In the first stage, the BS and each MS u design the analog precoding/combining vectors $\mathbf{a}_u$ and $\mathbf{w}_u$, to maximize the desired signal power for the uth MS without considering other users' interference. Note that this is the typical single-user analog precoding/combining design problem. Therefore, classical beam training algorithms developed for single-user systems as introduced in Section 5.4.2 can be used to design the analog precoding/combining vectors without explicit channel estimation.

In the second stage, each MS u first estimates its effective channel $\overline{\mathbf{h}}_u = \mathbf{w}_u^H \mathbf{H}_u \mathbf{A}$ for $u = 1, 2, ..., U$. Then, the effective channel $\overline{\mathbf{h}}_u$ is fed back to the BS. Note that the dimension of each effective channel $\overline{\mathbf{h}}_u$ is only $U \times 1$, which is much less than the original channel matrix. Therefore it can be expected that such feedback will enjoy low feedback overhead by utilizing the classical schemes [2,13]. Finally, the BS employs the classical ZF digital precoder based on the effective channel matrix $\overline{\mathbf{H}} = \left[\overline{\mathbf{h}}_1, \overline{\mathbf{h}}_2, ..., \overline{\mathbf{h}}_U\right]$. Thanks to the sparse mmWave massive MIMO channels, the effective MIMO channel is expected to be well conditioned [1,5], which makes adopting a simple multiuser digital precoding strategy like ZF capable of achieving near-optimal performance as proved by the following:

Theorem 5.2 *Let Algorithm 2 be used under the following assumptions:*

(1)   All channels are single path, that is, $L_u = 1$, $u = 1,2,...,U$.
(2)   The analog precoding vectors $\mathbf{a}_u$, $u = 1,2,...,U$ and the analog combining vectors $\mathbf{w}_u$, $u = 1,2,...,U$ are beam steering vectors with continuous angles.
(3)   Each MS u perfectly knows its channel $\mathbf{H}_u$, $u = 1,2,...,U$.
(4)   The BS perfectly knows the effective channels $\overline{\mathbf{h}}_u$, $u = 1,2,...,U$.

and define the $N_{BS} \times U$ matrix $\mathbf{A}_{BS}$ to gather the BS array response vectors associated with the U AoDs, that is, $\mathbf{A}_{BS} = [\mathbf{a}_{BS}(\phi_1), \mathbf{a}_{BS}(\phi_2),..., \mathbf{a}_{BS}(\phi_U)]$, with maximum and minimum singular values $\sigma_{max}(\mathbf{A}_{BS})$ and $\sigma_{min}(\mathbf{A}_{BS})$, respectively. Then, the achievable rate of user u is lower-bounded by

$$R_u \geq \log_2\left(1 + \frac{P}{\sigma_n^2 U} N_{BS} N_{MS} |\alpha_u|^2 G\left(\{\phi_u\}_{u=1}^U\right)\right), \tag{5.45}$$

where $G\left(\{\phi_u\}_{u=1}^U\right) = 4\left(\dfrac{\sigma_{max}^2(\mathbf{A}_{BS})}{\sigma_{min}^2(\mathbf{A}_{BS})} + \dfrac{\sigma_{min}^2(\mathbf{A}_{BS})}{\sigma_{max}^2(\mathbf{A}_{BS})} + 2\right)^{-1}.$

Proof See Ref. [17].   □

Theorem 5.3 *Denote the single-user rate as $\overline{R} = \log_2\left(1 + \dfrac{P}{\sigma_n^2 U} N_{BS} N_{MS} |\alpha_u|^2\right).$*

When Algorithm 2 is used with assumptions stated in Theorem 5.2, the relation between the achievable rate by MS u, and the single-user rate $\overline{R}$ satisfies

(1) $\mathbb{E}\left(\overline{R} - R_u\right) \leq K(N_{BS}, U)$
(2) $\lim_{N_{BS} \to \infty} R_u = \overline{R}$ almost surely

where $K(N_{BS}, U)$ is a constant whose value depends only on N_{BS} and U.

Proof See Ref. [17].   □

Theorem 5.3 indicates that the average achievable rate of MS *u* using the two-stage multiuser hybrid precoding grows with the same slope of the single-user rate at high SNR, and stays within a constant gap $K(N_{BS}, U)$ from it, which only depends on the number of MSs *U* and the number of BS antennas N_{BS}. More important, when N_{BS} increases, the matrix $\mathbf{A}_{BS}$ will be more well conditioned, and the ratio between its maximum and minimum singular values will approach one. As a result, the term $G\left(\{\phi_u\}_{u=1}^U\right)$ in Eq. (5.45) will be closer to one, and the gap between the achievable rate using Algorithm 2 and the single-user rate will decrease.

Performance evaluation
In this section, the performance of the two-stage hybrid precoding is evaluated in terms of achievable rate.

In Fig. 5.14, we consider the system model with a BS employing an 8×8 UPA with four MSs, each having a 4×4 UPA. The channel for each user is assumed to have single path. The azimuth AoAs/AoDs are assumed to be uniformly distributed in $[0, 2\pi]$, and the elevation AoAs/AoDs are uniformly distributed in $[-\pi/2, \pi/2]$. From Fig. 5.14, we observe that the performance of the two-stage hybrid precoding is very close to the single-user rate. Besides, the two-stage hybrid precoding can also achieve almost the same performance of the fully digital BD precoding, and enjoy an obvious gain over the fully analog beam steering solution, which increases with SNR as the beam steering solution starts to be interference limited. The tightness of the derived lower bound in Theorem 5.1 is also verified by Fig. 5.14.

In Fig. 5.15, the same setup in Fig. 5.14 is considered at SNR $= 0$ dB, but with different numbers of BS antennas N_{BS}. Fig. 5.15 shows that even at very large numbers of antennas, there is still a considerable gain of hybrid precoding over beam steering. We can also observe from Fig. 5.15 that the difference between the rate achieved by the two-stage hybrid precoding and the single-user rate decreases when the number of BS antennas increases, which verifies the conclusions in Theorem 5.3.

Summary of two-stage multiuser hybrid precoding

The basic idea of the two-stage multiuser hybrid precoding is to divide the design of the hybrid precoder into two stages. In the first stage, the BS RF precoder and the MS RF combiners are jointly designed to maximize the desired signal power of each user.

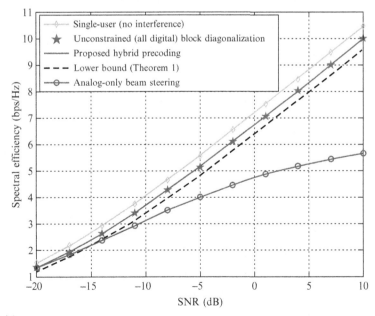

FIG. 5.14

Achievable rate comparison for an $N_{BS} = 64$, $N_{MS} = 16$, and $U = 4$ multiuser mmWave massive MIMO system.

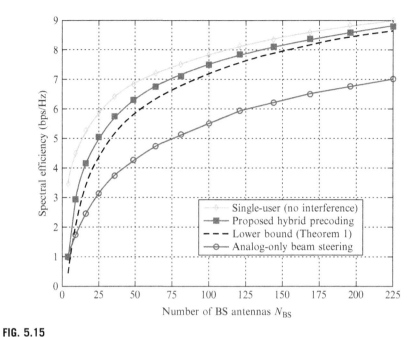

FIG. 5.15

Achievable rate comparison against the number of BS antennas N_{BS}.

In the second stage, the BS digital precoder is designed to manage the multiuser interference. Simulation results show that the two-stage hybrid precoding outperforms the fully analog beam steering solution and approaches the performance of the fully digital BD precoding in a typical multiuser mmWave massive MIMO system.

Besides the two-stage hybrid precoding, some multiuser hybrid precoding schemes have also been proposed. For example, Ref. [28] proposes a multiuser hybrid precoding scheme to minimize the Euclidean distance between the received signals of the system with hybrid precoding and the ones of the system with fully digital precoding. In Ref. [29], a similar idea called mean squared error (MSE)-based hybrid precoding is proposed, which aims to minimize the MSE between the original transmitted signals before precoding and the received signals after combining.

5.6 CONCLUSIONS

Precoding and beamforming design will be an important component for future mmWave massive MIMO systems, and the corresponding 5G standards. In this chapter, we first present the traditional precoding (beamforming) technologies, including digital precoding for low-frequency MIMO systems and analog beamforming for mmWave communications with large antenna arrays. Digital precoding aims to cancel interferences between different data streams, which can achieve satisfying performance. But, it usually requires a large number of RF chains when it is extended to mmWave massive MIMO systems, leading to unaffordable hardware cost and energy consumption. In contrast, analog beamforming tries to improve the antenna array gain by utilizing the analog phase shifter network, which only requires a small number of RF chains. However, it usually suffers from performance loss compared with digital precoding. What's more, it is also difficult to be extended to multiuser, multistream systems. After the traditional precoding (beamforming) technologies have been introduced, we investigate hybrid precoding, which combines the advantages of both digital precoding and analog beamforming together. Hybrid precoding can achieve performance quite close to digital precoding with only a small number of RF chains, which is in terms of the number of data streams instead of the number of antennas. As a result, hybrid precoding seems more appropriate for future mmWave massive MIMO systems.

It is worth pointing out that besides hybrid precoding, there are also some advanced technologies proposed recently to relieve the potential high energy consumption and hardware cost of mmWave massive MIMO systems. In Ref. [1], an mmWave massive MIMO architecture with low-resolution (e.g., 1-bit) analog-to-digital converters (ADCs) is proposed. Low-resolution ADCs can be implemented with simpler circuits and consume less power, which can be seen as an alternative to hybrid precoding. More important, it has been proved that at low SNR, it only incurs a little rate loss compared with high-resolution quantization [30]. In Ref. [31], beamspace MIMO is proposed. By fully utilizing the angular sparsity of the mmWave channel, beamspace MIMO employs a lens to transform the conventional MIMO channel into the sparse beamspace channel. Then, several beams of the obtained beamspace channel can be selected according to a specific criterion to reduce the effective dimension of the MIMO channel without obvious performance loss. After that, only a small-size digital precoder realized by a small number of RF chains will be required.

REFERENCES

[1] A. Alkhateeb, J. Mo, N. Gonzalez-Prelcic, R. Heath, MIMO precoding and combining solutions for millimeter-wave systems, IEEE Commun. Mag. 52 (12) (2014) 122–131.
[2] S. Hur, T. Kim, D. Love, J. Krogmeier, T. Thomas, A. Ghosh, Millimeter wave beamforming for wireless backhaul and access in small cell networks, IEEE Trans. Commun. 61 (10) (2013) 4391–4403.
[3] M.H. Costa, Writing on dirty paper (corresp.), IEEE Trans. Inf. Theory 29 (3) (1983) 439–441.
[4] O. El Ayach, S. Rajagopal, S. Abu-Surra, Z. Pi, R. Heath, Spatially sparse precoding in millimeter wave MIMO systems, IEEE Trans. Wirel. Commun. 13 (3) (2014) 1499–1513.
[5] T. Bai, A. Alkhateeb, R. Heath, Coverage and capacity of millimeter-wave cellular networks, IEEE Commun. Mag. 52 (9) (2014) 70–77.
[6] Z. Pi, F. Khan, An introduction to millimeter-wave mobile broadband systems, IEEE Commun. Mag. 49 (6) (2011) 101–107.

[7] A. Alkhateeb, O. El Ayach, G. Leus, R. Heath, Channel estimation and hybrid precoding for millimeter wave cellular systems, IEEE J. Sel. Top. Signal Process. 8 (5) (2014) 831–846.

[8] C.A. Balanis, Antenna Theory: Analysis and Design, Wiley, Hoboken, NJ, 2012.

[9] D. Gesbert, M. Shafi, D. Shiu, P.J. Smith, A. Naguib, From theory to practice: an overview of MIMO space-time coded wireless systems, IEEE J. Sel. Areas Commun. 21 (3) (2003) 281–302.

[10] D. Tse, P. Viswanath, Fundamentals of Wireless Communication, Cambridge University Press, Cambridge, UK, 2005.

[11] Z. Shen, R. Chen, J.G. Andrews, R. Heath, B.L. Evans, Low complexity user selection algorithms for multiuser MIMO systems with block diagonalization, IEEE Trans. Signal Process. 54 (9) (2006) 3658–3663.

[12] U.D. Wessel, Achievable rates for Tomlinson-Harashima precoding, IEEE Trans. Inf. Theory 44 (3) (1998) 824–830.

[13] J. Wang, Z. Lan, C.-W. Pyo, T. Baykas, C.-S. Sum, M.A. Rahman, J. Gao, R. Funada, F. Kojima, H. Harada, et al., Beam codebook based beamforming protocol for multi-Gbps millimeter-wave WPAN systems, IEEE J. Sel. Areas Commun. 27 (8) (2009) 1390–1399.

[14] O. El Ayach, R. Heath, S. Abu-Surra, S. Rajagopal, Z. Pi, The capacity optimality of beam steering in large millimeter wave MIMO systems, in: Proceedings of the 2012 IEEE International Workshop Signal Processing Advances Wireless Communications, 2012, pp. 100–104.

[15] C. Cordeiro, D. Akhmetov, M. Park, IEEE 802.11 ad: introduction and performance evaluation of the first multi-Gbps WiFi technology, in: Proceedings of the 2010 ACM International Workshop on mmWave Communications, 2010, pp. 3–8.

[16] X. Gao, L. Dai, S. Han, C.-L. I, R. Heath, Energy-efficient hybrid analog and digital precoding for mmWave MIMO systems with large antenna arrays, IEEE J. Sel. Areas Commun. 34 (4) (2016) 998–1009.

[17] A. Alkhateeb, G. Leus, R. Heath, Limited feedback hybrid precoding for multi-user millimeter wave systems, IEEE Trans. Wirel. Commun. 14 (11) (2015) 6481–6494.

[18] J. Tropp, A. Gilbert, Signal recovery from random measurements via orthogonal matching pursuit, IEEE Trans. Inf. Theory 53 (12) (2007) 4655–4666.

[19] S.F. Cotter, B.D. Rao, K. Engan, K. Kreutz-Delgado, Sparse solutions to linear inverse problems with multiple measurement vectors, IEEE Trans. Signal Process. 53 (7) (2005) 2477–2488.

[20] Y. Lee, C.-H. Wang, Y.-H. Huang, A hybrid RF/baseband precoding processor based on parallel-index-selection matrix-inversion-bypass simultaneous orthogonal matching pursuit for millimeter wave MIMO systems, IEEE Trans. Signal Process. 63 (2) (2015) 305–317.

[21] A. Alkhateeb, O. El Ayach, G. Leus, R. Heath, Hybrid precoding for millimeter wave cellular systems with partial channel knowledge, in: Proceedings of the IEEE Information Theory and Applications Workshop (ITA'13), 2013, pp. 1–5.

[22] S. Han, C.-L. I, Z. Xu, C. Rowell, Large-scale antenna systems with hybrid precoding analog and digital beamforming for millimeter wave 5G, IEEE Commun. Mag. 53 (1) (2015) 186–194.

[23] Y.-C. Liang, E.Y. Cheu, L. Bai, G. Pan, On the relationship between MMSE-SIC and BI-GDFE receivers for large multiple-input multiple-output channels, IEEE Trans. Signal Process. 56 (8) (2008) 3627–3637.

[24] O. El Ayach, R. Heath, S. Rajagopal, Z. Pi, Multimode precoding in millimeter wave MIMO transmitters with multiple antenna sub-arrays, in: Proceedings of the IEEE Global Communications Conference (GLOBECOM'13), December, 2013, pp. 3476–3480.

[25] S. Cui, A.J. Goldsmith, A. Bahai, Energy-constrained modulation optimization, IEEE Trans. Wirel. Commun. 4 (5) (2005) 2349–2360.

[26] P. Amadori, C. Masouros, Low RF-complexity millimeter-wave beamspace-MIMO systems by beam selection, IEEE Trans. Commun. 63 (6) (2015) 2212–2222.

[27] T.S. Rappaport, J.N. Murdock, F. Gutierrez, State of the art in 60-GHz integrated circuits and systems for wireless communications, Proc. IEEE 99 (8) (2011) 1390–1436.

[28] T.E. Bogale, L.B. Le, Beamforming for multiuser massive MIMO systems: digital versus hybrid analog-digital, in: Proceedings of the IEEE Global Communications Conference (GLOBECOM'14), December, 2014, pp. 10–12.

[29] M. Kim, Y.H. Lee, MSE-based hybrid RF/baseband processing for millimeter-wave communication systems in MIMO interference channels, IEEE Trans. Veh. Technol. 64 (6) (2015) 2714–2720.

[30] J. Singh, O. Dabeer, U. Madhow, On the limits of communication with low-precision analog-to-digital conversion at the receiver, IEEE Trans. Commun. 57 (12) (2009) 3629–3639.

[31] J. Brady, N. Behdad, A. Sayeed, Beamspace MIMO for millimeterwave communications: system architecture, modeling, analysis, and measurements, IEEE Trans. Antennas Propag. 61 (7) (2013) 3814–3827.

Channel estimation for mmWave massive MIMO systems

Z. Gao*, L. Dai[†], C. Hu[†], X. Gao[†], Z. Wang[†]

Beijing Institute of Technology, Beijing, China* Tsinghua University, Beijing, China[†]

CHAPTER OUTLINE

mmWave Massive MIMO. http://dx.doi.org/10.1016/B978-0-12-804418-6.00006-6

6.1 INTRODUCTION

The acquisition of accurate channel state information (CSI) is essential in millimeter-wave (mmWave) massive MIMO systems [1]. Hence, how to reliably and efficiently acquire CSI is a fundamental issue to fully exploit the potential advantages of mmWave massive MIMO systems. Compared with conventional mmWave communications and massive MIMO working at sub-3–6 GHz, channel estimation (CE) in mmWave massive MIMO systems is more challenging mainly due to the following reasons.

1. *The number of antennas in mmWave massive MIMO can be very large.* In mmWave massive MIMO systems, the number of antennas at both the base station (BS) and mobile station (MS) can be much larger than that in conventional massive MIMO working at sub-3–6 GHz due to the much smaller wavelength of mmWave signals [2]. This implies the challenge that CE in mmWave massive MIMO can be more difficult even when time division duplex (TDD) leveraging the channel reciprocity is considered, since the user side may also employ dozens of antennas. Hence, mmWave massive MIMO systems may suffer from the prohibitively high pilot overhead for CE. It should be also pointed out that even for TDD-based communication systems, synchronization, calibration error of radio frequency (RF) chains, and other issues to guarantee the channel reciprocity are not trivial, especially in mmWave communications [1].

2. *mmWave communications suffer from the special hardware constraints.* The hardware cost and energy consumption of transceivers, including high-speed analog-to-digital convertors (ADCs) and digital-to-analog convertors, synthesizers, mixers, and so on, in mmWave communications are much larger than those in conventional cellular communications. Hence, massive low-cost antennas but a limited number of expensive RF chains can be an appealing transceiver structure for mmWave massive MIMO systems [1–4], which will be further illustrated in Section 6.2.2. Such a transceiver structure makes the CE more challenging. Due to the much smaller number of RF chains than that of antennas, it is difficult to exploit the effective observations of small size to estimate the mmWave massive MIMO channels of large size.

3. *The signal-to-noise ratio (SNR) before beamforming is low in mmWave communications.* In mmWave communications, the bandwidth can be hundreds of megahertz or even multiple gigahertzes, which will introduce much more thermal noise [1]. Moreover, due to the strong signal directivity of mmWave, the

low SNR before beamforming also makes the CE more difficult [2]. Because reliable CE usually requires sufficient SNR, it indicates that at least partial channel state information at the transmitter (CSIT) may be required to ensure beamforming at the transmitter to match mmWave MIMO channels for reliable CE. Hence, CE in mmWave massive MIMO systems should consider not only signal processing in the digital baseband, but also the characteristics of analog phase-shift networks (PSNs) in the analog RF, which can make CE more complex. Finally, the Doppler shift in mmWave can be more obvious due to the high frequency band, and the blockage effect should also be considered [1].

4. *Channel feedback is required even for TDD mode*: Because single-antenna users are typically considered in conventional massive MIMO due to the limited form factor, only CSIT is required for precoding. However, in mmWave massive MIMO where both the BS and MS can be equipped with massive antennas, precoding in the uplink and combining in the downlink at MS are also necessary for improved link reliability. Therefore, CSI known by both the BS and MS is also required for mmWave massive MIMO systems, which indicates another challenge that CSI acquired in the uplink by leveraging the channel reciprocity should also be fed back to the MS. In the future 5G heterogeneous network (HetNet), it is the consensus that mmWave can be used to support the high-rate data plane, while the frequency band in conventional cellular networks can be used to support the control plane [2]. Hence, both the BS and MS can use the control plane to accomplish the CSI feedback.

In this chapter, we will provide a comprehensive survey of state-of-the-art CE solutions for mmWave massive MIMO systems. This chapter is organized as follows. In Section 6.2, we introduce the preparatory work, including the mmWave massive MIMO channel models in typical uniform linear array (ULA) scenario and two kinds of transceiver structures for mmWave massive MIMO systems. In Sections 6.3–6.6, we details four kinds of CE schemes for mmWave massive MIMO systems, and their pros and cons are also discussed. Finally, in Section 6.7, we briefly discuss the codebook-based CSI acquisition scheme and the potential that how the existing CE schemes initially proposed for conventional massive MIMO working at sub-3–6 GHz can be tailored to mmWave massive MIMO systems.

6.2 PREPARATORY WORK

In this section, we first briefly introduce the channel model in mmWave massive MIMO communications. Then, we further provide two promising transceiver structures for mmWave massive MIMO systems: the hybrid MIMO transceiver structure with analog PSN and the receiver with one-bit ADC. Both of the two transceiver structures are inspired by the hardware constraints in mmWave communications, while they can provide high data rates with negligible performance loss compared with the conventional MIMO systems, thanks to the sparsity of mmWave MIMO channels.

6.2.1 CHANNEL MODEL

Experiments have demonstrated that mmWave massive MIMO channels exhibit obvious sparsity in the spatial domain or angular domain. This is due to the high path loss for nonline-of-sight (NLOS) signals, where only a small number of dominated paths (typically, 3–5 paths in realistic environments) consist of the significant mmWave MIMO multipath components. If we consider the widely used ULA, the point-to-point mmWave massive MIMO channel can be modeled as [5]

$$\mathbf{H} = \sqrt{\frac{N_T N_R}{\rho}} \sum_{l=1}^{L} \alpha_l \mathbf{a}_R(\theta_l) \mathbf{a}_T^*(\varphi_l) = \sqrt{\frac{N_T N_R}{\rho}} \mathbf{A}_R \mathbf{H}_a \mathbf{A}_T^* \tag{6.1}$$

where N_T and N_R are the numbers of transmitter and receiver antennas, respectively, ρ is the average path loss, L is the number of multipath, α_l is the complex gain of the lth path, θ_l or $\varphi_l \in [0, 2\pi]$ are azimuth angles of arrival or departure (AoAs/AoDs). In (6.1),

$$\mathbf{a}_R(\theta_l) = \frac{1}{\sqrt{N_R}} \left[1, e^{j2\pi d \sin(\theta_l)/\lambda}, ..., e^{j2\pi(N_R-1)d \sin(\theta_l)/\lambda} \right]^T \tag{6.2}$$

and

$$\mathbf{a}_T(\varphi_l) = \frac{1}{\sqrt{N_T}} \left[1, e^{j2\pi d \sin(\varphi_l)/\lambda}, ..., e^{j2\pi(N_T-1)d \sin(\varphi_l)/\lambda} \right]^T \tag{6.3}$$

are steering vectors at the receiver and transmitter, respectively. Moreover,

$$\mathbf{A}_R = [\mathbf{a}_R(\theta_1), \mathbf{a}_R(\theta_2), ..., \mathbf{a}_R(\theta_L)] \tag{6.4}$$

$$\mathbf{A}_T = [\mathbf{a}_T(\varphi_1), \mathbf{a}_T(\varphi_2), ..., \mathbf{a}_T(\varphi_L)] \tag{6.5}$$

and the diagonal matrix $\mathbf{H}_a = \text{diag}\{\alpha_1, \alpha_2, ..., \alpha_L\}$, where λ and d are wavelength and antenna spacing, respectively.

The sparsity of mmWave channels in the spatial domain or angular domain (L is small, such as 3–5) and massive MIMO channel matrix (N_T and N_R can be large, such as dozens or even hundreds) implies that the mmWave massive MIMO channel matrix has the low-rank property. For instance, Fig. 6.1 provides the energy probability distribution of singular values of $\mathbf{H}$ with descending order against different L's, where $N_T = 512$, $N_R = 32$, and path gains follow the independent and identically distributed (i.i.d.) complex Gaussian distribution [2]. It can be observed that the mmWave massive MIMO channel matrix has the obvious low-rank property. The sparsity of mmWave channels in the spatial domain or angular domain as well as the low-rank property of the mmWave massive MIMO channel matrix will be exploited to design the cost- and energy-efficient transceiver for mmWave massive MIMO systems.

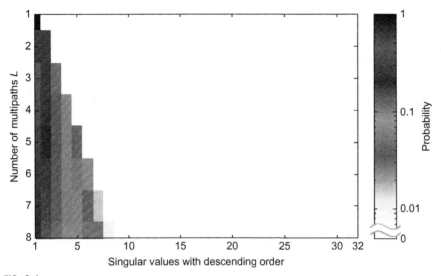

FIG. 6.1

Energy probability distribution of singular values with descending order of mmWave massive MIMO channel matrix versus different L's, where $N_R = 512$ and $N_T = 32$.

6.2.2 TRANSCEIVER STRUCTURE IN mmWave MASSIVE MIMO

In mmWave massive MIMO systems, the hardware cost and power consumption of the high-speed ADC, synthesizers, and so on, can be nonnegligible. Therefore, for mmWave massive MIMO systems, the transceiver structure that each antenna requires an associated RF chain as the conventional MIMO systems working at sub-3–6 GHz can be impossible. To solve this problem, the transceiver structure with the analog PSN has been emerging as a promising candidate for mmWave massive MIMO systems. This specific transceiver structure can be illustrated in Fig. 6.2 [4]. In Fig. 6.2, it can be observed that for the BS (MS), the number of RF chains N_{RF} (we assume that the MS also has N_{RF} RF chains except for Section 6.5) is much smaller than that of antennas N_{BS} (N_{MS}), which is essentially different from that in conventional MIMO systems with full digital precoding/combining. Consider such transceiver structure in the downlink: the beamforming processing at the BS can be considered the cascade of the analog RF beamforming matrix $\mathbf{F}_{RF} \in \mathbb{C}^{N_{BS} \times N_{RF}}$ and the digital baseband beamforming matrix $\mathbf{F}_{BB} \in \mathbb{C}^{N_{RF} \times N_s}$, where N_s is the number of independent data stream. Accordingly, the combining at the MS also has the analog RF combining matrix $\mathbf{W}_{RF} \in \mathbb{C}^{N_{MS} \times N_{RF}}$ and digital baseband combining matrix $\mathbf{W}_{BB} \in \mathbb{C}^{N_{RF} \times N_s}$. If $\mathbf{s} \in \mathbb{C}^{N_s \times 1}$ is the training signal used for CE at the BS, and $\mathbf{n} \in \mathbb{C}^{N_s \times 1}$ is the effective received noise at the MS, the received signal at the MS after combining can be expressed as

$$\mathbf{y} = \mathbf{W}_{BB}^H \mathbf{W}_{RF}^H \mathbf{H} \mathbf{F}_{RF} \mathbf{F}_{BB} \mathbf{s} + \mathbf{n} = \mathbf{W}^H \mathbf{H} \mathbf{F} \mathbf{s} + \mathbf{n} \tag{6.6}$$

where $\mathbf{W} = \mathbf{W}_{RF} \mathbf{W}_{BB}$ and $\mathbf{F} = \mathbf{F}_{RF} \mathbf{F}_{BB}$.

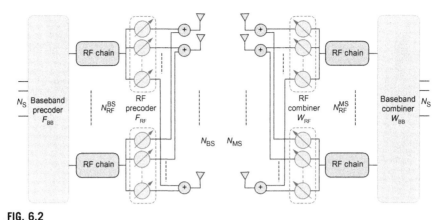

FIG. 6.2

Transceiver with the analog PSN for the reduced number of RF chains.

By far, most state-of-the-art CE schemes for mmWave massive MIMO systems are based on such a transceiver structure with analog PSN. In Sections 6.3, 6.4, and 6.6, we will provide more details on these CE schemes.

Low-resolution ADC at the receiver is another promising candidate to reduce the receiver's cost and power consumption for mmWave massive MIMO systems [1]. It has been demonstrated that, compared with high-resolution ADCs, low-resolution ADCs can be implemented with simpler circuits, consume less power, and only suffer from a slight performance loss at low SNR [4]. Fig. 6.3 illustrates the receiver structure employing each one-bit ADC to respectively receive the in-phase and quadrature baseband signal from each RF chain. In Section 6.4, we will also provide the specific CE scheme in mmWave massive MIMO systems with the one-bit ADC at the receiver.

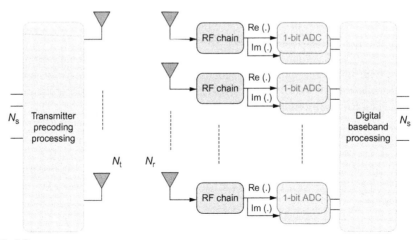

FIG. 6.3

Receiver structure with one-bit ADC.

6.3 COMPRESSIVE SENSING (CS)-BASED CHANNEL ESTIMATION SCHEMES

In this section, we will detail the CS-based CE schemes [4,6,7] for mmWave massive MIMO systems. We commence by briefly introducing the concept of compressive sensing (CS) theory. Then, we formulate the issue of CE as the standard CS problem. Moreover, the specific selection of the transformation matrix, the design of codebooks, and practical sparse channel reconstruction algorithms are also discussed.

6.3.1 CONCEPT OF CS THEORY

Naturally, most continuous signals from the real world have inherent redundancy or correlation, which implies that the effective information rate conveyed by such signals can be much smaller than their bandwidth. Equivalently, the number of effective degrees of freedom of the corresponding discrete signals can be much smaller than their dimensions, which indicates that these signals exhibit sparsity in some transformation domains. Against this background, CS theory has been developed, which shows that the sparsity of a signal can be exploited to recover the original signal from far fewer samples than those required by the classical Shannon-Nyquist sampling theorem [8].

To be specific, we consider the standard CS problem, that is,

$$\mathbf{y} = \boldsymbol{\Phi}\mathbf{x} \in \mathbb{C}^{M \times 1} \tag{6.7}$$

where the sparse signal $\mathbf{x} \in \mathbb{C}^{N \times 1}$ has the sparsity level k (i.e., $\mathbf{x}$ has k nonzero elements), $\boldsymbol{\Phi} \in \mathbb{C}^{M \times N}$ is the measurement matrix, and $\mathbf{y} = \boldsymbol{\Phi}\mathbf{x} \in \mathbb{C}^{M \times 1}$ is the measurement signal. In CS theory, what matters is how to acquire $\mathbf{x}$ by solving the underdetermined Eq. (6.7) with $M \ll N$ provided $\mathbf{y}$ and $\boldsymbol{\Phi}$. Generally, $\mathbf{x}$ may not exhibit the sparsity itself, but the sparsity appears in some transformation domains, that is, $\mathbf{x} = \boldsymbol{\Psi}\mathbf{s}$ where $\boldsymbol{\Psi} \in \mathbb{C}^{N \times N}$ is the transform matrix, and $\mathbf{s} \in \mathbb{C}^{N \times 1}$ is the sparse signal with the sparsity level k. In this way, (6.7) can be further expressed as

$$\mathbf{y} = \boldsymbol{\Phi}\boldsymbol{\Psi}\mathbf{s} = \boldsymbol{\Theta}\mathbf{s} \tag{6.8}$$

where $\boldsymbol{\Theta} = \boldsymbol{\Phi}\boldsymbol{\Psi}$. In CS theory, there are three fundamental issues, as follows [8]:

(1) *Sparse transformation domain*: It is essential for CS to find the suitable transform matrix $\boldsymbol{\Psi}$ to transform the original signal $\mathbf{x}$ into the sparse signal $\mathbf{s}$.

(2) *Sparse signal compression*: The design of $\boldsymbol{\Theta}$ should compress the dimension of measurements while minimizing the information loss, which can be evaluated by the coherence or restricted isometry property of $\boldsymbol{\Theta}$.

(3) *Sparse signal recovery*: It is important to design sparse signal recovery algorithms for the reliable reconstruction of $\mathbf{s}$ from the limited measurement signal $\mathbf{y}$.

6.3.2 FORMULATE CHANNEL ESTIMATION AS CS PROBLEM

We consider the downlink CE in mmWave massive MIMO systems with the hybrid MIMO transceiver structure for single-user scenarios, which has been shown in Fig. 6.2. In this scenario, the BS has N_{BS} antennas but $N_{RF} \ll N_{BS}$ RF chains, while the MS has N_{MS} antennas but $N_{RF} \ll N_{MS}$ RF chains. In the downlink CE, the BS transmits N_{BS}^{Tr} training beam patterns, denoted as $\{\mathbf{f}_p \in \mathbb{C}^{N_{BS} \times 1} : p = 1, 2, ..., N_{BS}^{Tr}\}$ with $N_{BS}^{T} = N_{RF} N_{BS}^{Slot}$, and the MS uses N_{MS}^{Tr} combining patterns, denoted as $\{\mathbf{w}_q \in \mathbb{C}^{N_{MS} \times 1} : q = 1, 2, ..., N_{MS}^{Tr}\}$, to receive signals, where $N_{MS}^{T} = N_{RF} N_{MS}^{Slot}$. During the stage of CE, the BS transmits N_{BS}^{Tr} different training beam patterns in N_{BS}^{Tr} successive time slots, while the MS employs N_{MS}^{Slot} combining patterns to receive signals for every training beam pattern in N_{MS}^{Slot} successive sub-time slots. To be specific, if the BS transmits the training beam pattern $\mathbf{f}_p \in \mathbb{C}^{N_{BS} \times 1}$ and the MS employs the receiver combining pattern $\mathbf{w}_q \in \mathbb{C}^{N_{MS} \times 1}$, the received signal can be written as

$$y_{q,p} = \mathbf{w}_q^H \mathbf{H} \mathbf{f}_p + \mathbf{w}_q^H \mathbf{n}_p \tag{6.9}$$

At the receiver, the MS employs N_{MS}^{Slot} successive sub-time slots to receive each training beam pattern transmitted by the BS. It should be pointed out that, due to the hybrid MIMO transceiver structure with $N_{RF} \ll N_{BS}$ RF chains at the MS, the MS can simultaneously employ N_{RF} combining patterns to acquire the signal $\mathbf{y}_p \in \mathbb{C}^{N_{RF} \times 1}$ in each sub-time slot. For the pth transmit training beam pattern, the received signal $\mathbf{y}_p \in \mathbb{C}^{N_{MS}^{Tr} \times 1}$ can be expressed as

$$\mathbf{y}_p = \mathbf{W}^H \mathbf{H} \mathbf{f}_p + \mathbf{W}^H \mathbf{n}_p \tag{6.10}$$

where $\mathbf{W} = \left[\mathbf{W}_1, \mathbf{W}_2, ..., \mathbf{W}_{N_{MS}^{Slot}} \right] \in \mathbb{C}^{N_{MS} \times N_{MS}^{Tr}}$, $\mathbf{W}_l = \left[\mathbf{w}_{(l-1)N_{RF}+1}, \mathbf{w}_{(l-1)N_{RF}+2}, ..., \mathbf{w}_{lN_{RF}} \right]$ $\in \mathbb{C}^{N_{MS} \times N_{RF}}$ for $1 \leq l \leq N_{MS}^{Slot}$, and $\mathbf{n}_p \in \mathbb{C}^{N_{MS}^{Tr} \times 1}$ is noise. Because the BS will transmit N_{BS}^{Tr} different training beam patterns in N_{BS}^{Tr} successive time slots, the aggregated receive signal at the MS can be further expressed as

$$\mathbf{Y} = \mathbf{W}^H \mathbf{H} \mathbf{F} + \mathbf{N} \tag{6.11}$$

where $\mathbf{Y} = \left[\mathbf{y}_1, \mathbf{y}_2, ..., \mathbf{y}_{N_{BS}^{Tr}} \right]$, $\mathbf{F} = \left[\mathbf{f}_1, \mathbf{f}_2, ..., \mathbf{f}_{N_{BS}^{Tr}} \right]$, and $\mathbf{N} = \mathbf{W}^H \left[\mathbf{n}_1, \mathbf{n}_2, ..., \mathbf{n}_{N_{BS}^{T}} \right]$. By vectorizing the received signal $\mathbf{Y}$, we can further obtain

$$\mathbf{y}_{MS} = \text{vec}(\mathbf{Y}) = \left(\mathbf{F}^T \otimes \mathbf{W}^H \right) \text{vec}(\mathbf{H}) + \mathbf{v} \tag{6.12}$$

where we use the equality $\text{vec}(\mathbf{ABC}) = \left(\mathbf{C}^T \otimes \mathbf{A} \right) \text{vec}(\mathbf{B})$ for matrices $\mathbf{A}$, $\mathbf{B}$, and $\mathbf{C}$, and $\mathbf{v} = \text{vec}(\mathbf{N})$. It is clear that we can use the conventional linear algorithms such as least squares (LS) estimator to acquire the estimation of $\text{vec}(\mathbf{H})$ from $\mathbf{y}_{MS}$, that is,

$$\text{vec}(\hat{\mathbf{H}}) = \left(\left(\mathbf{F}^T \otimes \mathbf{W}^H \right)^H \left(\mathbf{F}^T \otimes \mathbf{W}^H \right) \right)^{-1} \left(\mathbf{F}^T \otimes \mathbf{W}^H \right)^H \mathbf{y}_{MS} \tag{6.13}$$

However, it requires (6.13) to be an overdetermined equation, that is, $N_{BS}^{Tr} N_{MS}^{Tr} \geq N_{MS} N_{BS}$. In mmWave massive MIMO systems with hundreds of antennas, employing an LS algorithm to acquire $\hat{\mathbf{H}}$ will suffer from the prohibitively high computational complexity and time overhead for CE.

Fortunately, by exploiting the sparsity of mmWave massive MIMO channels, it is expected to exploit the CS theory to estimate mmWave massive MIMO channels of large size from the measurements of small size. As we discussed in Section 6.3.1, the first key issue in CS theory is to find the suitable transformation matrix to transform the vector vec($\mathbf{H}$) to a sparse signal. By leveraging the sparse AoAs/AoDs of mmWave massive MIMO channels in the angular domain [4,6,7], assumes that the AoAs/AoDs are taken from a uniform grid on G points with $G \geq L$, that is, θ_l and φ_l for $1 \leq l \leq L$ comes from the uniform angle set $\{0, 2\pi/G, \ldots, 2\pi(G-1)/G\}$. It should be pointed out that the actual AoAs/AoDs are continuous values with the range from 0 to 2π. By ignoring the grid quantization error, the channel matrix $\mathbf{H}$ can be expressed as

$$\mathbf{H} = \mathbf{A}_{R,G} \widetilde{\mathbf{H}}_a \mathbf{A}_{T,G}^H \tag{6.14}$$

where

$$\mathbf{A}_{R,G} = \left[\mathbf{a}_R\left(\overline{\theta}_0\right), \mathbf{a}_R\left(\overline{\theta}_1\right), \ldots, \mathbf{a}_R\left(\overline{\theta}_{G-1}\right) \right] \in \mathbb{C}^{N_{\text{MS}} \times G} \tag{6.15}$$

and

$$\mathbf{A}_{T,G} = \left[\mathbf{a}_T\left(\overline{\varphi}_0\right), \mathbf{a}_T\left(\overline{\varphi}_1\right), \ldots, \mathbf{a}_T\left(\overline{\varphi}_{G-1}\right) \right] \in \mathbb{C}^{N_{\text{MS}} \times G} \tag{6.16}$$

with $\overline{\theta}_g = 2\pi g/G$ and $\overline{\varphi}_g = 2\pi g/G$ for $0 \leq g \leq G-1$. Due to the sparsity of mmWave massive MIMO channels, $\widetilde{\mathbf{H}}_a$ appears the sparsity, that is, most elements of $\widetilde{\mathbf{H}}_a$ are zero, and only L elements associated with the AoA and AoD are nonzero. By inserting (6.14) into (6.12), we can further obtain

$$\begin{aligned}
\mathbf{y}_{\text{MS}} = \text{vec}(\mathbf{Y}) &= \left(\mathbf{F}^T \otimes \mathbf{W}^H\right) \text{vec}\left(\mathbf{A}_{R,G}^H \widetilde{\mathbf{H}}_a \mathbf{A}^H\right) + \mathbf{v} \\
&= \left(\mathbf{F}^T \otimes \mathbf{W}^H\right) \left(\mathbf{A}_{T,G}^T \otimes \mathbf{A}_{R,G}^H\right) \text{vec}\left(\widetilde{\mathbf{H}}_a\right) + \mathbf{v} \\
&= \left(\mathbf{F}^T \mathbf{A}_{T,G}^T\right) \otimes \left(\mathbf{W}^H \mathbf{A}_{R,G}^H\right) \text{vec}\left(\widetilde{\mathbf{H}}_a\right) + \mathbf{v}
\end{aligned} \tag{6.17}$$

where we use the equality $(\mathbf{A} \otimes \mathbf{B})(\mathbf{C} \otimes \mathbf{D}) = (\mathbf{AC}) \otimes (\mathbf{BD})$ for matrices $\mathbf{A}$, $\mathbf{B}$, $\mathbf{C}$, and $\mathbf{D}$. For simplicity, (6.17) can be further written as

$$\mathbf{y}_{\text{MS}} = \mathbf{Q} \mathbf{z}_a + \mathbf{v} \tag{6.18}$$

where $\mathbf{Q} = \left(\mathbf{F}^T \mathbf{A}_{T,G}^T\right) \otimes \left(\mathbf{W}^H \mathbf{A}_{R,G}^H\right) \in \mathbb{C}^{N_{\text{MS}}^{\text{Tr}} N_{\text{BS}}^{\text{Tr}} \times G^2}$ and $\mathbf{z}_a = \text{vec}\left(\widetilde{\mathbf{H}}_a\right) \in \mathbb{C}^{G^2 \times 1}$. In this way, the channels of mmWave massive MIMO systems can be estimated by solving the following optimization problem under the framework of CS [8], that is,

$$\begin{aligned}
\hat{\mathbf{z}}_a = \arg \min_{\mathbf{z}_a} \|\mathbf{y}_{\text{MS}} - \mathbf{Q}\mathbf{z}_a\|_2, \\
\text{s.t. } \|\mathbf{y}_{\text{MS}} - \mathbf{Q}\mathbf{z}_a\|_2 \leq \delta
\end{aligned} \tag{6.19}$$

where $\hat{\mathbf{z}}_a$ is the estimate of the sparse signal $\mathbf{z}_a$, and δ is a threshold related to the noise. Consequently, the final estimate of the channels is given by

$$\hat{\mathbf{H}} = \mathbf{A}_{R,G}^H \hat{\widetilde{\mathbf{H}}}_a \mathbf{A}_{T,G} \tag{6.20}$$

where $\hat{\mathbf{z}}_a = \text{vec}\left(\hat{\widetilde{\mathbf{H}}}_a\right)$.

6.3.3 SPARSE CHANNELS RECONSTRUCTION VIA CS

The second issue in CS theory is how to reconstruct the original sparse signals of high dimension from the measurements of low dimension. By far, most state-of-the-art sparse CE schemes in [4,6,7] are developed from the classical orthogonal matching pursuit (OMP) algorithm due to its low complexity and good performance.

To solve the optimization problem in (6.19), Lee et al. [6] has proposed the OMP-based sparse channel estimator, which is provided in Algorithm 1.

ALGORITHM 1 OMP-BASED SPARSE CHANNEL ESTIMATOR

Input: Measurement matrix $\mathbf{Q}$, measurement signal $\mathbf{y}_{MS}$, and a predefined threshold δ.

   Output: The estimated $\hat{\mathbf{z}}_a$.

   1: $\Omega_0 = \phi$; % Empty the support set

   2: $\mathbf{r}_{-1} = \mathbf{0}$; % Set the residual of the last iteration

   3: $\mathbf{r}_0 = \mathbf{y}_{MS}$; % Set the residual of the current iteration

   4: $t = 1$; % Set the index of iteration

   5: **While** $\|\mathbf{r}_{t-1} - \mathbf{r}_t\|_2^2 \geq \delta$ **do**

   6:    $i = \arg\max_{1 < \tilde{i} < G^2} \|\mathbf{q}_{\tilde{i}}^H \mathbf{r}_{t-1}\|$; % Find the optimal AoA/AoD pair

   7:    $\Omega_t = \Omega_{t-1} \cup \{i\}$; % Update the AoA/AoD pair

   8:    $(\hat{\mathbf{z}}_a)_{\Omega_t} = \mathbf{Q}_{\Omega_t}^\dagger \mathbf{y}_{MS}$; % Update the estimated channel

   9:    $\mathbf{r}_t = \mathbf{y}_{MS} - \mathbf{Q}\hat{\mathbf{z}}_a$; % Update the residual

  10:   $t = t + 1$ % Update the index of iterations

  11: **End While**

It should be pointed out that in Algorithm 1, $\mathbf{q}_i$ is the ith column of $\mathbf{Q}$, and $(\hat{\mathbf{z}}_a)_{\Omega_t}$ denotes the subvector whose elements come from $\hat{\mathbf{z}}_a$ with the indexes associated with Ω_t, $\mathbf{Q}_{\Omega_t}$ denotes the submatrix of $\mathbf{Q}$ with the indexes defined by Ω_t, and $\mathbf{Q}^\dagger$ is the pseudoinverse of $\mathbf{Q}$.

As we have discussed, the actual AoAs/AoDs are continuously distributed with the range from 0 to 2π. Hence, the larger G implies the finer resolution of the transformation matrix for improved AoA/AoD estimation performance. However, the large G will lead to prohibitively high computational complexity. To realize accurate CE with low complexity, Lee et al. [6] further proposed an improved OMP algorithm named multigrid OMP (MG-OMP) algorithm. This algorithm can adaptively refine the resolution of the transformation matrix according to the previous CE. The MG-OMP algorithm is provided in Algorithm 2.

ALGORITHM 2 MG-OMP-BASED SPARSE CHANNEL ESTIMATOR

Input: Measurement matrix $\mathbf{Q}_0$, measurement signal $\mathbf{y}_{MS}$, the number of total stages R (determined by the required resolution of AoA/AoD) and a predefined threshold δ.

   Output: The estimated $\hat{\mathbf{z}}_a$.

   1: $\Omega_0 = \phi$; % Empty the support set

   2: $\mathbf{r}_{-1} = \mathbf{0}$; % Set the residual of the last iteration

   3: $\mathbf{r}_0 = \mathbf{y}_{MS}$; % Set the residual of the current iteration

   4: $t = 1$; % Set the index of iteration

   5: **While** $\|\mathbf{r}_{t-1} - \mathbf{r}_t\|_2^2 \geq \delta$ **do**

6: $r = 1$; % Set the index of stage

7: **While** $r \leq R$ **do**

8: $i = \arg \max_{1 \leq i \leq G_r^2} \left\| \mathbf{q}_i^H \mathbf{r}_{t-1} \right\|$; % Find the optimal AoA/AoD pair with the
resolution in the rth stage

9: Obtain the indexes of AoD/AoA corresponding to i, where $g_{r-1} = \lceil i/G_{r-1} \rceil$ and
$g'_{r-1} = \mod(j-1, G_{r-1}) + 1$, respectively

10: Update the transformation matrix

$$\mathbf{A}_{R,G_r} = \left[\mathbf{a}_R(\overline{\theta}_1), \mathbf{a}_R(\overline{\theta}_2), \dots, \mathbf{a}_R(\overline{\theta}_{g'_r}), \dots, \mathbf{a}_R(\overline{\theta}_{G_r}) \right]$$

$$\mathbf{A}_{T,G_r} = \left[\mathbf{a}_T(\overline{\varphi}_1), \mathbf{a}_T(\overline{\varphi}_2), \dots, \mathbf{a}_T(\overline{\varphi}_{g_r}), \dots, \mathbf{a}_T(\overline{\varphi}_{G_r}) \right]$$

where $\overline{\theta}_{g_r} \in \left[\overline{\theta}_{g_{r-1}-1}, \overline{\theta}_{g_{r-1}+1} \right]$ with $g_r = 1, \dots, G_r$ and $\overline{\varphi}_{g'_r} \in \left[\overline{\varphi}_{g'_{r-1}-1}, \overline{\varphi}_{g'_{r-1}+1} \right]$ with
$g'_r = 1, \dots, G_r$.

11: Update the measurement matrix

$$\mathbf{Q}_{(t-1)R+r} = \left(\mathbf{F}^T \mathbf{A}_{T,G_r}^T \right) \otimes \left(\mathbf{W}^H \mathbf{A}_{R,G_r}^H \right)$$

12: $r = r + 1$

13: **End While**

14: $\Omega_t = \Omega_{t-1} \cup \{i\}$; % Update the AoA/AoD pair

15: $\left(\hat{\mathbf{z}}_{a,(t-1)R+r} \right)_{\Omega_t} = \left(\mathbf{Q}_{(t-1)R+r} \right)_{\Omega_t}^{\dagger} \mathbf{y}_{MS}$; % Update the estimated channel

16: $\mathbf{r}_t = \mathbf{y}_{MS} - \mathbf{Q}_{(t-1)R+r} \hat{\mathbf{z}}_{a,(t-1)R+r}$; % Update the residual

17: $t = t + 1$; % Update the index of iterations

18: **End While**

The MG-OMP algorithm starts with a coarse grid (or coarse resolution to estimate AoA/AoD) and only increases the resolution around the grid where the AoAs/AoDs are present. For the rth stage in the tth iteration, (6.18) in the MG-OMP algorithm can be rewritten as

$$\mathbf{y}_{MS} = \mathbf{Q}_{(t-1)R+r} \mathbf{z}_{a,(t-1)R+r} + \mathbf{v} \tag{6.21}$$

where $\mathbf{Q}_{(t-1)R+r} = \left(\mathbf{F}^T \mathbf{A}_{T,G_r}^T \right) \otimes \left(\mathbf{W}^H \mathbf{A}_{R,G_r}^H \right) \in \mathbb{C}^{N_{MS}^T N_{BS}^T \times G_r^2}$ is the measurement matrix at the rth stage in the tth iteration, $\mathbf{A}_{R,G_r} \in \mathbb{C}^{N_{MS} \times G_r}$, and $\mathbf{A}_{T,G_r} \in \mathbb{C}^{N_{BS} \times G_r}$. Note that columns in $\mathbf{A}_{R,G_r}$ and $\mathbf{A}_{T,G_r}$ are the array response vectors corresponding to the candidate grid of AoA/AoD.

6.3.4 DESIGN TRAINING BEAM AND COMBINING PATTERNS ACCORDING TO CS THEORY

The last but not the least issue in CS theory is how to reliably compress the original sparse signal and minimize the information loss. For (6.21), this problem can be converted to the design of $\mathbf{Q}_{(t-1)R+r} = \left(\mathbf{F}^T \mathbf{A}_{T,G_r}^T \right) \otimes \left(\mathbf{W}^H \mathbf{A}_{R,G_r}^H \right) \in \mathbb{C}^{N_{MS}^T N_{BS}^T \times G_r^2}$. Because the measurement matrix is the product of the projection/compressive matrix (**F** and **W**) and the transformation matrix ($\mathbf{A}_{R,G_r}$ and $\mathbf{A}_{T,G_r}$), and the transformation matrix is adaptive in the different stages and iterations, in this section we will elaborate the projection/compressive matrix $\mathbf{F}$ and $\mathbf{W}$.

For mmWave massive MIMO systems with the hybrid transceiver structure, both the beamforming at the transmitter and the combining at the receiver can be divided into the analog part and the digital part, that is, $\mathbf{F} = \mathbf{F}_{RF}\mathbf{F}_{BB}$ and $\mathbf{W} = \mathbf{W}_{RF}\mathbf{W}_{BB}$ as in (6.6), where $\mathbf{F}_{RF} \in \mathbb{C}^{N_T \times N_{BS}^{Tr}}$ and $\mathbf{W}_{RF} \in \mathbb{C}^{N_R \times N_{MS}^{Tr}}$ are the beamforming matrix and combining matrix in the RF, respectively, and $\mathbf{F}_{BB} \in \mathbb{C}^{N_{BS}^{Tr} \times N_{BS}^{Tr}}$ and $\mathbf{W}_{BB} \in \mathbb{C}^{N_{MS}^{Tr} \times N_{MS}^{Tr}}$ are the beamforming matrix and combining matrix in the digital baseband, respectively.

Due to the hardware constraint in mmWave massive MIMO systems, the RF beamforming matrix and combining matrix have the following properties:

(1) Elements of $\mathbf{F}_{RF}$ and $\mathbf{W}_{RF}$ are constant envelope.
(2) Columns of $\mathbf{F}_{RF}$ should cover all the AoDs and columns of $\mathbf{W}_{RF}$ should cover all the AoAs.
(3) Both $\mathbf{F}_{BB}$ and $\mathbf{W}_{BB}$ are block diagonal matrices given by

$$\mathbf{F}_{BB} = \text{diag}\left(\mathbf{F}_{BB,1}, \dots, \mathbf{F}_{BB,i}, \dots, \mathbf{F}_{BB,N_{BS}^{Slot}}\right) \tag{6.22}$$

and

$$\mathbf{W}_{BB} = \text{diag}\left(\mathbf{W}_{BB,1}, \dots, \mathbf{W}_{BB,i}, \dots, \mathbf{W}_{BB,N_{MS}^{Slot}}\right) \tag{6.23}$$

with $\mathbf{F}_{BB,i} \in \mathbb{C}^{N_{RF} \times N_{RF}}$ and $\mathbf{W}_{BB,i} \in \mathbb{C}^{N_{RF} \times N_{RF}}$.

With these properties, [6] has suggested to use the discrete Fourier transform (DFT) beams for RF beamforming $\mathbf{F}_{RF}$ at the transmitter, where columns of $\mathbf{F}_{RF}$ are selected from the DFT matrix with the size of $N_{BS}^{Tr} \times N_{BS}^{Tr}$. Similarly, RF combining patterns $\mathbf{W}_{RF}$ at the receiver use the DFT matrix, where columns of $\mathbf{W}_{RF}$ are selected from the DFT matrix with the size of $N_{MS}^{Tr} \times N_{MS}^{Tr}$.

Based on $\mathbf{F}_{RF}$ and $\mathbf{W}_{RF}$, we further elaborate $\mathbf{F}_{BB}$ and $\mathbf{W}_{BB}$ under the framework of CS theory. In CS theory, the coherence of the matrix can be considered the quality of the measurement matrix on the compressive performance. The coherence of the measurement is defined as

$$\mu(\mathbf{Q}) = \max_{1 \leq m < n \leq G^2} \frac{\langle \mathbf{q}_m, \mathbf{q}_n \rangle}{\langle \mathbf{q}_n, \mathbf{q}_n \rangle \langle \mathbf{q}_m, \mathbf{q}_m \rangle} \tag{6.24}$$

where we omit r, t and G for convenience, and $\mathbf{q}_m$ denotes the mth column of the matrix $\mathbf{Q}$. In CS theory, the smaller $\mu(\mathbf{Q})$ can lead to better sparse signal recovery performance. Because $\mu(\mathbf{Q}) = \max\left\{\mu(\mathbf{F}^T\mathbf{A}_{T,G}^T), \mu(\mathbf{W}^H\mathbf{A}_{R,G}^H)\right\}$ and $\mathbf{F}^T\mathbf{A}_{T,G}^T$ and $\mathbf{W}^H\mathbf{A}_{R,G}^H$ are respectively processed at the BS and MS, a suitable digital beamforming or combining matrix design can be converted as the following problems:

$$\mathbf{A}_{T,G} = \min_{\mathbf{A}_{T,G}} \mu\left(\mathbf{F}_{BB}^T\mathbf{F}_{RF}^T\mathbf{A}_{T,G}^T\right) \tag{6.25}$$

$$\mathbf{A}^{R,G} = \min_{\mathbf{A}_{R,G}} \mu\left(\mathbf{W}_{BB}^H\mathbf{W}_{RF}^H\mathbf{A}_{R,G}^H\right) \tag{6.26}$$

In [6], it was shown that the optimal solution to (6.25) can be given as

$$\mathbf{F}_{\text{BB},i} = \mathbf{U}_{\text{BB},i}\left(\mathbf{\Lambda}_{\text{BB},i}^{-1/2}\right)^{H}, 1 \le i \le N_{\text{BS}}^{\text{Slot}} \qquad (6.27)$$

where $\mathbf{U}_{\text{BB},i}$ and $\mathbf{\Lambda}_{\text{BB},i}$ are the matrices of the eigenvectors and eigenvalues, respectively, that is,

$$\mathbf{F}_{\text{RF},i}^{H}\mathbf{A}_{T,G}\mathbf{A}_{T,G}^{H}\mathbf{F}_{\text{RF},i} = \mathbf{U}_{\text{BB},i}\mathbf{\Lambda}_{\text{BB},i}\mathbf{U}_{\text{BB},i}^{H} \qquad (6.28)$$

Similarly, the optimal solution to (6.26) can be given as

$$\mathbf{W}_{\text{BB},i} = \mathbf{U}_{\text{BB},i}\left(\mathbf{\Lambda}_{\text{BB},i}^{-1/2}\right)^{H}, 1 \le i \le N_{\text{MS}}^{\text{Slot}} \qquad (6.29)$$

where $\mathbf{U}_{\text{BB},i}$ and $\mathbf{\Lambda}_{\text{BB},i}$ are the matrices of the eigenvectors and eigenvalues, respectively, that is,

$$\mathbf{W}_{\text{RF},i}^{H}\mathbf{A}_{R,G}\mathbf{A}_{R,G}^{H}\mathbf{W}_{\text{RF},i} = \mathbf{U}_{\text{BB},i}\mathbf{\Lambda}_{\text{BB},i}\mathbf{U}_{\text{BB},i}^{H} \qquad (6.30)$$

6.3.5 REMARK

In this section, we have presented the CS-based CE for mmWave massive MIMO systems proposed in [6], which is similar to those proposed in [4, 7]. The CE scheme proposed in [7] mainly considers the mmWave massive MIMO channels with LOS. The CE scheme proposed in [4] can handle the mmWave massive MIMO channels with multiple paths, and an adaptive multi-resolution codebook is elaborated, which is similar to [6].

However, these are still some key issues for the practical implementation of the CS-based CE schemes. The first issue is the optimal beamforming or combining pattern design. Although [6] has provided the optimal digital baseband beamforming and combining patterns, the optimal analog beamforming and combining pattern design is still a problem to be further investigated. The second issue is the trade-off between the quantization error for the continuous AoA/AoD in practice and the CE performance with the assumption of the discrete AoA/AoD. Hence, a theoretical analysis is desired to enlighten the practical algorithm design given the requirement of the accuracy of CE performance. Finally, if the mmWave massive MIMO is not sparse in some special scenarios (indoor office, for instance), an adaptive CS-based CE scheme should be provided to handle this situation.

6.4 CHANNEL ESTIMATION WITH ONE-BIT ADCs AT THE RECEIVER

In this section, we will introduce the CE scheme for mmWave massive MIMO systems with one-bit ADCs at the receiver [9]. Similar to the hybrid MIMO systems with analog PSN, the receiver structure with one-bit ADCs also exploits the sparsity of mmWave massive MIMO channels with slight performance loss compared to the conventional MIMO transceiver structure.

6.4.1 VIRTUAL CHANNEL REPRESENTATION OF mmWave MASSIVE MIMO CHANNELS

We begin by introducing the virtual channel representation of the mmWave massive MIMO channels as follows:

$$\mathbf{H} = \mathbf{A}_R \widetilde{\mathbf{H}}_a \mathbf{A}_T^H \tag{6.31}$$

where $\mathbf{A}_R \in \mathbb{C}^{N_R \times N_R}$ and $\mathbf{A}_T \in \mathbb{C}^{N_T \times N_T}$ are unitary DFT matrices, and $\widetilde{\mathbf{H}}_a$ is the virtual channel matrix. It should be pointed out that (6.31) is a special case of (6.14), where the resolution of AoA at the MS is $G = N_R$ and the resolution of AoD at the BS is $G = N_T$. By using the Fourier transformation in (6.31), elements of the original mmWave massive MIMO channel matrix in the antenna domain are one-to-one mapped to the angular domain, and the mth row and nth column element in $\widetilde{\mathbf{H}}_a$ can be interpreted as the channel gains associated with the mth receiver AoA and nth transmitter AoD. The virtual channel of $\mathbf{H}$ with $N_T = 256$, $N_R = 32$ and two paths can be illustrated in Fig. 6.4.

For convenience, [9] assumes that the transmitter/receiver array response vectors are selected from the columns of the DFT matrix $\mathbf{A}_T/\mathbf{A}_R$. Under this assumption, $\widetilde{\mathbf{H}}_a$ is a pure sparse matrix. That is to say, only L elements in $\widetilde{\mathbf{H}}_a$ are nonzero. To simplify the analysis, we assume that each element in $\widetilde{\mathbf{H}}_a$ follows the Bernoulli-Gaussian distribution, that is,

$$P_r\left(\left[\widetilde{\mathbf{H}}_a\right]_{m,n} = 0\right) = 1 - \eta \tag{6.32}$$

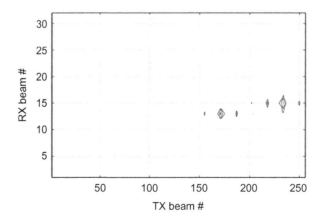

FIG. 6.4

The energy of elements in the virtual channel representation of $\widetilde{\mathbf{H}}_a$ with $N_T = 256$, $N_R = 32$, and two paths.

$$P_r\left(\left[\widetilde{\mathbf{H}}_a\right]_{m,n} = h\right) = \frac{\eta}{\pi\sigma_L^2}e^{-\frac{|h|^2}{\sigma_L^2}}, |h| > 0 \tag{6.33}$$

where $\left[\widetilde{\mathbf{H}}_a\right]_{m,n}$ is the mth row and nth column element in $\widetilde{\mathbf{H}}_a$, $\eta = L/N_tN_r$ is the probability that $\left[\widetilde{\mathbf{H}}_a\right]_{m,n} \neq 0$, and σ_L^2 is the average channel gain.

6.4.2 THE MAXIMAL LIKELIHOOD (ML) ESTIMATOR

We consider the received signal after ADC can be expressed as

$$\begin{aligned}\mathbf{Y} &= \text{sign}(\mathbf{HF} + \mathbf{N})\\ &= \text{sign}\left(\mathbf{A}_R\widetilde{\mathbf{H}}_a\mathbf{A}_T^H\mathbf{A}_T\mathbf{S} + \mathbf{N}\right)\\ &= \text{sign}\left(\mathbf{A}_R\widetilde{\mathbf{H}}_a\mathbf{S} + \mathbf{N}\right)\end{aligned} \tag{6.34}$$

where $\mathbf{Y} \in \mathbb{C}^{N_R \times T_o}$, T_o is the required training overhead, sign() is signum function, $\mathbf{F} = \mathbf{A}_T\mathbf{S} \in \mathbb{C}^{N_T \times T_o}$ is the training signal at the transmitter, and $\mathbf{N} \in \mathbb{C}^{N_T \times T_o}$ is the noise. By vectorizing $\mathbf{Y}$, we can obtain

$$\begin{aligned}\mathbf{y} &= \text{vec}(\mathbf{Y})\\ &= \text{sign}\left(\text{vec}\left(\mathbf{A}_R\widetilde{\mathbf{H}}_a\mathbf{S} + \mathbf{N}\right)\right)\\ &= \text{sign}\left(\left(\mathbf{S}^T \otimes \mathbf{A}_R\right)\text{vec}\left(\widetilde{\mathbf{H}}_a\right) + \text{vec}(\mathbf{N})\right)\end{aligned} \tag{6.35}$$

For mmWave massive MIMO systems with one-bit ADC at the receiver as shown in Fig. 6.3, each RF chain has two ADCs; thus the receiver signal at the digital baseband can be expressed as

$$\bar{\mathbf{y}} = \text{sign}(\Xi\bar{\mathbf{h}} + \bar{\mathbf{n}}) \tag{6.36}$$

where

$$\bar{\mathbf{y}} = \begin{bmatrix}\text{Re}(\mathbf{y})\\ \text{Im}(\mathbf{y})\end{bmatrix} \tag{6.37}$$

$$\bar{\mathbf{h}} = \begin{bmatrix}\text{Re}\left(\text{vec}\left(\widetilde{\mathbf{H}}_a\right)\right)\\ \text{Im}\left(\text{vec}\left(\widetilde{\mathbf{H}}_a\right)\right)\end{bmatrix} \tag{6.38}$$

$$\Xi = \begin{bmatrix}\text{Re}\left(\mathbf{S}^T \otimes \mathbf{A}_R\right) & -\text{Im}\left(\mathbf{S}^T \otimes \mathbf{A}_R\right)\\ \text{Im}\left(\mathbf{S}^T \otimes \mathbf{A}_R\right) & \text{Re}\left(\mathbf{S}^T \otimes \mathbf{A}_R\right)\end{bmatrix} \tag{6.39}$$

$$\bar{\mathbf{n}} = \begin{bmatrix}\text{Re}(\text{vec}(\mathbf{N}))\\ \text{Im}(\text{vec}(\mathbf{N}))\end{bmatrix} \tag{6.40}$$

To this end, this CE problem has been converted as the ML estimation problem; that is, estimate $\bar{\mathbf{h}}$ provided $\bar{\mathbf{y}}$ and Ξ, that is,

$$p(\bar{\mathbf{y}}|\bar{\mathbf{h}}) = \prod_{i=1}^{2N_R} \Phi\left(\frac{\bar{y}_i \psi_i^T \bar{\mathbf{h}}}{\sigma}\right) \tag{6.41}$$

where $\bar{y}_i$ is the ith element of $\bar{\mathbf{y}}$, ψ_i is the ith row of Ξ, σ^2 is the variance of the real part or imagery part of the AWGN, $\Phi(\cdot)$ is the normal cumulative distribution function.

It is clear that the optimal estimation of $\bar{\mathbf{h}}$ is the ML estimator, that is,

$$\frac{\partial \log p(\bar{\mathbf{y}}|\bar{\mathbf{h}})}{\partial \bar{\mathbf{h}}} = 0 \tag{6.42}$$

However, finding the closed-form estimation of $\bar{\mathbf{h}}$ can be difficult. In Section 6.4.3, we will use the iterative approach to estimate $\bar{\mathbf{h}}$ meeting (6.42).

6.4.3 ESTIMATE CHANNELS WITH ITERATIVE APPROACH

In this subsection, we will briefly introduce two CE algorithms with the iterative approach to acquire the estimation of $\bar{\mathbf{h}}$ meeting (6.42). The first CE algorithm is provided in Algorithm 3, named the expectation-maximization (EM) algorithm.

ALGORITHM 3 EM ALGORITHM

Input: The received signal $\bar{\mathbf{y}}$ and the measurement matrix Ξ.
    Output: The estimation of $\bar{\mathbf{h}}$
    1: **Repeat**
    2: **Expectation Step**:
        Compute the posterior covariance Σ and mean μ of $p\left(\hat{\mathbf{h}}|\Xi, \bar{\mathbf{y}}\right)$
    3: **Maximization Step**:
        Compute the estimation of σ_L^2 associated with each element in $\bar{\mathbf{h}}$
    4: $\bar{\mathbf{h}} = \mu$
    5: **Until** the estimations of $\bar{\mathbf{h}}$ converge.

In addition to the EM algorithm in Algorithm 3, Mo et al. [9] also proposed to combine the generalized approximate message passing (GAMP) algorithm and EM algorithm as EM-GAMP algorithm. For EM-GAMP algorithm, the sparsity property of $\bar{\mathbf{h}}$ is exploited, and the parameters η and σ_L^2 are learned because they are unknown in practice.

6.4.4 REMARK

This section reviews the CE scheme for the mmWave massive MIMO systems with the one-bit receiver [9]. In this scenario, the CE problem is converted to the ML estimation problem, where EM algorithm, approximate message passing algorithm, or their variants are employed or proposed to estimate the channels. However, existing CE algorithms only consider the discrete AoA/AoD, which is not practical. Actually, for the continuous AoA/AoD in practice, each path will suffer from the angular spread in the virtual angular domain due to the leakage effect. Hence, how to effectively estimate the channels with the actually continuous AoA/AoD is still challenging and will be a hot topic in the near future.

6.5 PARAMETRIC CHANNEL ESTIMATION SCHEMES FOR mmWave MASSIVE MIMO SYSTEMS

For mmWave massive MIMO systems with N_T transmit antennas and N_R receive antennas, the size of $\mathbf{H}$ can be large, and there can be $N_T N_R$ unknown elements to be estimated. However, thanks to the sparsity of mmWave massive MIMO channels as shown in (6.1), there are only $3L \ll N_T N_R$ parameters that totally determine $\mathbf{H}$, that is, $\{\alpha_l, \theta_l, \varphi_l, 1 \leq l \leq L\}$. Hence, if we can acquire the estimation of only $3L$ parameters and know a priori information of the array manifold, we can reconstruct $\mathbf{H}$ of large size. Compared with the previous CE schemes in Sections 6.3 and 6.4, which assume the discrete AoA/AoD, in this section we propose a parametric CE scheme [2], which can acquire the super-resolution estimation of the parameters AoA/AoD and the associated path gains.

The parametric CE scheme mainly consists of the coarse CE to acquire the sufficient SNR at the receiver and the fine CE to estimate the parameters dominating the channels. Besides, based on this CE scheme, we will also present an associated hybrid beamforming scheme, which can effectively support multiuser and multistream for each user. Finally, we also provide the simulation results to verify the proposed scheme.

6.5.1 SUPER-RESOLUTION SPARSE CHANNEL ESTIMATION

The super-resolution sparse CE scheme consists of the following four stages.

(1) The first stage is coarse CE. This stage can be illustrated in Fig. 6.5A, which aims to acquire partial CSIT to generate the appropriate beamforming patterns for the following fine CE with the improved received signal power. Specifically, the BS sequentially broadcasts L_{BS} predefined beamforming patterns in L_{BS} successive time slots, while in every time slot, each user sequentially receives the signal with L_{MS} combining patterns in L_{MS} successive

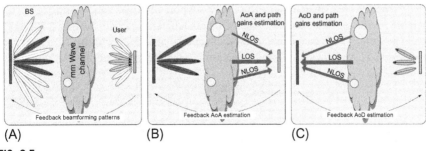

FIG. 6.5

Super-resolution sparse channel estimation (CE) for mmWave massive MIMO systems: (A) Coarse CE; (B) AoA and path gains estimation at the MS; and (C) AoD and path gains estimation at the BS.

sub-time slots. Then each MS feeds back the indexes of several optimal beamforming/combining patterns to the BS.

(2) The second stage is the fine CE at the MS. This process can be shown in Fig. 6.5B, which aims to estimate AoA and the associated path gains at each MS. The BS performs beamforming according to the feedback, while the kth MS estimates AoA and path gains by exploiting the finite rate of innovation theory (analog CS) [8]. With the aid of the predefined training signals $\mathbf{F} \in \mathbb{C}^{N_{BS} \times T_{BS}}$ with the time overhead T_{BS}, according to (6.10), the received signals at the MS can be expressed as follows:

$$\mathbf{Y} = \mathbf{W}^H \mathbf{A}_R \mathbf{H}_a \mathbf{A}_T^H \mathbf{F} + \mathbf{N} \tag{6.43}$$

Furthermore, (6.43) can be further written as

$$\widetilde{\mathbf{Y}} = \mathbf{A}_R \widetilde{\mathbf{S}} + \mathbf{N} \tag{6.44}$$

where $\widetilde{\mathbf{Y}} = \left(\mathbf{W}^H\right)^{\dagger} \mathbf{Y}, \widetilde{\mathbf{S}} = \mathbf{H}_a \mathbf{A}_T^H \mathbf{F}$. It is clear that $\mathbf{A}_R$ is a Vandermonde matrix, which indicates that we can use the estimation of signal parameters via rational invariance technique (ESPRIT) algorithm to estimate AoA and the associated path gains, that is, θ_l and α_l for $1 \le l \le L$. Then the MS feeds back the estimated parameters to the BS.

(3) The third stage is the fine CE at the BS. This stage can be shown in Fig. 6.5C, which aims to estimate AoD and path gains at the BS. The specific procedure is similar to the second stage, where the kth MS transmits training signals while the BS estimates channels.

(4) Finally, according to the estimated AoA/AoD and the corresponding path gains, both the BS and the MS can reconstruct the high-dimensional channel matrix. In this way, we can realize both CSIT and CSI at the receiver (CSIR).

6.5.2 MULTIUSER AND MULTISTREAM (MU-MS) HYBRID BEAMFORMING/COMBINING

To verify the validity of the proposed super-resolution sparse CE, we further propose the MU-MS beamforming/combining scheme based on the hybrid MIMO transceiver structure [2], which can effectively support multiuser and multistream for each MS. To be specific, according to the estimated parameters in Section 6.5.1, both the BS and the kth MS can reconstruct the channel matrix $\mathbf{H}_k$ according to (1). $\mathbf{H}_k$ can be expressed as follows according to singular value decomposition (SVD):

$$\mathbf{H}_k = \left[\mathbf{U}_k^1, \mathbf{U}_k^2\right] \begin{bmatrix} \mathbf{\Sigma}_k^1 & \mathbf{0} & \mathbf{0} \\ \mathbf{0} & \mathbf{\Sigma}_k^2 & \mathbf{0} \end{bmatrix} \begin{bmatrix} \left(\mathbf{V}_k^1\right)^H \\ \left(\mathbf{V}_k^2\right)^H \end{bmatrix} \approx \mathbf{U}_k^1 \mathbf{\Sigma}_k^1 \left(\mathbf{V}_k^1\right)^H \tag{6.45}$$

where both $\left[\mathbf{U}_k^1, \mathbf{U}_k^2\right] \in \mathbb{C}^{N_{MS} \times N_{MS}}$ and $\left[\mathbf{V}_k^1, \mathbf{V}_k^2\right]^H \in \mathbb{C}^{N_{BS} \times N_{BS}}$ are unitary matrices, $\mathbf{\Sigma}_k^1 \in \mathbb{C}^{R_k \times R_k}$ and $\mathbf{\Sigma}_k^2 \in \mathbb{C}^{(N_{MS}-R_k) \times (N_{MS}-R_k)}$ are diagonal matrices whose diagonal elements are singular values of $\mathbf{H}_k$, and R_k is the effective rank of $\mathbf{H}_k$. The approximation in (6.45) is due to the low-rank property of $\mathbf{H}_k$ with $\mathbf{\Sigma}_k^2 \approx \mathbf{0}$, so that $\mathbf{U}_k^1 \in \mathbb{C}^{N_{MS} \times R_k}$ and $\left(\mathbf{V}_k^1\right)^H \in \mathbb{C}^{R_k \times N_{BS}}$.

Eq. (6.45) indicates that N_{RF} can be reduced to R_k in single-user (SU)-MIMO due to only $N_s = R_k$ effective independent streams. Moreover, we can use the precoding matrix $\mathbf{F}_k = \mathbf{V}_k^1 \in \mathbb{C}^{N_{BS} \times R_k}$ and the combining matrix $\mathbf{W}_k = \left(\mathbf{U}_k^1\right)^H \in \mathbb{C}^{R_k \times N_{MS}}$ to effectively realize the multistream transmission. Moreover, with the cascade of the digital precoding matrix $\mathbf{F}_{BB,k} \in \mathbb{C}^{R_k \times R_k}$ (or combining matrix $\mathbf{W}_{BB,k} \in \mathbb{C}^{R_k \times R_k}$) and analog precoding matrix $\mathbf{F}_{RF,k} \in \mathbb{C}^{N_{BS} \times R_k}$ (or combining matrix $\mathbf{W}_{RF,k} \in \mathbb{C}^{N_{MS} \times R_k}$), we can use $\mathbf{F}_{BB,k}\mathbf{F}_{RF,k}$ (or $(\mathbf{W}_{BB,k}\mathbf{W}_{RF,k})^H$) to approximate $\mathbf{F}_k$ (or $\mathbf{W}_k$).

Considering the precoding, for instance, we can use the following iterative approach to acquire $\mathbf{F}_{BB,k}$ and $\mathbf{F}_{RF,k}$ that can minimize $\|\mathbf{F}_k - \mathbf{F}_{RF,k}\mathbf{F}_{BB,k}\|_F$, that is,

$$\left(\mathbf{F}_{RF,k}, \mathbf{F}_{BB,k}\right) = \arg \min_{\tilde{\mathbf{F}}_{RF,k}, \tilde{\mathbf{F}}_{BB,k}} \left\|\mathbf{F}_k - \tilde{\mathbf{F}}_{RF,k}\tilde{\mathbf{F}}_{BB,k}\right\|_F \tag{6.46}$$

and elements in $\mathbf{F}_{RF,k}$ are constant modulus. To find the suitable $\mathbf{F}_{BB,k}$ and $\mathbf{F}_{RF,k}$, we propose the following heuristic Algorithm 4.

ALGORITHM 4 PROPOSED HEURISTIC ALGORITHM THAT CAN ACQUIRE THE SOLUTION TO (6.46)

1: $\tilde{\mathbf{F}}_k \leftarrow \mathbf{F}_k$
2: **Repeat**
3: Set every element of $\mathbf{F}_{RF,k}$ to have the same phase with the corresponding element in $\tilde{\mathbf{F}}_k$, and elements in $\mathbf{F}_{RF,k}$ are constant modulus.
4: $\mathbf{F}_{BB,k} \leftarrow \left(\mathbf{F}_{RF,k}\right)^{\dagger}\mathbf{F}_k$
5: $\tilde{\mathbf{F}}_k \leftarrow \mathbf{F}_k\left(\mathbf{F}_{BB,k}\right)^{\dagger}$
6: **Until** $\mathbf{F}_{RF,k}$ and $\mathbf{F}_{BB,k}$ converge.

Similarly, we can acquire the analog combining matrix $\mathbf{F}_{RF,k}$ and digital combining matrix $\mathbf{F}_{BB,k}$ with the same approach.

Besides, some power allocation strategies such as waterfilling can be integrated in the digital baseband precoding/combining to further improve the achievable capacity. Furthermore, consider the downlink MU-MIMO, where the channel matrix between BS and K MSs can be denoted as $\mathbf{H} \in \mathbb{C}^{KN_{MS} \times N_{BS}}$, and it can be represented as $\mathbf{H} = \left[\mathbf{H}_1^T | \mathbf{H}_2^T | \cdots | \mathbf{H}_K^T\right]^T$ with $\mathbf{H}_k \approx \mathbf{U}_k^1 \mathbf{\Sigma}_k^1 \left(\mathbf{V}_k^1\right)^*$ for $1 \leq k \leq K$ according to (6.45). Hence, we can further obtain

$$\mathbf{H} \approx \text{diag}\left\{\mathbf{U}_1^1, \mathbf{U}_2^1, ..., \mathbf{U}_K^1\right\}\text{diag}\left\{\mathbf{\Sigma}_1^1, \mathbf{\Sigma}_2^1, ..., \mathbf{\Sigma}_K^1\right\}\begin{bmatrix}\left(\mathbf{V}_1^1\right)^H \\ \left(\mathbf{V}_2^1\right)^H \\ \vdots \\ \left(\mathbf{V}_K^1\right)^H\end{bmatrix} \tag{6.47}$$

where $\mathbf{H}_k$ for $1 \leq k \leq K$ are assumed to share the same effective rank $R_k = R$. For precoding/combining in the proposed MU-MIMO systems, the analog precoding matrix at the BS with KR RF chains is $\mathbf{F}_{RF} = \left[\mathbf{F}_{RF,1}, \mathbf{F}_{RF,2}, ..., \mathbf{F}_{RF,K}\right] \in \mathbb{C}^{N_{BS} \times KR}$, and the analog and digital combining matrices for the kth MS with R RF chains can be $\mathbf{W}_{RF,k}$ and $\mathbf{W}_{BB,k}$, respectively.

To further eliminate the multiuser interference, digital precoding $\mathbf{F}_{BB} = (\widetilde{\mathbf{V}}\mathbf{F}_{RF}\widetilde{\mathbf{F}}_{BB})^{-1}$ is proposed at the BS, where $\widetilde{\mathbf{F}}_{BB} = \text{diag}\{\mathbf{F}_{BB,1}, \mathbf{F}_{BB,2}, ..., \mathbf{F}_{BB,K}\}$ and $\widetilde{\mathbf{V}} = \left[(\mathbf{V}_1^1)^*, (\mathbf{V}_2^1)^*, ..., (\mathbf{V}_K^1)^*\right]^T$. The precoding/combining in the uplink is similar to that in the downlink

The proposed precoding/combining scheme can diagonalize the equivalent channel

$$\text{diag}\{\mathbf{W}_{BB,1}^H \mathbf{W}_{RF,1}^H, \mathbf{W}_{BB,2}^H \mathbf{W}_{RF,2}^H, ..., \mathbf{W}_{BB,K}^H \mathbf{W}_{RF,K}^H\}\mathbf{H}\mathbf{F}_{RF}\mathbf{F}_{BB} \qquad (6.48)$$

to realize multiuser and multistream transmission, which is essentially different from existing hybrid precoding/combining schemes. Moreover, thanks to the obvious low-rank property of the mmWave massive MIMO channel matrix as shown in Fig. 6.2, the proposed precoding/combining with the reduced number of RF chains only suffers from a negligible performance loss, which will be shown in Section 6.5.3.

6.5.3 NUMERICAL SIMULATIONS

Fig. 6.6 compares the capacity (bits per channel use (bpcu)) of the proposed MU-MS hybrid precoding/combining scheme and the optimal full digital precoding/combining scheme in the downlink, where both the waterfilling power allocation and equal-power allocation are investigated. In simulations, ULA is considered at both BS and MS, the working frequency is 60 GHz, $K = 4$, $N_{BS} = 512$, and $N_{MS} = 32$. For the optimal full digital scheme, we assume the number of RF chains is equal to that of antennas, where the ideal CSIT and CSIR are assumed as the upper bound of capacity. In the proposed scheme, the numbers of RF chains at each MS and BS are 4 and 16, respectively, where cases of ideal CSI known by transceiver and non-ideal CSI acquired by the proposed parametric CE scheme are considered. For mmWave massive MIMO channels, L in simulations follows the discrete uniform distribution $U_d[2,6]$, and AoA/AoD follow the continuous uniform distribution $U_c[0,2\pi)$. For path gains, we consider Rican fading consisting of one line-of-sight (LOS) path and $L-1$ equal-power nonline-of-light (NLOS) paths, where path gains follow the mutually independent complex Gaussian distribution with zero means, and K_{factor} denotes the ratio between the power of LOS paths and the power of NLOS paths.

Fig. 6.6 shows that the proposed MU-MS hybrid scheme with ideal CSIT and CSIR suffers from a negligible capacity loss compared with the optimal full digital scheme, although the proposed scheme only uses a much smaller number of RF chains. This is because the proposed scheme exploits the low-rank property of the mmWave massive MIMO channel matrix, where the capacity exhibits a ceiling effect when the number of RF chains is sufficiently large [4]. Moreover, with the increased overhead for CE, the capacity of the proposed scheme with the parametric CE scheme approaches that with the ideal CSI. This is because the increased number of measurements can improve the CE performance. Besides, schemes with

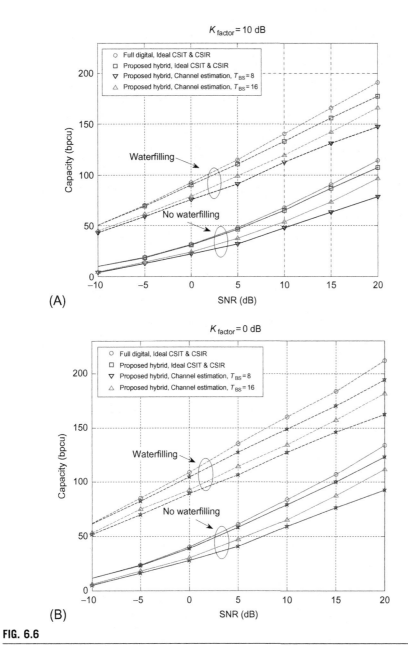

FIG. 6.6

Capacity comparison between the proposed hybrid precoding/combining scheme and the optimal full digital one: (A) $K_{factor} = 0$ dB; (B) $K_{factor} = 10$ dB.

waterfilling power allocation outperform those with equal-power allocation, which indicates that waterfilling or other power allocations should be considered in practical system design for improved capacity.

6.5.4 REMARK

In this section, we have proposed a parametric CE scheme for mmWave massive MIMO systems [2]. The proposed scheme removes the unpractical assumption of the discrete AoA/AoD discussed in Sections 6.5.2 and 6.5.3, and it can estimate the super-resolution AoA/AoD. To further verify the validity of the proposed parametric CE scheme, we also propose an associated MU-MS hybrid precoding/combining scheme, which can support multiuser and multistream for each MS. Simulation results also demonstrate that the channel capacity of the proposed MU-MS hybrid precoding/combining scheme with the proposed parametric CE scheme approaches that of the optimal full digital precoding/combining scheme, with a slight performance loss.

6.6 SUBSPACE ESTIMATION AND DECOMPOSITION (SED)-BASED CHANNEL ESTIMATION

In mmWave massive MIMO systems, due to the sparsity of mmWave massive MIMO channels, the mmWave massive MIMO channel matrix appears to be the low-rank property. This implies that only a few eigenmodes can be used for transmission, especially for mmWave massive MIMO systems where the majority of eigenmodes have negligible power [2]. In the parametric CE scheme, the complete channel matrix is first reconstructed according to the dominated channel parameters, and then the dominated eigenmodes are obtained according to the SVD of the reconstructed complete channel matrix. However, this scheme requires a priori information of the array manifold. In this section, the SED-based CE scheme proposed in [10] can directly estimate the dominated eigenmodes rather than the entire channel.

6.6.1 SUBSPACE ESTIMATION IN TRADITIONAL MIMO SYSTEMS

As we have discussed in Section 6.5, (6.45), only R (the index k is omitted for simplicity in this section) eigenmodes associated with the dominated singular values are effective for transmission. Hence, our goal is to directly estimate the dominated singular vectors $\mathbf{U}^1$ and $(\mathbf{V}^1)^H$. For convenience, we first illustrate how to acquire the subspace estimation in traditional MIMO systems with a full digital precoding/combining scheme.

In conventional MIMO systems, the subspace estimation can be realized by the multiple amplify-and-forward (AF) operations between the BS and MS. To further specifically illustrate this process, we consider how to estimate the right dominated singular vectors at the BS in a noiseless single-stream transmission. First, the BS selects a random unit-norm beamforming vector $\mathbf{q}_1$, and then transmits it to the MS. The received signal $\mathbf{s}_1 = \mathbf{H}\mathbf{f}_1$ is then fed back to the MS. This procedure is done iteratively. Specifically, we illustrate this procedure in Algorithm 5.

ALGORITHM 5 SUBSPACE ESTIMATION USING ARNOLDI ITERATION (SE-ARN)

Initiation: Set the number of iterations $m \leq N_T$ and random unit norm $\mathbf{q}_1$ with $\mathbf{Q} = [\mathbf{q}_1]$.

1: $l = 1$;
2: **While** $l \leq m$ do
    % BS-initiate echoing: estimate $\mathbf{H}^H\mathbf{H}\mathbf{f}_l$
3: $\mathbf{s}_l = \mathbf{H}\mathbf{f}_1$; % The MS receives the training signal from BS
4: $\mathbf{p}_l = \mathbf{H}^H\mathbf{s}_l + \mathbf{v}_l$; % The MS feeds back the received signal to BS
    % Gram-Schmidt orthogonalization at the BS
5: $t_{m,l} = \mathbf{q}_m^H\mathbf{p}_l, \forall m = 1,...,l$
6: $\mathbf{r}_l = \mathbf{p}_l - \Sigma_{m=1}^l \mathbf{q}_m t_{m,l}$
7: $t_{l+1,l} = \|\mathbf{r}_l\|_2$
8: $\mathbf{Q} = [\mathbf{Q}, \mathbf{q}_{l+1} = \mathbf{r}_l/t_{l+1,l}]$; % Update $\mathbf{Q}$
9: **End While**
    % Compute $\mathbf{V}^1$
10: $\mathbf{T}_m = \tilde{\boldsymbol{\Theta}}\tilde{\boldsymbol{\Lambda}}\tilde{\boldsymbol{\Theta}}^{-1}$
11: $\mathbf{V}^1 = \mathbf{Q}_m\tilde{\boldsymbol{\Theta}}_{1:R}$; % Columns in $\tilde{\boldsymbol{\Theta}}$ associated with R largest singular values consist of $\tilde{\boldsymbol{\Theta}}_{1:R}$

The same procedure can be also applied using the MS-initiated echoing to estimate $\mathbf{U}^1$, which is the R dominated eigenvectors of $\mathbf{H}\mathbf{H}^H$.

6.6.2 EXTEND TO HYBRID MIMO TRANSCEIVER STRUCTURE

In this subsection, we further extend the SED-based CE scheme to mmWave massive MIMO systems with the hybrid transceiver structure.

It is clear that how to obtain the estimation of $\{\mathbf{H}^H\mathbf{H}\mathbf{q}_l\}_{l=1}^m$ at the BS is essential to the SED-based CE scheme. However, for the hybrid MIMO transceiver structure with the analog and digital beamforming/combining at the transmitter and receiver, there exist several issues that can challenge the application of the SE-ARN algorithm to mmWave massive MIMO systems. First, the digital beamforming vector $\mathbf{q}_l$ needs to be approximated by the product of the analog and digital beamforming matrices. Moreover, the BS-initiated echoing relies on the AF operation at the MS, which can be impossible with the hybrid MIMO transceiver structure. In addition, both the BS and MS can only process the received digital baseband signal after the analog RF combining other than the received signal in the RF.

We consider the estimation of the dominated right singular vectors. In Algorithm 5, the training signal $\mathbf{q}_l$ transmitted by the BS can be realized by the cascade of the analog precoding matrix $\mathbf{F}_{RF,l}$ and the digital precoding matrix $\mathbf{F}_{BB,l}$, that is, $\mathbf{q}_l \approx \mathbf{F}_{RF,l}\mathbf{F}_{BB,l} = \mathbf{f}_l$. At the receiver, the MS obtains the received signal $\mathbf{s}_l = \mathbf{W}_{BB,l}^H\mathbf{W}_{RF,l}^H\mathbf{H}\mathbf{f}_l$, and then the MS feeds back the received signal using the hybrid precoding. Finally, the BS receives the feedback signal with the hybrid combining. As a result, the received signal at the BS in the lth iteration can be expressed as

$$\mathbf{p}_l = \mathbf{F}_{l,r}^H \mathbf{H}^H \mathbf{W}_{l,t} \mathbf{W}_{l,r}^H \mathbf{H} \mathbf{f}_l = \mathbf{F}_{l,r}^H \mathbf{H}^H \mathbf{W}_{l,t} \mathbf{W}_{l,r}^H \mathbf{H} (\mathbf{q}_l - \mathbf{e}_l) \qquad (6.49)$$

where $\mathbf{F}_l = \mathbf{F}_{\mathrm{RF},l} \mathbf{F}_{\mathrm{BB},l}$, $\mathbf{W}_l = \mathbf{W}_{\mathrm{RF},l} \mathbf{W}_{\mathrm{BB},l}$, $\mathbf{q}_l = \mathbf{f}_l + \mathbf{e}_l$, and precoding and combining are denoted by the subscript t and r, respectively.

According to (6.49), we can find that there are two imperfect factors due to the hybrid MIMO transceiver structure.

(1) *Analog-Processing Impairments*: Processing the signal at the MS with the hybrid combining $\mathbf{W}_{l,r}^H$ and precoding $\mathbf{W}_{l,t}$ can degrade the estimation performance of the dominated singular vectors. Moreover, processing the received signal at the BS with the hybrid combining $\mathbf{F}_{l,r}$ indicates that the effective received signal at the BS is a low-dimensional observation.

(2) *Decomposition-Induced Distortions*: The error from decomposing $\mathbf{q}_l$ at the BS can also degrade the estimation performance.

The two imperfect factors are due to the by-product of transferring the digital baseband precoding/combining burden to the analog RF. In the following, we will further illustrate how to resolve these two imperfect factors.

First, the SE-ARN algorithm requires the random unit norm $\mathbf{q}_1$, and this can be easily implemented with the hybrid MIMO transceiver structure with sufficient accuracy.

Second, the actual received signal at the MS is $\mathbf{W}_{l,r}^H \mathbf{H} \mathbf{f}_l$ and the feedback signal from the MS is $\mathbf{W}_{l,t} \mathbf{W}_{l,r}^H \mathbf{H} \mathbf{f}_l$ other than $\mathbf{H} \mathbf{f}_l$ as in conventional MIMO systems with full digital precoding/combining. However, this distortion can be resolved by using multiple time slots with the DFT analog combining or precoding matrix. Particularly, we consider that $\mathbf{W}_{l,t} = \mathbf{W}_{l,r} \in \mathbb{C}^{(T_o N_{\mathrm{RF}}) \times (T_o N_{\mathrm{RF}})}$ is the $T_o N_{\mathrm{RF}} \times T_o N_{\mathrm{RF}}$ DFT matrix, and $\mathbf{W}_{l,r} = [\mathbf{W}_{l,r}^1 \mathbf{W}_{l,r}^2 \ldots \mathbf{W}_{l,r}^{T_o}]$ with $\mathbf{W}_{l,r}^l \in \mathbb{C}^{(N_{\mathrm{RF}}) \times (T_o N_{\mathrm{RF}})}$, thus $\mathbf{W}_{l,t} \mathbf{W}_{l,r}^H = \mathbf{I}$. Similarly, the BS can also use the DFT analog precoding/combining to realize the AF without distortion.

6.6.3 REMARK

This section introduces SED-based CE for mmWave massive MIMO systems. Compared with CS-based CE in Section 6.3 and the CE scheme for the one-bit ADC receiver in Section 6.4, SED-based CE does not require the assumption of the discrete AoA/AoD. Moreover, compared with our proposed parametric CE scheme in Section 6.5, the SED-based CE scheme does not require a priori information of the array manifold, and it can directly estimate the dominated singular vectors rather than reconstructing the complete channel matrix. However, the mmWave communications suffer from low SNR before beamforming, and the multiple AF operations between the BS and MS can introduce much noise and degrade the accuracy of the CE.

6.7 OTHER CHANNEL ESTIMATION SCHEMES

In this section, we will briefly discuss the potential of how the existing CE schemes initially proposed for conventional massive MIMO systems working at sub-3–6 GHz can be tailored to mmWave massive MIMO systems. The codebook-based CSI acquisition scheme is also briefly presented.

6.7.1 CAN CHANNEL ESTIMATION SCHEMES IN MASSIVE MIMO BE TAILORED TO mmWave MASSIVE MIMO?

Compared with conventional massive MIMO systems working at sub-3–6 GHz, the mmWave massive MIMO are more attractive due to the larger bandwidth and compact antenna array. However, mmWave massive MIMO systems suffer from low SNR before beamforming. Moreover, the hybrid MIMO transceiver structure prevents the conventional CE schemes proposed for conventional massive MIMO from being directly used for mmWave massive MIMO systems. By far, there have been extensive studies on CE for conventional massive MIMO systems [11–13]. Meanwhile, the standardization of conventional massive MIMO, including pilot design and CE, is in progress. Hence, it is significant to design a CE scheme for mmWave massive MIMO compatible with that in conventional massive MIMO for the reduced hardware cost and extra system overheads. One viable approach is to design a suitable analog precoding/combining that can convert the CE problem in hybrid MIMO systems to that in conventional MIMO systems. This can be realized by the DFT analog precoding/combining using multiple time slots as in the SED-based CE scheme. From the other approach, the CS-based CE schemes [11–13] initially used for conventional massive MIMO by exploiting the sparsity of massive MIMO channels can be also used for mmWave massive MIMO systems with some slight modifications, thanks to the much sparser property of mmWave massive MIMO channels.

6.7.2 CODEBOOK-BASED CHANNEL ESTIMATION SCHEMES

The codebook-based CE schemes have been widely researched in many standards, including IEEE 802.15.3c (TG3c) for WPAN and IEEE 802.11.ad for WLAN [1]. More specifically, in IEEE 802.15.3c, a multilevel codebook is designed together with a ping-pong searching scheme. The basic idea of this method is to divide the whole beamspace into several parts with low resolution at first, and then the iterative ping-pong searching scheme is employed to sound the best part of the beamspace. After that, the searched best part will be divided again with higher resolution for the next round of searching. Such procedure will be continued until some criterion is satisfied. In IEEE 802.11ad, a single-sided searching scheme is adopted to find the optimal beam by two steps: the combiner is fixed to exhaustively search the best precoder at first, and then this best precoder is fixed to exhaustively search the best combiner.

For codebook-based CE, the acquired CSI is implicit other than explicit, as in the CE schemes discussed previously. Moreover, once the codebook-based CE is finished, both the BS and the MS also finish the beamforming and combining. Hence, this kind of CSI acquisition scheme can also be considered beamforming/combining, which will be further detailed in Chapter 7.

6.8 SUMMARY

The main challenges of CE in mmWave massive MIMO systems come from the large antenna array, low SNR before beamforming, and the special transceiver structures (hybrid MIMO transceiver structure and the receiver with one-bit ADCs). To solve these issues, in this chapter we have discussed various CE schemes for mmWave massive MIMO systems. We have detailed four kinds of state-of-the-art CE schemes: the CS-based CE scheme [4,6,7], the CE scheme with the one-bit ADC at the receiver [9], the parametric CE scheme [2], and the SED-based CE scheme [10]. Except for the third scheme, the other three kinds of schemes are based on the hybrid MIMO transceiver structure. Compared with the CS-based CE scheme and the CE scheme with the one-bit ADC at the receiver, the parametric CE scheme removes the assumption of the discrete AoA/AoD. The parametric CE scheme can acquire the super-resolution AoA/AoD estimation, but it requires the array manifold as a priori information, like the first two schemes. SED-based CE can directly acquire the dominated eigenmodes rather than the entire channel for transmission. But the ping-pong operation between the BS and MS can introduce too much noise and degrade the final estimation performance, especially in mmWave with low SNR before beamforming. Besides, we briefly discussed the potential approach of how to tailor the existing CE schemes [11–13] initially proposed for conventional massive MIMO systems working at sub-3–6 GHz to mmWave massive MIMO systems. Finally, we also briefly introduced the codebook-based CE scheme. Because it can simultaneously accomplish CSI acquisition and beamforming, we which will be detailed in Chapter 7.

REFERENCES

[1] L. Wei, R.Q. Hu, Y. Qian, G. Wu, Key elements to enable millimeter wave communications for 5G wireless systems, IEEE Wirel. Commun. 21 (6) (2014) 136–143.

[2] Z. Gao, L. Dai, D. Mi, Z. Wang, M.A. Imran, M.Z. Shakir, MmWave massive MIMO based wireless backhaul for 5G ultra-dense network, IEEE Wirel. Commun. 22 (5) (2015) 13–21.

[3] S. Han, I. Chih-Lin, Z. Xu, C. Rowell, Large-scale antenna systems with hybrid precoding analog and digital beamforming for millimeter wave 5G, IEEE Commun. Mag. 53 (1) (2015) 186–194.

[4] A. Alkhateeb, J. Mo, N. Gonzalez-Prelcic, R.W. Heath, MIMO precoding and combining solutions for millimter-wave systems, IEEE Commun. Mag. 52 (12) (2014) 122–131.

[5] A. Alkhateeb, O.E. Ayach, G. Leus, R.W. Heath, Channel estimation and hybrid precoding for millimeter wave cellular systems, IEEE J. Sel. Top. Sign. Proces. 8 (5) (2014) 831–846.

[6] J. Lee, G.T. Gil, Y.H. Lee, Exploiting spatial sparsity for estimating channels of hybrid MIMO systems in millimeter wave communications, IEEE Globecom 2014 (2014) 3326–3331.

[7] A. Alkhateeb, G. Leus, R.W. Heath, Compressed sensing based multi-user millimeter wave systems: How many measurements are needed? in: Proceedings of the IEEE International Conference on Acoustics, Speech and Signal Processing (ICASSP), Brisbane, Australia, 2015. http://arxiv.org/abs/1505.00299.

[8] Y.C. Eldar, G. Kutyniok, Compressed Sensing: Theory and Applications, Cambridge University Press, Cambridge, 2012.

[9] J. Mo, P. Schniter, N.G. Prelcic, R.W. Heath, Channel estimation in millimeter wave MIMO system with one-bit quantization, IEEE ACSSC 2014 (2014) 957–961.

[10] H. Ghauch, T. Kim, M. Bengtsson, M. Skoglund, Subspace estimation and decomposition for large millimeter-wave MIMO systems, IEEE J. Sel. Top. Sign. Proces. 10 (3) (2016) 528–542.

[11] Z. Gao, L. Dai, Z. Wang, S. Chen, Spatially common sparsity based adaptive channel estimation and feedback for FDD massive MIMO, IEEE Trans. Signal Process. 63 (23) (2015) 6169–6183.

[12] Z. Gao, L. Dai, Z. Wang, Structured compressive sensing based superimposed pilot design for large-scale MIMO systems, Electron. Lett. 50 (12) (2014) 896–898.

[13] Z. Gao, L. Dai, Z. Lu, C. Yuen, Z. Wang, Super-resolution sparse MIMO-OFDM channel estimation based on spatial and temporal correlations, IEEE Commun. Lett. 18 (7) (2014) 1266–1269.

Channel feedback for mmWave massive MIMO

7

P.-H. Kuo*, B. Su†, C.-P. Yen*

Industrial Technology Research Institute (ITRI), Hsinchu, Taiwan*
National Taiwan University, Taipei, Taiwan†

CHAPTER OUTLINE HEAD

7.1 INTRODUCTION

It is well known that the potential benefits of massive antenna arrays can be fully harvested if accurate channel state information (CSI) is available at the transmitter. Specifically, a base station with a large-scale antenna array can carry out beamforming that achieves very high array gain with appropriate downlink precoding [1]. This is fulfilled by matching the precoder with the instantaneous channel response, so obviously the optimal precoder is derived from the timely multiple-input multiple-output (MIMO) channel status. In many studies, the channel reciprocity nature of time division duplexing (TDD) has been assumed for wireless transceiver systems equipped with large antenna arrays [2]. With channel reciprocity, the downlink channel can

mmWave Massive MIMO. http://dx.doi.org/10.1016/B978-0-12-804418-6.00007-8

be postulated to be equivalent to the Hermitian conjugate of the uplink channel. Thus, TDD is commonly presumed in research relating to massive MIMO systems as ideally the base station can acquire downlink CSI simply by estimating uplink pilot signals. Nevertheless, there are some deficiencies of TDD-based CSI acquisition. For instance, the CSI measured by the base station itself with uplink pilot signals is unable to capture the downlink interferences that potentially exist (e.g., intercell interferences). Also, any mismatch between transmitter and receiver circuit responses of user equipment (UE) due to hardware implementations may result in inaccurate CSI, and additional calibration processes may be needed to facilitate CSI acquisition in such situations [3]. Moreover, the issue of *pilot contamination* [4] with TDD channel acquisition has been identified as the main hurdle that limits the performance of massive MIMO systems.

Nowadays, frequency division duplex (FDD) remains a dominant mode of cellular communications around the globe. Apparently, FDD is more suitable for systems with symmetric traffic between downlink and uplink. Moreover, FDD is less sensitive to latency, as both downlink and uplink resources are always available; this is an important characteristic for mission-critical applications that have been highlighted as a prospective trend of 5G [5]. Unlike TDD, in FDD different frequency bands are used for downlink and uplink transmissions, so channel reciprocity cannot be exploited for CSI acquisition. In other words, the base station is unable to acquire the detailed downlink CSI by simply measuring the uplink pilot symbols transmitted from the UE. The knowledge of CSI at the transmitter side is essential for massive MIMO schemes such as closed-loop beamforming, so CSI acquisition mechanism design is particularly important for massive MIMO communication systems operating in FDD mode.

A dedicated feedback mechanism is employed in modern cellular systems such as LTE-A to facilitate downlink MIMO precoding. With the feedback mechanism, the UE reports the measured downlink channel information to the base station using dedicated uplink resources. Compared to reciprocity-based CSI acquisition in TDD, the CSI derived and reported by the UE may be more accurate, as it is able to capture the impacts of downlink interferences. In order to undertake CSI feedback in a more efficient manner, the amplitude and phase information of MIMO channel responses is quantized using a predefined set of vectors/matrices dubbed "codebook" in practical systems. Thus channel feedback can be achieved by merely using a limited number of bits representing the indication of one codebook entry, which could be, for example, a quantized version of the ground-true channel response or the eigenvectors of the MIMO channel matrix. In 3GPP LTE specifications, such a CSI component is termed the precoding matrix indicator (PMI). Based on this codebook approach with limited feedback, the required radio resource for CSI reporting is significantly reduced and hence more tangible.

As the number of transmitter antennas grows, the codebook size may have to be extensively expanded in bid to accommodate fine-grain spatial channel structures. However, if we presume that the radio resource dedicated for CSI reporting remains unchanged, using a larger codebook (more feedback bits) may affect the feedback reliability due to a lower coding rate. Such a dilemma has motivated numerous research efforts on CSI feedback mechanisms for massive MIMO systems. For example, the authors of [6] have proposed a novel codebook design technique for massive MIMO, which allows a low-complexity encoding process that is more scalable than the conventional look-up table approach. Many other alternatives are also available in the literature.

To further investigate the prospect schemes that allow downlink precoding operations for massive MIMO base stations, this chapter examines two alternative methods that allow more efficient CSI feedback mechanisms for massive MIMO systems in FDD mode, plus a downlink precoding method that does not need CSI feedback from the users. The rest of this chapter discusses the following schemes:

- *CSI compression based on compressive sensing*: By utilizing the sparse representation of a MIMO channel matrix with spatially correlated antenna arrays, we develop a CSI compression method-based compressive sensing, as well as its enhanced version with adaptive sparsifying basis.
- *CSI feedback based on multistage beamforming*: Instead of letting the users measure the channel responses corresponding to each of the transmitter antennas, we examine an alternative scheme based on beamformed pilot symbols, which enables more efficient CSI measurements and reporting.
- *Downlink precoding without CSI knowledge*: We investigate a downlink precoding technique that allows the base station to leverage only the knowledge of angles of arrival (AoAs) of transmission paths, in which the burdens of CSI feedback can be dispensed.

7.2 CHANNEL FEEDBACK WITH COMPRESSIVE SENSING

During the last decade, a novel mathematical tool dubbed *compressive sensing* (also known as *compressive sampling*) has been intensively studied and applied in various scientific and engineering research areas, such as image processing and pattern recognition [7]. According to the theory of compressive sensing, signals can be reconstructed via optimization with much fewer samples than that required by the traditional Shannon-Nyquist sampling theorem, as long as the signal itself is sparse in a certain domain. Hence, this technique is particularly useful for engineering problems dealing with signals with potential sparse representations.

For large antenna arrays as in massive MIMO systems, the antenna spacing is generally small due to the limitation of equipment form factors. For example, consider the infrastructure nodes, such as small cells' base stations and remote radio heads with restricted physical sizes. A very compact antenna array has to be installed on these radio nodes if the massive MIMO feature with numerous antennas is to be supported. Thus, a strong spatial correlation among the corresponding MIMO spatial channel matrix entries can be anticipated [8], which may result in a sparse representation in a certain domain other than space. Therefore, compressive sensing techniques can be applied to compress the channel response, which effectively reduces the required feedback overhead for CSI reporting.

Prior to delving into the details of CSI feedback methods based on compressive sensing, we will first review the nuts and bolts of compressive sensing and its general algorithms for sparse signal compressions and recovery.

7.2.1 ALGORITHMIC FRAMEWORK OF COMPRESSIVE SENSING

Compressive sensing is useful to recover sparse signals from a small number of random measurements of the original signal, as well as to compress sparse signals in an efficient manner. To describe the algorithmic procedures of compressive sensing, we first consider an $N \times 1$ sparse signal vector denoted as x, wherein only $K \ll N$ elements of x have nonzero values, while all the remaining $N - K$ entries are zero. Compressive sensing encoding is capable of capturing these K nonzero components by compressing the signal into $M \ll N$ measurements via random projections. To be specific, the $N \times 1$ sparse signal x can be compressed into an $M \times 1$ vector denoted as y by

$$y = \Phi x \tag{7.1}$$

where Φ is an $M \times N$ matrix with independent and identically distributed (i.i.d.) random entries, the entries of which can be simply generated in accordance to distributions such as Gaussian or Bernoulli. Because $M \ll N$, Eq. (7.1) is obviously an underdetermined linear system (there are fewer equations than unknowns) that may have infinitely many solutions if one intends to solve x from y. Fortunately, the concept of compressive sensing has stated that, as long as signal vector x is sparse and sufficiently incoherent sampling (through Φ) has applied, x can be recovered from y with a very high probability via an optimization procedure.

In many practical cases, the $N \times 1$ target signal x does not exhibit any sparsity, which makes compressive sensing an infeasible approach. Nevertheless, if the correlations among the signal samples (the elements of x) are sufficiently high, we may be able to obtain a sparse representation of x if it undergoes a certain sparsifying transformation:

$$S = \Psi x \tag{7.2}$$

where S is an $N \times 1$ vector that represents x in another domain, and Ψ is an $N \times N$ transformation basis that sparsifies x. The quintessential examples of Ψ include discrete Fourier transform (DFT) and discrete cosine transform (DCT), and these transformations are usually orthogonal, which means the reverse transformation is simply

$$x = \Psi^{\mathrm{T}} S \tag{7.3}$$

In fact, sparse representation has been widely applied in conventional data compression schemes, wherein information on locations and values of nonzero elements in S are encoded for purposes of data restoration. Compressive sensing eliminates the necessity of such overheads because S is blindly encoded to form y via random projection as in Eq. (7.1):

$$y = \Phi x = \Phi \Psi^{\mathrm{T}} S \tag{7.4}$$

As mentioned previously, reconstruction of x from y can be formulated as an optimization process. To be more specific, the decoder should first estimate S with an l_1-norm minimization:

$$\min \|S\|_1, \text{ subject to } y = \Phi\Psi^T S \qquad (7.5)$$

Then, x can be recovered based on Eq. (7.3). Typically, the optimization problem shown in Eq. (7.5) can be solved by the well-known algorithms including linear programming, basic pursuit, and orthogonal matching pursuit. The details of these optimization algorithms are omitted here as they are available in the vast literature (e.g., Ref. [9]). Note that S can be solved accurately with a high probability as long as M is sufficiently large and $\Phi\Psi^T$ satisfies restricted isometry property as described in [10]. Hence, the value of M should be carefully selected. Theoretically, the size of y should satisfy $M \geq cK \log(N/K)$, where c is a small constant. A rule of thumb is to set M as two to four times K.

7.2.2 APPLICATIONS OF COMPRESSIVE SENSING TO CSI FEEDBACK SCHEMES

We consider a massive MIMO point-to-point wireless communication system with $N_t \gg 1$ transmitter antennas at the base station and $N_r \geq 1$ receiver antennas at the UE. The downlink signal model can be written as

$$r = Hd + n \qquad (7.6)$$

where d and r are $N_t \times 1$ transmitter signal vector and $N_r \times 1$ receiver signal vector, respectively and H is an $N_r \times N_t$ MIMO channel matrix with flat-fading entries. Finally, n is an $N_r \times 1$ vector representing additive Gaussian white noise. Assuming that H is perfectly estimated at the UE via downlink pilot signals, the base station has to acquire at least partial, if not full, information relating to the estimated channel status in order to adaptively configure the physical layer parameters such as MIMO precoder weights. For instance, the Hermitian conjugate or pseudoinverse of the MIMO matrix H can be used for precoder construction of massive MIMO systems (namely conjugate beamforming and zero-forcing beamforming) [11]. Alternatively, singular value decomposition (SVD) can be applied on H to extract the beamforming vector, which corresponds to the principal singular vector of H.

With massive MIMO systems, feedback of H can be an arduous task in FDD mode. As aforementioned, in conventional cellular communication systems, the vector/matrix quantization approach is adopted so the channel response H can be quantized by a predefined codebook with finite entries, and feedback from the UE can be implemented simply by reporting a codebook entry index. Due to a large number of transmitter antennas in massive MIMO, the codebook size has to be expanded to capture H properly. In other words, the feedback payload has to be increased. This is obviously undesirable in terms of uplink radio resource usage.

Compressive sensing techniques delineated here could be employed to compress the required feedback payload by reducing the dimension of H via random projection. For the sake of implementation convenience, the real and imaginary parts of the MIMO channel matrix $\mathbf{H}$ are processed separately in all proposed schemes with a common operation, and the notation $\widehat{\mathbf{H}}$ is used to represent either the real part or imaginary part of $\mathbf{H}$. As the initial settings, the target signal $\widehat{\mathbf{H}}$ should first be vectorized into an $(N_r \times N_t) \times 1$ vector:

$$\boldsymbol{h} = \mathrm{vec}\left(\widehat{\mathbf{H}}\right) \tag{7.7}$$

Note that this step can be omitted for a multiple-input single-output channel, as the channel has a vector form in the first place. To compress $\boldsymbol{h}$, random projection in Eq. (7.1) can be applied, so we have

$$\boldsymbol{y} = \boldsymbol{\Phi}\boldsymbol{h} \tag{7.8}$$

Because $\boldsymbol{\Phi}$ is an $M \times (N_r \times N_t)$ matrix consisting of independent random entries and the value of M is smaller than $(N_r \times N_t)$, the channel $\boldsymbol{h}$ is compressed and encoded as an $M \times 1$ vector $\boldsymbol{y}$ in Eq. (7.8), and feedback load is thereby reduced by a compression ratio of $\eta = M/(N_r \times N_t)$. As we have assumed that in a massive MIMO system, antenna arrays at both transmitter and receiver are closely packed, the resultant high spatial correlations among the elements of $\boldsymbol{h}$ may lead to a sparse representation in a certain domain, as in Eq. (7.2). Hence, based on compressive sensing theory, the channel information can be accurately recovered from $\boldsymbol{y}$ even if the value of M is set to be much smaller than $(N_r \times N_t)$, so a low compression ratio can be achieved to allow an efficient CSI feedback scheme.

It is worth noting that the sparsifying transformation matrix $\boldsymbol{\Psi}$ and the elements of $\boldsymbol{\Phi}$ can be determined offline (for example, fixed by specification of the system); therefore both the base station and the UE are aware of $\boldsymbol{\Psi}$ as well as the contents of $\boldsymbol{\Phi}$ based on preconfigurations. Thus the base station is able to undertake the optimization procedures in Eq. (7.5) to estimate $\boldsymbol{S}$ (the sparse form of $\boldsymbol{h}$), and eventually recover $\boldsymbol{h}$ by applying inverse sparsifying transformation:

$$\boldsymbol{h} = \boldsymbol{\Psi}^{\mathrm{T}}\boldsymbol{S} \tag{7.9}$$

The overall channel feedback scheme is illustrated in Fig. 7.1.

7.2.3 SPARSIFYING BASIS

The choice of the sparsifying basis, $\boldsymbol{\Psi}$, plays a key role in performance of the recovery. Under a given compression ratio, it is desirable to select a sparsifying basis that provides a more sparse representation (fewer nonzero elements in $\boldsymbol{S}$, or smaller K) of $\boldsymbol{h}$, as it leads to signal recovery with higher accuracy. In this chapter, two such bases are considered: two-dimensional discrete cosine transformation (2D-DCT) and Karhunen-Loeve transformation (KLT).

In the context of digital image processing, 2D-DCT has been widely applied for image compression, as it is useful in exploiting spatial correlation among the pixels to reduce the image dimensions. With massive MIMO systems, an array consisting of a large number of antennas could be very compact due to the constrained form factor

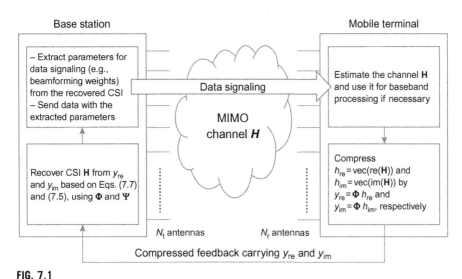

FIG. 7.1

CSI feedback framework based on compressive sensing.

of the equipment, which leads to small antenna spacings. Based on such conjecture, the elements of either real or imaginary MIMO channel matrix $\widehat{\mathbf{H}}$ are expected to be strongly correlated in at least one dimension due to small antenna spacing at the transmitter side. Therefore, analogous to exploiting the spatial correlation among pixels of a digital image, 2D-DCT can be leveraged as the sparsifying basis to realize a sparse representation of $\widehat{\mathbf{H}}$ in the spatial frequency domain by fully exploiting such correlation structure. The matrix operation of applying 2D-DCT on $\widehat{\mathbf{H}}$ can be written as $\boldsymbol{C}_{N_t}^{\mathrm{T}}\widehat{\mathbf{H}}\boldsymbol{C}_{N_r}$, where $\boldsymbol{C}_L$ is an $L \times L$ DCT matrix. Because DCT matrix is a 2D-separable basis [12], the sparse representation of $\boldsymbol{h}$ can be expressed as

$$S = (\boldsymbol{C}_{N_t} \otimes \boldsymbol{C}_{N_r})^{\mathrm{T}}\mathrm{vec}\left(\widehat{\mathbf{H}}\right) = (\boldsymbol{C}_{N_t} \otimes \boldsymbol{C}_{N_r})^{\mathrm{T}}\boldsymbol{h} \tag{7.10}$$

where $\otimes$ denotes Kronecker product. Hence, it is apparent that the sparsifying basis $\boldsymbol{\Psi}$ associating to 2D-DCT is

$$\boldsymbol{\Psi}_{\mathrm{2D-DCT}} = (\boldsymbol{C}_{N_t} \otimes \boldsymbol{C}_{N_r})^{\mathrm{T}} \tag{7.11}$$

The sparsity that 2D-DCT can achieve is heavily dependent on the correlation structure of $\boldsymbol{h}$. In contrast, KLT is capable of providing the optimal sparse representation with only one nonzero element ($K = 1$) regardless of the correlation structure [13], which promises accurate channel recovery even if only a small number of measurements are available. That is, a very low compression ratio can be guaranteed. Nevertheless, application of KLT is rather difficult in practice, as its sparsifying basis is signal dependent and has to be derived from the target signal itself. To be specific, given that we have the instantaneous $(N_r \times N_t) \times (N_r \times N_t)$ correlation matrix of $\boldsymbol{h}$:

$$\boldsymbol{W} = \boldsymbol{h}\boldsymbol{h}^{\mathrm{T}} \tag{7.12}$$

the sparsifying basis of KLT, Ψ_{KLT}, can be computed as the eigenvectors of W:

$$W = \Psi_{KLT} \Lambda \Psi_{KLT}{}^T \tag{7.13}$$

where Λ is a diagonal matrix, the nonzero entries of which are eigenvalues of W. Thus, in spite of the optimal sparsity that KLT can reach, application of KLT is impractical as this is essentially a chicken-and-egg paradox, wherein the sparsifying basis is derived from the target signal that we wish to recover. From this point of view, a fixed-value sparsifying basis such as 2D-DCT is more feasible if the CSI feedback mechanism based on compressive sensing is adopted for massive MIMO systems. However, if the channel is relatively static (for instance, in low-mobility scenarios), the validity of Ψ_{KLT} obtained earlier could be sustained over several transmission time intervals. In the next section, we elaborate a CSI feedback protocol that employs 2D-DCT and KLT in a hybrid manner.

7.2.4 CSI FEEDBACK PROTOCOL WITH ADAPTIVE SPARSIFYING BASIS

It is known that the optimal sparse representation of the MIMO channel matrix delivered by KLT has only one nonzero value (denoted as s), which is presumably the first entry of the sparse vector (S) due to eigenstructure properties. Thus, the feedback burden can be significantly reduced because the value of s alone is sufficient for CSI recovery at the base station. Although KLT is signal dependent, as mentioned earlier, it could be useful if the wireless channel does not change rapidly. In particular, a KLT basis derived in one time instance could still be effective in several subsequent time instances. Here we propose a compressive sensing-based CSI feedback protocol that utilizes both 2D-DCT and KLT, which targets low-mobility scenarios with slow-varying channels. The scheme allows the feedback content to be switched between CS measurements y (for initial CSI construction based on 2D-DCT) and s value (for subsequent CSI recovery based on KLT) to improve efficiency. The flow chart of this proposed scheme is illustrated in Fig. 7.2.

The operational procedure of the proposed scheme is elucidated as follows:

1. The UE estimates the instantaneous $N_r \times N_t$ MIMO channel matrix **H** based on the pilot symbols transmitted by the base station, and then compresses the measured channel into an $M \times 1$ vector y via random projection described by Eq. (7.8). The compressed CSI, y, is then sent to the base station using an uplink radio resource dedicated for CSI feedback. Additionally, the UE emulates the channel recovery procedure by undertaking l_1-norm minimization as described by Eq. (7.5) using 2D-DCT basis, in order to predict the recovered channel that is to be perceived by the base station. It is worth noting that, due to inevitable recovery errors, the recovered channel matrix (denoted as $\mathbf{H}_{est}$) could be slightly deviated from the ground-truth channel **H**. The UE should further perform eigendecomposition of $\mathbf{H}_{est}$ to compute its KLT basis, denoted as Q_{est}. In summary, apart from y, the UE also derives and stores Q_{est} for future use.

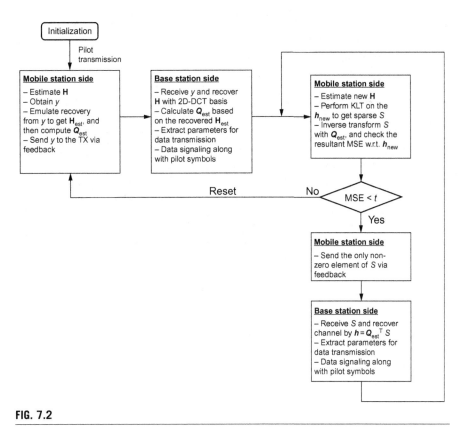

FIG. 7.2

The flowchart of CSI feedback protocol based on adaptive sparsifying basis.

2. Once the compressed CSI **y** is received by the base station via feedback, the massive MIMO channel can be recovered by l_1-norm minimization by the base station using 2D-DCT basis. This is identical to the emulation process that the UE carried out in Step 1. Here we assume that the same version of $\mathbf{H}_{est}$ is obtained by the base station via channel reconstruction due to the identical recovery algorithm. Based on the recovered channel matrix $\mathbf{H}_{est}$, the base station further computes and stores the correspondent KLT basis, $\boldsymbol{Q}_{est}$. Because the CSI is recovered, appropriate beamforming weights can be applied for downlink data transmission. Meanwhile, new pilot symbols are also transmitted in downlink to track channel variation.

3. The UE detects the downlink data as well as estimating the latest channel status denoted as $\mathbf{H}_{new}$ (the vectorized version is written as $\boldsymbol{h}_{new}$). Instead of compressing the estimated channel $\mathbf{H}_{new}$ by random projection as before, KLT is applied to obtain the optimal sparse representation of $\mathbf{H}_{new}$, which is a sparse vector with only one nonzero element (denoted as S) located in the first entry. Ideally, the feedback load can be reduced significantly as the UE only has to feed back S, which is presumably sufficient for channel recovery at the base station given that $\boldsymbol{Q}_{est}$

(derived in Step 1) is still valid due to slow channel variation. However, prior to feedback of S, the UE should check the validity of $\mathbf{Q}_{\text{est}}$ by using it to emulate CSI recovery based on KLT:

$$h_{\text{try}} = \mathbf{Q}_{\text{est}}S \qquad (7.14)$$

and then inspect the mean squared error (MSE) between h_{try} and the latest channel h_{new}. If the MSE is smaller than a preconfigured threshold level t, then the UE can simply report S for efficient feedback. Otherwise, the UE should repeat Step1 and report the compressed CSI **y** instead.

4. Upon the reception of S, the base station could simply recover the channel required by massive MIMO operations using Eq. (7.9) because $\mathbf{Q}_{\text{est}}$ is already available (obtained in Step 2) and its validity has been tested by the UE in Step 3.

With the protocol described previously, CSI reporting for massive MIMO becomes very efficient, as the payload comprises only one scalar value (S) in most feedback sessions, with occasional feedback of compressed channel vector (**y**), as long as the channel does not change too rapidly. Both 2D-DCT and KLT basis are applied throughout the operation. The simulation results in Fig. 7.3 show a performance comparison between the two feedback methods based on compressive sensing. Apparently, when only 2D-DCT basis is employed and the compression ratio is fixed as 40%, the resultant performance is almost aligned with the cases where perfect CSI is available at the transmitter. However, the compressed channel vector **y** has to be

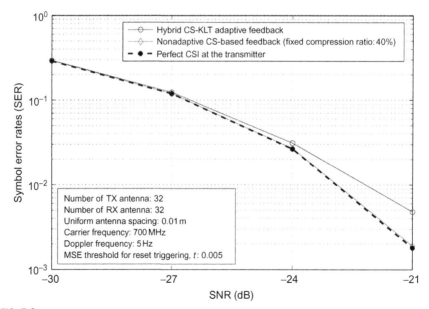

FIG. 7.3

Performance comparison between compressive sensing-based CSI feedback schemes with fixed and adaptive sparsifying basis.

reported by the UE in every feedback session. On the other hand, when the scheme adaptive sparsifying basis is used, the burden of CSI feedback is relaxed in spite of marginal performance degradation, as shown in Fig. 7.3.

The compressive sensing approach in this section provides a potential way to reduce the required CSI feedback payload for massive MIMO systems. However, in order to get a sparse representation of the channel, it is essential to assume high spatial correlation among the antennas. In practice, such an assumption might become invalid, especially when the carrier frequency is increased to millimeter-wave (mmWave) ranges. In order to accommodate systems operating in higher frequencies, we examine CSI acquisition techniques based on beamforming in the next section.

7.3 CSI ACQUISITION WITH ANGULAR-DOMAIN BEAMFORMING

In this section, we discuss the applications of angular-domain beamforming to CSI acquisition mechanisms for massive MIMO systems. Fig. 7.4 illustrates a downlink pilot signals transmission scheme with angular-domain beamforming, where the UE can only perceive signal energy and conduct channel estimation in a few beamforming directions. Based on such a setting, a feedback scheme with multistage beamforming (e.g., two- and three-stage) is presented in this section.

7.3.1 MULTISTAGE BEAMFORMING AND FEEDBACK

Instead of measuring and reporting CSI corresponding to a large number of antennas for massive MIMO systems as the conventional approach, it is possible to reduce the dimensionality of the desired channel via beamforming. In particular, the number of propagation paths of a beamformed signal is expected to be much lower; the user can

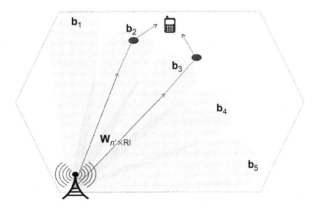

FIG. 7.4

Illustration of angular-domain beamforming in a cellular network.

focus on these "beams" rather than measuring channel responses corresponding to all transmitter antennas of the base station. From this perspective, the feedback load can be relented as the user only has to consider the channel status in the beamspace. Such a CSI acquisition scheme can be fulfilled with a sequential procedure consisting of multiple stages [14]. Here we specifically look at a two-stage scheme and a three-stage scheme.

CSI acquisition scheme with two-stage beamforming

In the two-stage beamforming scheme, CSI acquisition involves two different phases, namely beam selection and beam-based CSI reporting, as described as follows:

- In the first stage, the base station transmits beams in multiple fixed directions, and then the user selects preferred beams and feedback in the first stage. The flowchart of the scheme is shown in Fig. 7.5. The selection criterion can be based on the received energy level or signal-to-interference-plus-noise ratio (SINR) of the measured pilot signals on each of the beams. After the measurement, the user reports the index (indexes) of the selected $N \geq 1$ beams. Based on the beam(s) selection report received from the user, the base station configures the user with a set of certain beam directions it plans to apply for downlink transmission. For beamforming-based communication schemes, accurate beam tracking is the key to high system performance. If beam tracking is not precisely performed, it will not only significantly reduce the received SINR, but also cause interference. Thus, to decide beam configurations for the user, the base station should also consider some other available information, such as the user's mobility, directions of motion and potential interferences from the beams configured for the other users, in order to improve the system performance.
- In the second stage, the base station beamforms the downlink pilot signals on the configured directions, so the user can further measure and report the CSI of the configured beamformed channels. Thus the base station can transmit downlink data using beamforming as well as the reported CSI.

In the beginning of the second stage, the base station uses the configured beamforming weight vectors $\{\mathbf{b}_{I_i}\}$ for the user to transmit downlink pilot symbols; that is the transmitted signal vector can be written as

$$\mathbf{x} = \mathbf{Bs} \qquad (7.15)$$

where $\mathbf{B}$ consists of stacking column vectors of $\{\mathbf{b}_{I_i}\}$, that is, $\mathbf{B} = [\mathbf{b}_{I_1} \ \mathbf{b}_{I_2} \ \cdots \ \mathbf{b}_{I_M}]$ in which $\{I; i = 1, 2, ..., M, I_i \in \{1, 2, ..., N\}\}$ is the beam index set selected by the user. Thus the user measures the beamformed channel $\mathbf{H}_{bf} = \mathbf{HB}$ and evaluates the corresponding CSI. Therefore the size of the beamformed channel is reduced from $N_r \times N_t$ to $N_r \times M$. Because the number of beams (M) directed to a specific UE is usually much smaller than the number of transmitter antennas (N_t), the dimension of the MIMO channel is reduced significantly.

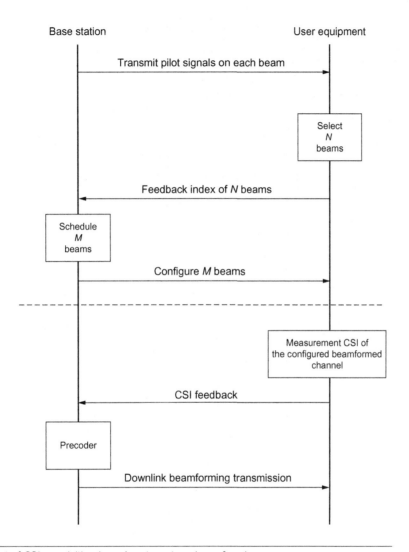

FIG. 7.5

Flow chart of CSI acquisition based on two-stage beamforming.

For data transmission, the base station applies the precoder $\mathbf{W}$ from the CSI reported by the user for downlink data transmission; that is, the received signal at the UE can be written as $\mathbf{x} = \mathbf{BWs}$. Apparently, the signal undergoes two precoding processes, and this scheme is therefore referred to as two-stage beamforming.

CSI acquisition scheme with three-stage beamforming

It is shown that the dimension of the effective channel after beamforming is drastically reduced in the two-stage Beamforming scheme. Remarkably, the dimension can be further reduced if the channel matrix's eigenstructure is known. This is due to the fact that the transmitter antennas are correlated to each other and/or there is only a small number of dominant paths that exist between the base station and the user. Both factors will yield the beamformed channel matrix $\mathbf{H}_{bf} = \mathbf{HB}$ to be rank deficient. By using this property, it is possible for the base station to concentrate its transmission power to the dominant eigenvectors only, thereby improving the energy efficiency of the base station, and further curtailing the feedback overhead for the UE.

To accomplish this, the user has to provide more information to the base station. In particular, the base station needs the information to facilitate power allocation among the eigenvectors. It is known that the channel matrix can be factored as $\mathbf{H}_{bf} = \mathbf{R}_{R\_bf}^{1/2} \mathbf{H}_{iid} \mathbf{R}_{T\_bf}^{1/2}$, where $\mathbf{R}_{T\_bf} = E\{\mathbf{H}_{bf}^H \mathbf{H}_{bf}\}$ is the transmitter correlation matrix, $\mathbf{R}_{R_bf} = E\{\mathbf{H}_{bf} \mathbf{H}_{bf}^H\}$ is the receiver correlation matrix and $\mathbf{H}_{iid}$ is a matrix composed of i.i.d. random variables. The SVD of $\mathbf{R}_{R\_bf}$ can be written as $\mathbf{R}_{R\_bf} = \mathbf{UDU}^H$, in which the columns of $\mathbf{U}$ are its eigenvectors and the diagonal elements of the diagonal matrix $\mathbf{D}$ are the associated eigenvalues. Similarly, we have $\mathbf{R}_{T\_bf}^{1/2} = \mathbf{UD}^{1/2}\mathbf{U}^H$. Note that the rank of $\mathbf{R}_{T\_bf}^{1/2}$, that is, R is less than or equal to the number of configured beams M, hence further dimensional reduction is possible. More specifically, if the base station transmits only on the eigenvectors, that is, through the precoder $\mathbf{U}$ of dimension $M \times R$, then the effective channel becomes $\mathbf{H}_{bf}\mathbf{U} = \mathbf{HBU}$, the dimension of which is $N_r \times R$. Recall that $R \le M$; thus the dimension of the beamformed channel is reduced. Finally, after the CSI of the channel $\mathbf{H}_{bf}\mathbf{U}$ is reported to the base station, the downlink data signal can be precoded as $\mathbf{x} = \mathbf{BUWs}$.

The flowchart of this three-stage beamforming scheme is illustrated in Fig. 7.6. Note that the first and the third-stage are identical to the two stages shown in Fig. 7.5. As for the second stage, the user measures the transmitter correlation matrix and feeds back its eigenvectors to the base station. Compared to the two-stage beamforming in Section "CSI acquisition scheme with two-stage beamforming," additional feedback is needed. Fortunately, thanks to the fact that the transmitter correlation matrix is a long-term statistic, the reporting periodicity can be set to be longer than that of the third-stage CSI report. Thus, if the $R \ll M$, it is possible to achieve an even lower feedback overhead than in two-stage beamforming.

7.4 DOWNLINK PRECODING IN FDD BASED ON ANGLE OF ARRIVAL

In contrast to preceding sections that focus on more efficient CSI feedback, this section investigates alternative CSI acquisition mechanisms for FDD massive MIMO systems using beamforming techniques. The base stations equipped with a large number of antennas have the capabilities of transmit and receive beamforming, given that the antennas are arranged as a uniform linear array (ULA). Using beamforming in a massive MIMO system is known to have great potential in increasing sum rate

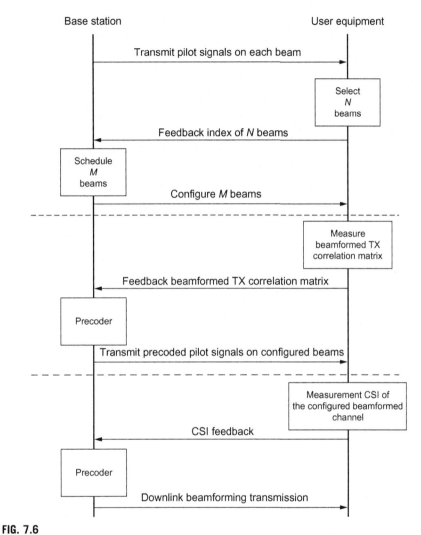

FIG. 7.6

Flow chart of CSI acquisition based on three-stage beamforming.

performance in multiuser scenarios, assuming that the CSI information is perfectly known. It has been reported that beamforming is also helpful for acquisition of partial CSI information that enables downlink precoding in FDD massive MIMO systems. In this section, we introduce some of these techniques [15].

The main differentiating feature of the approach introduced in this section is that it abandons any efforts on downlink CSI estimation for the purpose of downlink precoding. Subsequently, no explicit CSI feedback will be provided to the base station in the reverse channel either, because the UE does not have any CSI information prior to the downlink transmission anyway. Instead, the base station relies on its own capability of receive beamforming to obtain partial CSI information for a UE via a previous uplink signaling from the UE. Examples of such uplink signals initiated by a UE include a random access request, acknowledgment of a previous downlink transmission, an uplink data transmission, and so on. All these kinds of signaling do not contain information on downlink CSI. However, a base station equipped with a large-scale antenna array is able to obtain estimates of arriving angles of propagation paths through which the uplink signal was transmitted. A major advantage of this approach is that it does not require any allocation of transmission resources for downlink precoding purposes. Although in FDD the property of channel reciprocity cannot be applied like in TDD, it is noted that (e.g., Ref. [16]) the arriving angles of uplink propagation paths from a UE have strong correlations with departing angles of propagation paths in the downlink transmission to the same UE. Exploiting this property, precoding coefficients for downlink transmission can be designed without extra cost in bandwidth. Thus the intrinsic benefits of easier CSI acquisition in TDD could also be exploited by FDD.

The basic idea of the downlink precoder designs introduced here is to apply beamforming techniques and steer the transmit power to a direction through which we determine the target UE receives the most power, and at the same time minimize the power directed to other UEs accessing the same transmission resource. Although a UE may receive power from the base station through multiple angles, we would choose only one of the directions in order to avoid potential destructive interference. A simple way to implement this idea is to apply the beamspace division method where the angular-domain space is evenly divided into M mutually orthogonal beams. However, the beamspace division method may have the following disadvantages. First, when the number of antennas on the base station is small, this method would cause interference to other UEs due to large beamforming sidelobe power. Second, when the number of antennas on the base station is large, the resulting beamwidth is narrow and thus the performance of downlink transmission would be very sensitive to errors in the estimation of arriving angles of the uplink signal. In order to mitigate these two problems, some precoder designs have been proposed to boost transmission energy in the direction of the targeted UE and suppress interference caused to other UEs located at different directions.

Based on the previous idea, we introduce here two methods for multiuser precoder design without CSI feedback. The first takes advantage of the techniques of diagonal loading (DL) [17] widely used in the radar literature; the other one treats the beamforming problem as a spatial-domain filter design problem and uses the classical Parks-McClellan (PM) algorithm [18]. It is observed that the proposed methods have only a slight performance degradation compared to the ideal block diagonalization (BD) method, which requires full, accurate CSI feedback.

7.4.1 BACKGROUND AND ASSUMPTIONS

Prior to delving into the details of the proposed scheme, we first present the channel model and the background required for analysis.

Channel model

A base station equipped with a ULA with M antennas is considered. Assume that the base station serves K single-antenna UEs using the same time-frequency resources ($K \ll M$). The base station employs FDD with uplink carrier frequency f_u and downlink carrier frequency f_d. The variables λ_u and λ_d denote the corresponding uplink and downlink wavelengths, respectively; that is, $\lambda_u = c/f_u$ and $\lambda_d = c/f_d$ where c is the speed of light. We denote D as the antenna spacing of the ULA, and assume that the interantenna spacing is a half wavelength of the downlink carrier frequency, that is, $\lambda_u \approx \lambda_d = 2D$.

The radio propagation channel from the kth UE to the base station consists of P paths and is characterized as [19]

$$c_k(t; \theta) = \sum_{p=1}^{P} \alpha_{kp} \delta(t - \tau_{kp}) \delta(\theta - \theta_{kp}) \tag{7.16}$$

where the pth path has a complex gain α_{kp}, a time delay τ_{kp} with an AoA θ_{kp}. The AoA of a path is defined as the angle between the direction of the arriving path and a line parallel to the antenna array. Without loss of generality, we have $-\pi/2 \leq \theta_{kp} < \pi/2$. Suppose that these propagation paths can be grouped as separate clusters, each consisting of paths that have close AoAs. According to the grouping, the channel can be rewritten as

$$c_k(t; \theta) = \sum_{l=1}^{L} \sum_{p=1}^{P} \alpha_{klp} \delta(t - \tau_{klp}) \delta(\theta - \theta_{klp}) \tag{7.17}$$

where L denotes the number of clusters, P_l denotes the number of paths of the lth cluster, and α_{klp}, τ_{klp}, and θ_{klp} are the corresponding parameters for the pth path in the lth cluster. A representative angle ϑ_{kl} can be defined for each cluster to enable any path in this cluster to satisfy the following inequality with a predetermined threshold θ_{di}:

$$\left| \theta_{klp} - \vartheta_{kl} \right| < \theta_{di}, \ \forall p \in \{1, ..., P_l\} \ \text{and} \ \forall l \in \{1, ..., L\} \tag{7.18}$$

An example of such a cluster is shown as the dotted arrows in the circle labeled "Cluster of Paths" in Fig. 7.7.

Similarly, the downlink channel can be modeled as

$$
\begin{aligned}
h_k(t; \theta) \quad &= \sum_{p=1}^{P} \beta_{kp} \delta(t - \tau_{kp}') \delta(\theta - \theta_{kp}') \\
&= \sum_{l=1}^{L} \sum_{p=1}^{P_l'} \beta_{klp} \delta(t - \tau_{klp}') \delta(\theta - \theta_{klp}')
\end{aligned}
\tag{7.19}
$$

where P_l' denotes the number of paths in the lth path, β_{klp} is a complex gain, τ_{klp}' represents the time delay, and θ_{klp}' is the angle of departure (AoD) in the downlink.

The downlink AoD of an individual path θ_{kp}' can be equal to the AoA of its uplink paths when the propagation paths are purely line-of-sight or reflections. Such reciprocity no longer holds when the path proceeds through refraction because

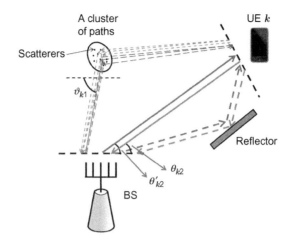

FIG. 7.7

Radio propagation paths between an FDD-based base station and UE k.

dispersion of electromagnetic waves causes divergence in the traveling paths of different radio frequencies. However, by holding that $\lambda_u \approx \lambda_d$, such divergence in a traveling direction can be bounded, and we can assume that

$$\left|\theta_{klp}' - \vartheta_{kl}\right| < \theta_{di}, \ \forall p\left\{1, ..., P_l'\right\} \text{ and } \forall l \in \{1, ..., L\} \tag{7.20}$$

We further assume that an orthogonal frequency division multiplexing modulation with cyclic prefixes is applied in the baseband transceivers so that for each subcarrier, the uplink channel can be assumed to be frequency flat. The steering vector in the downlink can be defined as

$$\boldsymbol{a}_d(\theta) = \frac{1}{\sqrt{M}}\left[1, e^{-j2\pi(D/\lambda_d)\sin\theta}, ..., e^{-j2\pi(M-1)(D/\lambda_d)\sin\theta}\right]^T \tag{7.21}$$

Let $x[n]$ denote the transmitted vector signal; subsequently, the received signal at the kth UE is

$$y_k[n] = \mathbf{h}_k^T x[n] + v_k[n] \tag{7.22}$$

where $v_k[n]$ is a zero-mean complex Gaussian random variable with variance σ_v^2, and

$$\mathbf{h}_k = \sum_{l=1}^{L}\sum_{p=1}^{P_l} \beta_{klp} e^{-j2\pi f_d \tau_{klp}'} \boldsymbol{a}_d\left(\theta_{klp}'\right) \tag{7.23}$$

In the downlink transmission, we use the term $s_k[n]$ to denote the information-bearing signal dedicated to the kth UE. The signal $s_k[n]$ is precoded by a vector $\boldsymbol{w}_k$, resulting in a transmitted signal as

$$x[n] = \sum_{k=1}^{K} \boldsymbol{w}_k^* s_k[n] \tag{7.24}$$

Consequently, the received signal at the kth UE in Eq. (7.22) is

$$y_k[n] = \mathbf{h}_k^{\mathrm{T}} \left(\mathbf{w}_k^* s_k[n] + \sum_{j \neq k} \mathbf{w}_j^* s_j[n] \right) + v_k[n] \tag{7.25}$$

Multiuser downlink precoding

The goal of this section is to develop precoding methods to enable downlink transmission that can simultaneously serve multiple UEs by designing the vector (i.e., $\mathbf{w}_k$, $k=1, 2, ..., K$) in Eq. (7.24). The design is based on the following assumptions:

- The base station can obtain an estimate of ϑ_{kl}, denoted as $\hat{\vartheta}_{kl}$, for all $k \in \{1, 2, ..., K\}$ and $l \in \{1, 2, ..., L\}$ through a previous uplink signaling. The estimation error of the representative angle is bounded by a predetermined value θ_{er}; in other words,

$$\left| \vartheta_{kl} - \hat{\vartheta}_{kl} \right| < \theta_{\mathrm{er}} \tag{7.26}$$

- Define $\nu_{kl} = |\mathbf{a}_{\mathrm{d}}^{\dagger}(\vartheta'_{kl})\mathbf{h}_k|$, which roughly represents the energy transmitted through the direction of ϑ_{kl} in the downlink. Similarly, define μ_{kl} as the uplink counterpart. We assume that the base station knows μ_{kl} but is not aware of ν_{kl}. Nevertheless, ν_{kl} is supposed to be highly correlated with μ_{kl} if the uplink and downlink carrier frequencies are sufficiently close to each other. Specifically, it is assumed that ν_{kl} has a large value with a high probability whenever μ_{kl} is large, as suggested in a previous study [20].
- The angle difference between clusters satisfies

$$\left| \vartheta_{kl} - \vartheta_{km} \right| > \theta_{\mathrm{th}}, l \neq m, \forall l, m \in \{1, 2, ..., L\} \tag{7.27}$$

where θ_{th} denotes an angle spacing threshold.

- Any pair of UEs with paths close to each other can be assigned to different time-frequency resources. Therefore, for the UEs accessing the same time-frequency resources, we can assume that

$$\left| \vartheta_{kl} - \vartheta_{jm} \right| > \theta_{\mathrm{th}}, j \neq k, \forall j, k \in \{1, 2, ..., K\} \text{ and } l, m \in \{1, 2, ..., L\} \tag{7.28}$$

- Finally,

$$\left| \theta_{kp} \right| \leq \theta_{\mathrm{ext}} < \pi, \ \forall k \in \{1, 2, ..., K\} \text{ and } \forall p \in \{1, 2, ..., P\} \tag{7.29}$$

where θ_{ext} denotes the extreme value of θ, suggesting that the angles of UEs served by this antenna array cannot be in the regions of $[-\pi, \theta_{\mathrm{ext}})$ and $(\theta_{\mathrm{ext}}, \pi)$. This assumption is justified by the following argument: if there are UEs located in those regions, then they can be served by another perpendicular antenna array of the base station.

7.4.2 DOWNLINK PRECODING DESIGN USING BEAMFORMING-BASED PARTIAL CSI

MVDR method

The minimum variance distortionless response (MVDR) beamforming is well known for mitigating the interferences with AoAs that are different from those of the desired signal. If the AoA of the desired signal is not accurately known, a DL method [21] can be used to avoid unexpected suppression of the desired signal. The MVDR-DL method uses a spatial-domain correlation matrix. In the application of interest, the spatial correlation matrix can be expressed as

$$\mathbf{R} = E_s \sum_{k=1}^{K} \sum_{l=1}^{L} |\mu_{kl}|^2 \boldsymbol{a}_d(\hat{\vartheta}_{kl}) \boldsymbol{a}_d^\dagger(\hat{\vartheta}_{kl}) + \frac{\sigma_v^2}{M} \mathbf{I}_M \qquad (7.30)$$

where E_s denotes the average energy of $s_k[n]$. The DL beamformer solves the following optimization problem:

$$\min_{\mathbf{w}_k} \mathbf{w}_k^\dagger (\mathbf{R} + \gamma_k \mathbf{I}_M) \mathbf{w}_k \text{ subject to } \boldsymbol{a}_d^\dagger(\theta_k) \mathbf{w}_k = 1 \qquad (7.31)$$

where $\theta_k = \hat{\vartheta}_{k\hat{l}_k}$ and γ_k denote the DL factor of UE k. The solution for this problem is $\mathbf{w}_k = \mathbf{u}_k / \|\mathbf{u}_k\|$ where

$$\mathbf{u}_k = \frac{(\mathbf{R} + \gamma_k \mathbf{I}_M)^{-1} \boldsymbol{a}_d(\theta_k)}{\boldsymbol{a}_d^\dagger(\theta_k)(\mathbf{R} + \gamma_k \mathbf{I}_M)^{-1} \boldsymbol{a}_d(\theta_k)} \qquad (7.32)$$

The DL factor γ_k for UE k is chosen as

$$\gamma_k = -\left(\frac{\sigma_v^2}{M} + |\mu_{k\hat{l}_k}|^2 E_s\right) \qquad (7.33)$$

to maximize SINR. The DL precoder is effective when the estimated direction of the desired signal (i.e., θ_k) is accurate. Nevertheless, the angle θ_k might be mismatched because of inaccurate AoA estimates. In such cases, mismatched angles would diminish the performance of the downlink transmission. Therefore, we propose another precoding method to improve the robustness against an angle mismatch.

Spatial-domain filter design approach

Here we present another precoder design method on the basis of viewing the beamforming design problem as a problem with designing a finite impulse response (FIR) filter in the spatial-domain. We employ the generalized PM algorithm [22], which extends the renowned PM FIR algorithm [18] to the case of allowing complex-value parameters.

Suppose we are designing an FIR filter, denoted as $w[n], n = 0, 1, \ldots, M-1$, with its Fourier transform expressed as

$$W(e^{j\omega}) = \sum_{n=0}^{M-1} w[n] e^{-j\omega n} \qquad (7.34)$$

We define $\mathbf{w} \triangleq [w[0] \, w[1] \cdots w[M-1]]^{\mathrm{T}}$. The optimum filter design problem involves minimizing the maximal value of the weighted approximation error of the frequency response [22]; in other words,

$$\min_{\mathbf{w}} \left\{ \max_{w \in \mathbf{B}} \left[G(e^{j\omega}) \left| F(e^{j\omega}) - W(e^{j\omega}) \right| \right] \right\} \tag{7.35}$$

where $F(e^{j\omega})$ denotes the desired frequency response, $G(e^{j\omega})$ is a nonnegative weighting function, and $\mathbf{B}$ represents a compact subset in $[-\pi, \pi]$.

For the precoding vector design problem, the angular-domain corresponds to the frequency domain. Specifically, we considered $\omega \triangleq \pi \sin \theta$. For a specific UE, we aimed to boost the power to its desired direction and eliminate interference to the AoA clusters of other UEs. Therefore, the ideal precoding vector $\mathbf{w}_k$ for UE k should have the property that its angular-domain response $W_k(e^{j\omega})$, $\omega \in \mathbf{B}$ be unity in the neighborhood of $\pi \sin \theta$ and 0 in the neighborhood of $\pi \sin \theta_j$ for $j \neq k$.

Let δ_θ denote the maximum allowable range of angle mismatch. We first define

$$C_r = [\pi \sin(\theta_r - \delta_\theta), \, \pi \sin(\theta_r + \delta_\theta)], \, r \in \{0, 1, ..., KL\} \tag{7.36}$$

Note that $C_r \cap C_q = \emptyset, q \neq r$ because the AoA clusters of different UEs are separated, as expressed in (7.28). Let $q(k)$ denote the index of the desired cluster for UE k (i.e., $q(k) = \hat{l}_k$); subsequently, a compact subset can be defined as

$$\mathbf{D}_{kr} = [a, b] \tag{7.37}$$

where

$$a = \begin{cases} \pi \sin(\theta_r + \beta \delta_\theta), & \text{if } r = q(k) \\ -\pi/2, & \text{if } r = 0 \\ \pi \sin(\theta_r + \delta_\theta), & \text{otherwise.} \end{cases} \tag{7.38}$$

and

$$b = \begin{cases} \pi \sin(\theta_r - \beta \delta_\theta), & \text{if } r + 1 = q(k) \\ -\pi/2, & \text{if } r = KL \\ \pi \sin(\theta_r - \delta_\theta), & \text{otherwise.} \end{cases} \tag{7.39}$$

We refer to this compact subset as a wild compact set it contains no UE's AoA clusters. Consequently, the compact subset can be written as

$$\mathbf{B}_k = \left(\cup_{r=1}^{KL} C_r \right) \cup \left(\cup_{r=0}^{KL} D_{kr} \right) \tag{7.40}$$

Based on these definitions, the desired angular-domain response for UE k is

$$F_k(e^{j\omega}) = \begin{cases} 1, & \omega \in C_{q(k)} \\ 0, & \omega \in \mathbf{B}_k \, C_{q(k)} \end{cases} \tag{7.41}$$

The weighting function $G_k(e^{j\omega})$ is designed as

$$G_k(e^{j\omega}) = \begin{cases} \rho_{\mathrm{d}}, & \omega \in C_{q(k)} \\ \rho_{\mathrm{i}}, & \omega \in C_r, r \neq q(k) \\ \rho_{\mathrm{v}}, & \omega \in D_{kr} \end{cases} \tag{7.42}$$

where ρ_d, ρ_i, and ρ_v denote the weights for the desired, interference and wild directions, respectively. To attain a high SINR and energy-efficient downlink transmission, we selected $\rho_d \approx \rho_i \gg \rho_v$. By substituting Eqs. (7.40)–(7.42) into Eq. (7.35), we obtain the optimal precoding vector for UE k, denoted as $\mathbf{w}_k$.

7.4.3 SIMULATION RESULTS

In our simulation, the average sum rate that was used as the performance metric can be calculated by

$$C_{\text{avg}} = \frac{1}{T} \sum_{t=1}^{T} \sum_{k=1}^{K} \log_2(1 + \text{SINR}_k(t)) \tag{7.43}$$

where T is the number of trials, and the SINR of UE k during trial t is calculated by

$$\text{SINR}_k(t) = \frac{\left| \mathbf{w}_k^\dagger(t)\mathbf{h}_k(t) \right|^2 E_s}{\sum_{j \neq k} \left| \mathbf{w}_j^\dagger(t)\mathbf{h}_k(t) \right|^2 E_s + \sigma_v^2} \tag{7.44}$$

The number of trials in all the simulation plots was set to $T = 1000$.

Fig. 7.8 illustrates the precoder pattern of the proposed methods and beamspace division method. In this figure, the targeted UE is located at an angle of 50 degrees, and two other UEs are located at 40 and 60 degrees. The pattern of the proposed precoder, which uses the DL method, concentrates the transmission power in the desired direction with a narrow allowable width of angle mismatch. By contrast, the pattern of the proposed precoder, which employs the PM method, has an approximately 4 dB loss in magnitude because the transmission power is distributed in the broadened beamwidth to mitigate the angle mismatch problem. At the angles of other UEs, the precoder has a favorable attenuation of only -70 dB. This property is conductive to gaining a favorable SINR.

Fig. 7.9 displays the sum rate performance of a system comprising four UEs. Here, perfect knowledge of representative angles is considered (i.e., $\hat{\vartheta}_{kl} = \vartheta_{kl}$, $\forall k = 1, 2, ..., K$ and $l = 1, 2, ..., L$). This figure illustrates that the proposed methods and beamspace division method differ slightly in performance. The performance of the proposed PM method is slightly worse than that of the proposed DL method because of the magnitude loss in the desired angles.

In the following plots we present the simulation results of mismatched cases (i.e., the estimated representative angles are mismatched). Specifically, the distribution of an estimated representative angle is expressed as

$$\hat{\vartheta}_{kl} \sim U(\vartheta_{kl} - \theta_{\text{er}}, \vartheta_{kl} + \theta_{\text{er}}) \tag{7.45}$$

where θ_{er} denotes the bound of the angle mismatch defined in Eq. (7.26). The bound θ_{er} are set to 0.5 degrees and 2 degrees in Figs. 7.10 and 7.11, respectively. The

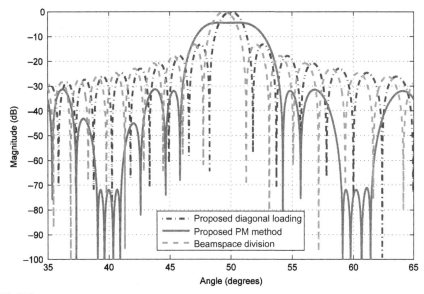

FIG. 7.8

Precoding vector pattern. The served UE is located at 50 degrees; two UEs causing interference are located at 40 and 60 degrees.

simulation plots are based on averaging the results for different mismatched values generated from the given distributions.

Figs. 7.10 and 7.11 display the simulated performance of each method with mismatched ϑ_{kl}, demonstrating that the performance of the proposed DL method is slightly more favorable than that of the proposed PM method. On the other hand, as exhibited in Fig. 7.11, when the value of θ_{er} is increased to 2 degrees, the proposed PM method outperformed the other methods and attained similar performance with the BD method because it has a broader beamwidth to alleviate the performance degradation caused by angle mismatching. These observations have shown that the proposed DL method outperforms the proposed PM method when the mismatched angle is smaller than 0.5 degrees, but its superiority vanishes when the mismatched angle is larger.

7.5 SUMMARY

CSI acquisition is an important and yet challenging issue for massive MIMO systems. Correspondingly, this chapter gives an overview on a few different directions in this context. Compressive sensing has been suggested to reduce the burden of CSI feedback, which relies on the potential sparse representation of MIMO channels

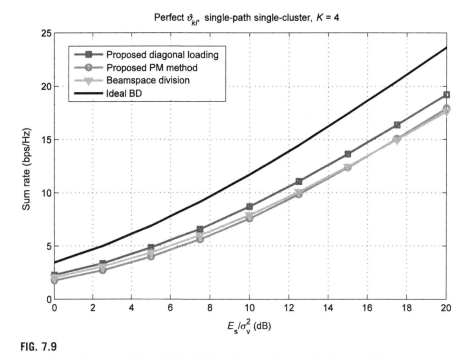

FIG. 7.9

Sum rate versus E_s/σ_v^2 with angles without mismatching in single-path single-cluster scenarios.

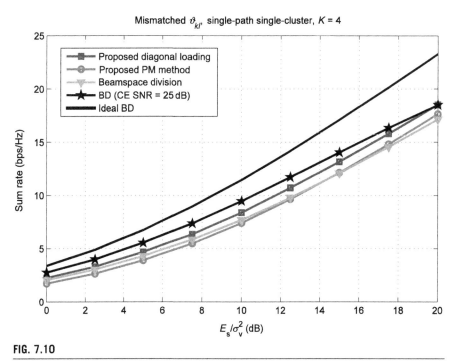

FIG. 7.10

Sum rate versus E_s/σ_v^2 with mismatched angles of 0.5 degrees in single-path single-cluster scenarios.

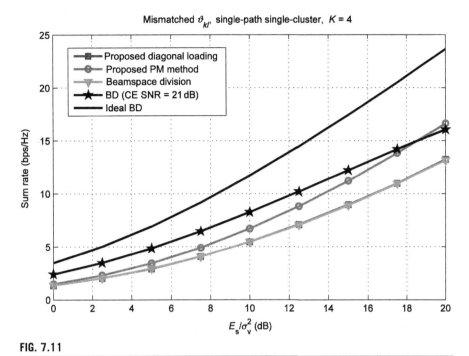

FIG. 7.11

Sum rate versus E_s/σ_v^2 with mismatched angles of 2 degrees in single-path single-cluster scenarios.

under spatial correlation. For a slowly varying propagation environment, a hybrid feedback compression method utilizing two different sparsifying basis (DCT and KLT) can further strike a balance between feedback load and CSI recovery performance. Nevertheless, because massive MIMO is likely to be deployed in mmWave frequency ranges in next-generation cellular networks, a compressive sensing-based approach might not be feasible, as the occurrence of strong spatial correlation is not likely with high carrier frequencies such as mmWave. Instead, due to a limited number of propagation paths, the beamforming-based approach is probably more practical, in which the user is able to focus on only a few angular beams for CSI measurements and reporting. Moreover, by utilizing the facts that AoAs of uplink signals could be aligned with departing angles of downlink, this chapter also discussed a downlink precoding technique that does not require CSI feedback from the users, which is very efficient in terms of resource and power saving at the user side. The schemes examined in this chapter have provided a foundation for future developments of mmWave massive MIMO technologies in 5G systems.

REFERENCES

[1] T.L. Marzetta, Noncooperative cellular wireless with unlimited numbers of base station antennas, IEEE Trans. Wirel. Commun. 9 (2010) 3590–3600.

[2] J. Jose, A. Ashikhmin, P. Whiting, S. Vishwanath, Channel estimation and linear precoding in multiuser multiple-antenna TDD systems, IEEE Trans. Veh. Technol. 60 (2011) 2102–2116.

[3] F. Kaltenberger, H. Jiang, M. Guillaud, R. Knopp, Relative channel reciprocity calibration in MIMO/TDD systems, in: Proceedings of the Future Network and Mobile Summit, 2010.

[4] A. Ashikhmin, T.L. Marzetta, Pilot contamination precoding in multi-cell large scale antenna systems, in: Proceedings of the IEEE International Symposium on Information Theory (ISIT), 2012, pp. 1137–1141.

[5] Technology Vision 2020—Reducing network latency to milliseconds, Nokia Network White Paper, 2015.

[6] J. Choi, Z. Chance, D.J. Love, U. Madhow, Noncoherent trellis coded quantization: a practical limited feedback technique for massive MIMO systems, IEEE Trans. Commun. 61 (2013) 5016–5029.

[7] R.G. Baraniuk, Compressive sensing, IEEE Signal Process. Mag. 24 (2007) 118–124.

[8] P.-H. Kuo, H.T. Kung, P.-A. Ting, Compressive sensing based channel feedback protocols for spatially-correlated massive antenna arrays, in: Proceedings of IEEE Wireless Communication and Networking Conference (WCNC), Paris, France, 2012, pp. 492–497.

[9] J.A. Tropp, A.C. Gilbert, Signal recovery from random measurements via orthogonal matching pursuit, IEEE Trans. Inf. Theory 53 (2007) 4655–4666.

[10] E. Candes, J. Romberg, T. Tao, Robust uncertainty principles: exact signal reconstruction from highly incomplete frequency information, IEEE Trans. Inf. Theory 52 (2006) 489–509.

[11] H. Yang, T.L. Marzetta, Performance of conjugate and zero-forcing beamforming in large-scale antenna systems, IEEE J. Sel. Areas Commun. 31 (2013) 172–179.

[12] Y. Rivenson, A. Stern, Compressed imaging with a separable sensing operator, IEEE Signal Process. Lett. 16 (2009) 449–452.

[13] J.E. Fowler, Compressive-projection principal component analysis, IEEE Trans. Image Process. 18 (2009) 2230–2242.

[14] J. Nam, J.-Y. Ahn, G. Caire, A. Adhikari, Joint spatial division and multiplexing—the large-scale array regime, IEEE Trans. Inf. Theory 59 (2013) 6441–6463.

[15] M.-F. Tang, B. Su, Downlink precoding for multiple users in FDD massive MIMO without CSI feedback, J. Signal Process. Syst. 83 (2) (2016) 151–163.

[16] Y. Li, M.S. Rahman, Y.-H. Nam, Full-dimension MIMO cellular systems realizing potential of massive-MIMO, IEEE Comsoc MMTC E-Lett. 9 (6) (2014) 6–10.

[17] B. Carlson, Covariance matrix estimation errors and diagonal loading in adaptive arrays, IEEE Trans. Aerosp. Electron. Syst. 24 (4) (1988) 397–401.

[18] T. Parks, J. McClellan, Chebyshev approximation for nonrecursive digital filters with linear phase, IEEE Trans. Circuit Theory 19 (2) (1972) 189–194.

[19] R. Ertel, P. Cardieri, K. Sowerby, T. Rappaport, J. Reed, Overview of spatial channel models for antenna array communication systems, IEEE Pers. Commun. 5 (1) (1998) 10–22.

[20] J. Capon, High-resolution frequency-wavenumber spectrum analysis, Proc. IEEE 57 (8) (1969) 1408–1418.

[21] F. Vincent, O. Besson, Steering vector errors and diagonal loading, IEE Proc. Radar Son Nav. 151 (6) (2004) 337–343.

[22] L. Karam, J. McClellan, Complex Chebyshev approximation for FIR filter design, IEEE Trans. Circuits Syst. II, Analog Digit. Signal Process. 42 (3) (1995) 207–216.

mmWave massive MIMO channel modeling

B. Ai*, K. Guan*, G. Li*, S. Mumtaz[†]

Beijing Jiaotong University, Beijing, China * Instituto de Telecomunicações, Aveiro, Portugal*[†]

CHAPTER OUTLINE

8.1 INTRODUCTION

MmWave massive multiple-input multiple-output (MIMO) has the potential to dramatically improve wireless access and throughput, as this combination enjoys both huge signal bandwidth at mmWave frequencies and spatial multiplexing capability of large antenna arrays [1,2]. Because these potential system benefits are directly related to the propagation environment and antenna arrays, realistic and reliable propagation, and channel models are of great importance for the mmWave massive MIMO access technologies' design. As so far there is no official or widely acknowledged mmWave massive MIMO channel model, either in academia or in the industry, we first point out the main specific features of the mmWave massive MIMO channel in terms of three layers: propagation mechanisms, static channel, and dynamic channel as follows:

(1) *Propagation mechanisms*: In massive MIMO systems, the primary source of channel state information (CSI) error is considered to be pilot contamination.

mmWave Massive MIMO. http://dx.doi.org/10.1016/B978-0-12-804418-6.00008-X

But, for mmWave frequencies, some specific propagation phenomena, such as very high propagation loss and near-line-of-sight (LOS) propagation, would give a chance to mitigate the pilot contamination effect. Thus it is of importance to establish comprehensive understanding of the internal relations between propagation mechanisms and various impact factors in the mmWave band. Apart from the propagation characteristics at each frequency point, the frequency dispersion of the huge bandwidths in the mmWave band is critical as well. Because one important motivation for the future broadband wireless communications to jump in the mmWave band is to obtain sufficient bandwidth, the mmWave massive MIMO channel is naturally an ultra-broadband channel. In this context, accurately revealing the frequency dispersion of the wave propagation and its interaction with various structures plays a key role in understanding how the propagation characteristics change along the bandwidth in the mmWave band. Moreover, besides this frequency-dependent behavior, in the sub-mmWave band, the propagation characteristics even change along the distance; this indeed impacts the design of the physical and link layers of mmWave massive MIMO systems.

In particular, the very strong distance-dependent behavior of the available bandwidth will even develop different modulations for different applications based on the targeted transmission distance for future communications.

As shown in Fig. 8.1 (from [2]), path loss in the sub-mmWave band for different transmission distances have different available bandwidths, even though in Fig. 8.1, the first transmission window covers the whole mmWave band when the distance is 100 m. It can be expected that such window will become narrower when the distance further increases, and finally this effect cannot be ignored in the mmWave massive MIMO channel characterization and modeling.

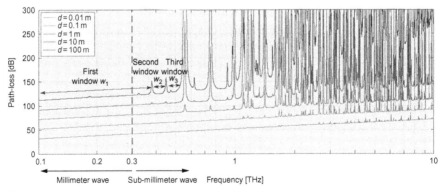

FIG. 8.1

Path loss in the mmWave and sub-mmWave bands with different transmission distances, from Fig. 8.2 in [4].

(2) *Statistic channel*: When the frequency band of the massive MIMO channel increases to mmWave band, some assumptions that are valid in the conventional bands do not hold anymore. One important assumption is the asymptotic pairwise orthogonality between the channel vectors under an independent and identically distributed (i.i.d.) Rayleigh fading channel. This assumption enables multiple spatially multiplexed data streams to be transmitted to multiple user equipment simultaneously with very simple linear precoding techniques. Furthermore, the mutual orthogonal channels also provide the optimal performance in terms of transmit power consumption or sum-rate channel capacity [4]. However, when it comes to the realistic design of multiuser (MU)-MIMO systems operating at mmWave frequencies, many practical considerations need to be taken into account, and in fact this orthogonality does not stand anymore [1]. Thus, the nonorthogonality of the mmWave massive MIMO channel should be investigated under a realistic propagation environment, including the factors that decide the spatial correlation and the overall system performance. Another feature is spatial nonstationarity, which exists in the conventional-band massive MIMO channel, but becomes more severe in the mmWave band. Because the far-field condition is more and more difficult to be fulfilled when the frequency goes up, the original widely used plane wavefront assumption should be replaced by the spherical wavefront (see Fig. 8.2). This poses new challenges both on the channel modeling and simulation approaches [3].

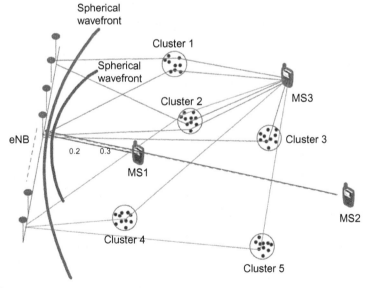

FIG. 8.2

Illustration of nonstationary phenomenon and spherical wavefront assumption, from Fig. 1 in [13].

(3) *Dynamic channel*: It was traditionally considered that the mmWave massive MIMO is only feasible for static channels. However, more and more evidence shows the strong potential of the mmWave bands for mobile application. Thus, channel dynamics introduced by moving objects and transceivers are the features that the mmWave massive MIMO channel modeling cannot avoid. Usually, picocell and femtocell applications typically do not involve high-velocity users, but utilizing mmWave frequencies means that a given velocity will lead to a Doppler shift that is proportionately higher. Thinking of an order of magnitude lower peak user velocity in a small cell and an order of magnitude increase in the carrier frequency, one could expect to see peak Doppler shifts of the mmWave massive MIMO channels similar to current macrocell applications (e.g., around 200 Hz). Thus, for pilot-based channel estimation, a rough estimate is that CSI for mmWave channels would have to be updated at least as often as currently deployed MIMO systems. Of course, in a massive MIMO implementation there is more CSI to update, so the required computational rate of mmWave massive MIMO systems would then increase as well [1]. Thus, lack of insight into such dynamics inhibits the realization of the mmWave massive communications in any dynamic environment.

In the following sections of this chapter, the state-of-the-art research on the channel features offered previously will be discussed. The latest progresses on the modeling and sounding for mmWave massive MIMO channels will be introduced, respectively.

8.2 SPECIFIC CHARACTERISTICS OF mmWave MASSIVE MIMO CHANNELS

This section will introduce the specific characteristics of mmWave massive MIMO channels in detail. Corresponding efforts and the state-of-the-art of the research on these features will be reviewed and discussed in terms of the propagation mechanisms, static channel, and dynamic channel, respectively.

8.2.1 PROPAGATION MECHANISMS OF mmWave AND SUB-mmWave

Through extensive research efforts on propagation mechanisms, the following key aspects of mmWave and sub-mmWave propagation have been identified:

(1) *LOS propagation*: For the mmWave band, as found in [5], the path loss increases in its absolute value with increasing frequency, but the exponent of distance power laws does not follow a monotone increase with frequency. For the sub-mm band, the free-space propagation property of electromagnetic waves at the 300 GHz band is determined by the spreading loss and mainly the molecular absorption loss [6]. The resulting path loss is highly frequency-selective and can easily go above 100 dB for distances just above a few meters. In addition, molecular absorption by water vapor molecules defines several transmission windows, whose position and width depend on the distance and molecular composition of the medium [4].

(2) *Reflections*: Specular reflection attenuations from smooth surfaces can be modeled with the well-known Fresnel equations, given that they are sufficiently thick with respect to the wavelength. Otherwise, reflections within the material or multiple reflections at the interfaces of layered media have to be taken into account, causing a highly frequency-dependent reflection behavior [7].

(3) *Scattering*: For the reason of the wavelength ranging in the order magnitude of typical building material surface height deviations, diffuse scattering from walls covered with rough materials such as ingrain wallpaper or plaster becomes highly relevant [8,9]. The scattering of a considerable amount of power out of the specular reflection direction can cause additional reflection attenuations of several 10 dB and induce multipath propagation through the diffuse components. Such effects can be described analytically with the Kirchhoff theory [9] and have additionally been modeled stochastically in [10].

(4) *Diffraction*: Very high diffraction attenuations of easily 30 dB and more make diffraction effects in the shadowing region behind objects negligible in the mmWave band. However, diffraction could still be proven suitable and helpful to model dynamic ray shadowing effects caused by human movement, where the body itself is approximated as a diffracting object [11].

(5) Frequency dispersion: Under consideration of the huge bandwidths from 1 GHz to beyond 10 GHz, the propagation phenomena themselves have to be treated as frequency-dependent. Each frequency component in an ultra-broadband signal experiences different attenuation and delay. This frequency-dispersion effect, or equivalent distortion in the time domain, needs to be characterized in mmWave and sub-mmWave band propagation models [4,12].

Although certain progress has been made, the material and geometry of considered objects are limited. This makes the current insight insufficient to interpret the propagation mechanism constitution of mmWave or sub-mmWave in many real scenarios. More measurements on propagation mechanisms should be made to accurately understand the propagation characteristics of mmWave, which is fundamental for mmWave massive MIMO channel establishment and physical layer key techniques. For instance, utilizing the distance-dependent transmission bandwidth mechanism, we can think of a new communication scheme in which a transceiver dynamically adapts the transmitted waveform based on the transmission distance in order to match the transmission window shape [4].

8.2.2 mmWave MASSIVE MIMO STATIC CHANNEL

(1) *Frequency selective fading*: Many research teams all over the world have done extensive static mmWave channel indoor and outdoor measurements. The team at New York University (NYU) is very representative. In their outdoor channel measurements, taken at 28 and 38 GHz [13], a wide range of delay spreads were observed depending on the measurement location. Note that these measurements were carried out without spatial selectivity; in an LOS or near-LOS environment that usually the mmWave propagation requires, a massive MIMO array is able to form a narrow beam that eliminates much of the multipath, in turn considerably reducing the delay spread and potentially reducing the need for equalization. However, even in this context, because employing mmWave frequencies can allow dramatic increases in user bandwidth to hundreds of MHz or even a few GHz (and hence symbol periods on the order of 1–10 ns or less), frequency selective fading may need to be addressed through either equalization or modulation, even with a massive MIMO array [1].

(2) *Mutual orthogonality*: As introduced in the previous section, the number of independent multipath components (MPCs) at mmWave frequencies is virtually limited, so the channel vectors are not i.i.d. Rayleigh but rather correlated fading [1]. Consequently, mmWave massive MIMO does not necessarily guarantee orthogonal channels between the users, especially when the users are in close proximity. Another consideration is that the number of transmitter (Tx) antennas in reality is finite as it is bounded by the compact form of the antenna array size. Increasing the number of Tx antennas without increasing the array dimensions virtually creates the mutual antenna coupling as well. To sum up, because neither the channels are i.i.d. nor the number of Tx antennas is infinite or made very large, the mutual orthogonal condition cannot be satisfied in reality, and therefore the optimal system performance cannot be always achieved. Here, we refer to a very representative work done by the authors of [2]. They study the nonorthogonality of the mmWave massive MIMO channel under a realistic outdoor propagation environment. First, we introduce the basic knowledge of channel mutual orthogonality. Let $\mathbf{H} = [\mathbf{h}_1, \mathbf{h}_2, \ldots, \mathbf{h}_K]$ denote the $M \times K$ composite MU-MIMO channel matrix. A mutual orthogonal channel requires that every pair of column vectors of $\mathbf{H}$ satisfies

$$\mathbf{h}_i{}^H \mathbf{h}_j = 0 \quad \text{for all} \quad i \neq j \tag{8.1}$$

Under the i.i.d. Rayleigh channels assumption, condition (8.1) can be asymptotically achieved with very large Tx antenna numbers, by the law of large numbers

$$\frac{1}{M} \mathbf{h}_i{}^H \mathbf{h}_j \to 0 \quad \text{as} \quad M \to \infty \quad \text{for} \quad i \neq j$$

Denote $\kappa(\mathbf{H})$ be the condition number of $\mathbf{H}$, which is defined as the ratio between the largest singular value $\sigma_{\max}$ and smallest singular value $\sigma_{\min}$ of $\mathbf{H}$, that is,

$$\kappa(\mathbf{H}) = \frac{\sigma_{\max}}{\sigma_{\min}}$$

If all the user channel vectors are normalized to have the same norm, for example, $\|h_K\|^2 = M$, then the power imbalance due to large-scale fading between the user channels is eliminated, and the condition number of the i.i.d. MU-MIMO channels is equal to 1. Such an ideal condition provides optimal performance for MU-MIMO channels: downlink spatial streaming can achieve maximum sum rate, by using only simple linear precoding methods at Tx to decompose the channel to multiple independent spatial interference-free subchannels. However, in reality, the mutual orthogonal condition cannot be satisfied, particularly in our interested mmWave massive MIMO propagation channel, because of many practical reasons previously discussed, including spatial correlation and compact form of the Tx arrays. In the experiments [2], the spatial correlation between user channel vectors occurs when the users are close to each other, and therefore their MPCs belong to the same cluster of scatters. In other cases, spatial correlation can be introduced by common scatterers (usually with large geometry, such as walls, buildings, and so on) seen by different users even when they are not well colocated. Moreover, other factors including the number of Tx antennas, the antenna spacing and the interuser distance can affect the mutual orthogonality of **H** as well.

Examining the condition number $\kappa(\mathbf{H})$ that quantifies the spread in magnitude of the singular values of **H** is one way of measuring the amount of lack of orthogonality or spatial correlation. The smaller and closer to 1 the condition number is, the better suited the channel is to support spatial multiplexing, and the better performance of the precoder that results in higher capacity. Assuming there are only two users, the pairwise orthogonality between two channel vectors h_i and h_j can be measured by the normalized scalar product

$$\rho_{ij} = \frac{\left|h_i^H h_j\right|}{\|h_i\| \|h_j\|} = \frac{1}{M} \left|h_i^H h_j\right| \tag{8.2}$$

The smaller and closer to zero this product is, the smaller the downlink interuser interference between user i and j is created employing the MF precoder.

In [2], the authors make extensive ray-tracing simulations in an open square in downtown Helsinki, Finland. Based on these simulations, the mutual orthogonality of mmWave massive MU-MIMO channels and its impact on the spatial multiplexing capability have been studied. As shown in Fig. 8.3, the downtown square is approximately 100×100 m^2. Besides the buildings and the ground, the propagation environment is affected by lamp posts and people. The people can block the LOS to some extent (which is also called obstructed-line-of-sight) in some cases.

For the precoding matrix design, two suboptimal but more practical linear precoding methods are considered: matched filtering (MF) and zero-forcing (ZF). Fig. 8.4 shows the average achievable sum rate of two users when Tx is equipped with 10×10 uniform planar array (UPA), $D = 2\lambda$, versus the interuser distance. The total power $P = 10$ dB and the noise power

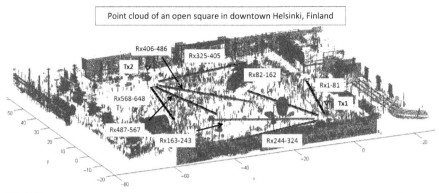

FIG. 8.3

The point cloud of an open square in downtown Helsinki, Finland, and the illustration of the simulated 2 Tx locations and 648 Rx locations, from Fig. 8.2 in [2].

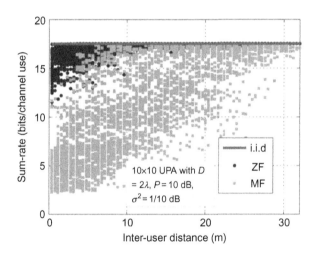

FIG. 8.4

Sum rate versus interuser distance, from Fig. 8.5 in [2].

$\sigma = 1/10$ dB [high signal-to-noise ratio (SNR) regime]. It can be found from Fig. 8.4 that in practice, the sum rate is far from the asymptotic i.i.d. sum capacity, especially when the users are close to each other. Specifically, the larger the distance between the two users, due to the decreasing of the correlation, the closer the sum rate of the MF precoder to that of the ZF precoder.

Now, let us extend the situation from two users to more users. When $K \geq 2$, each Tx with a 10×10 UPA with $D = 0.5\lambda$ chooses to serve $4K_r$ receivers (Rxs) from four routes (K_r Rx locations with a maximum separation on each route). Fig. 8.5 illustrates the cumulative distribution function (CDF) of the condition number of $\mathbf{H}$ for $K = \{4, 8, 12, 16\}$ users. It can be found that the four-user channel is well conditioned with a very small condition number, which means

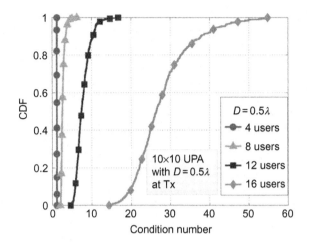

FIG. 8.5

CDF of condition number of **H** for different number of users, from Fig. 8.6 in [2].

that the channel is capable of supporting four parallel paths for spatial multiplexing. However, the channel becomes more and more ill-conditioned as K increases. In Fig. 8.6, because the channel matrix is nearly orthogonal, the sum rate with four users at average total transmission power $P=5$ dB and $\sigma^2=1$, using MF is close to that using ZF and of i.i.d. case. At $K=16$, both ZF and MF precoders have much smaller sum capacity compared with the i.i.d. case because the channel in this case is ill-conditioned indicated by large condition number $\kappa(\mathbf{H})$. In this low SNR regime, MF performs poorly as the signal energy at each user is low.

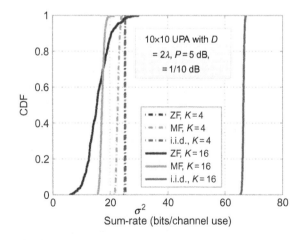

FIG. 8.6

CDF of sum rate for different number of users, from Fig. 8.7 in [2].

The above results obtained from [2] invalidate the i.i.d. channels assumption as well as the asymptotic mutual orthogonal channels of mmWave massive MU-MIMO in practical cases. Thus we can conclude that in real scenarios, the mmWave massive MIMO system capacity highly depends on many practical factors, such as the channel mutual orthogonality, the interuser distance, number of active users, antenna array dimensions, and so on. In order to achieve the sum rate that is closer to the ideal condition, the users in an mmWave system should be fewer than in microwave systems. Moreover, the users should be well separated, in a range of at least two times of correlation distance of shadowing, for an efficient spatial multiplexing with linear processing precoders.

(3) *Spherical wavefront assumption*: The plane wavefront assumption for conventional MIMO channel models does not hold for massive MIMO channels because of the very large antenna arrays. Thus in the latest studies, for instance, in [14], many efforts on the impacts of spherical wavefront assumption on both the LOS component and nonline-of-sight (NLOS) components have been studied. For mmWave massive MIMO channel models, the far-field assumption for conventional MIMO channel models, which is equivalent to the plane wavefront assumption, is not fulfilled for two reasons: first, the very short wavelength considerably lengthens the far-field distance; second, the dimension of the antenna array cannot be ignored when the number of antennas is massive. As shown in Fig. 8.2, the wavefront emitted either from scatter/cluster of scatters to the receive array is assumed to be spherical. As a result, the angle of arrival (AoA) is no longer linear along the array, and it needs to be computed based on geometrical relationships. As shown in Fig. 8.7 (which is from Fig. 8.9 in [14]), where for a 32-element antenna array there are in total 30 window positions, it can be observed that several estimated AoAs gradually differ along the massive MIMO antenna array axis. This phenomenon is mainly caused by the spherical wavefront assumption when the distances between their corresponding scatterers/clusters of scatterers and the antenna array do not meet the far-field assumption.

This point can be verified by some other channel parameters as well. For instance, in [14], the researchers found that the correlation function of massive MIMO channels does not only depend on the absolute difference between antenna indices, but also on the indexes of reference antennas. This virtually means that the wide-sense stationary properties on the antenna array axis are no longer valid under the spherical wavefront assumption. As shown in Fig. 8.8, the absolute space cross-correlation function (CCF) of the massive MIMO channel model under spherical assumption gradually drops when the normalized antenna spacing increases. This is entirely different from the traditional MIMO channels. Thus, getting a deeper insight into utilizing the spherical frontwave to replace the frontwave in the mmWave massive MIMO channel modeling is of great importance.

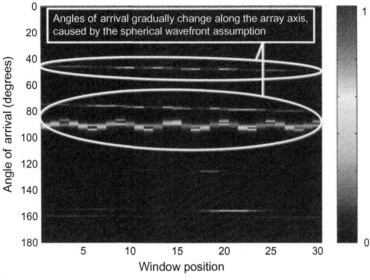

FIG. 8.7

A snapshot example of the normalized angle power spectrum of angle of arrival of the wideband ellipse model, from Fig. 8.9 in [14].

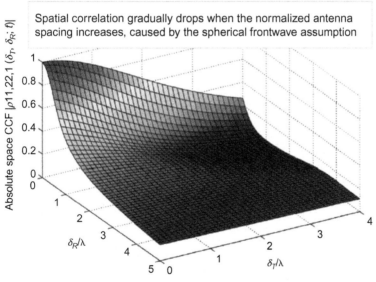

FIG. 8.8

Absolute space CCF of the ellipse model, from Fig. 8.6 in [14].

8.2.3 mmWAVE MASSIVE MIMO DYNAMIC CHANNELS

Most existing measurements were done for the static radio channel. In 2014, Samsung presented the world's first demonstration for 1.2 Gbps transmission at 28 GHz at 110 km/h cruising speed using dual-beamforming. This shows the strong potential of the mmWave band for mobile application, but the understanding of channel dynamics—spatial-temporal Doppler properties—is still absent. The lack of insight into such dynamics inhibits the realization of the communications using the mmWave band in any dynamic environment. Thus, dynamic channel measurements shall be designed and performed with various configurations, such as different moving transceivers and scatterers with variations of speed and motion direction. Similarly, dynamic simulations shall be made by adding the same mobility to the corresponding scenarios. Based on both the measurement and simulation results, the correlations between the mobility and channel dynamics will be revealed. For instance, it is critical to reveal the law of how the mobility and the variation of mobility influence the first- and second-order statistics of the channel dynamic parameters for mmWave massive MIMO channels.

8.3 STATE-OF-THE-ART OF MILLIMETER-WAVE MASSIVE MIMO CHANNEL STUDY

This section will introduce the state-of-the-art of mmWave massive MIMO channel study in terms of channel sounding and channel modeling, respectively, in detail.

8.3.1 MILLIMETER-WAVE MASSIVE MIMO CHANNEL MODELING

Here, the modeling for massive MIMO channels will be established by recommending a nonstationary wideband multiconfocal ellipse two-dimensional channel model, in which an assumption of spherical wavefront and birth-death process are employed [15]. This channel model is developed to acquire the effects of the spherical wavefront and the channel properties of cluster evolution on both array axis and time axis.

The assumption of spherical wavefront and birth-death process

The conventional MIMO channel models simply assume that the Rayleigh distance is far less than the distance between a scatterer and an antenna array; that is, the assumption of plane wavefront was generally employed to simplifying the MIMO channel model [16,17]. However, the researchers in [18,19] figure that the assumption of plane wavefront used in conventional MIMO channel models will not satisfactorily model massive MIMO channels while the number of antenna elements is massive. Moreover, the authors in [20] state that the plane wavefront assumption will underestimate the rank of the channel matrix [15]. Hence, it is imperative to consider a spherical wavefront channel model. Therefore the wavefront that is radiated from clusters will be received in Rx array with spherical assumption.

Then, the birth-death process (or Markov processes) is applied on both the array and time axis for the cluster evolution algorithm [15,21]. On the array axis, as studied in [18], the birth-death process is employed to capture the cluster appearance and disappearance on the array axis of antenna elements. On the time axis, the wideband massive MIMO channel model can be further developed to capture channel dynamic properties by combining the birth-death process for modeling cluster time evolution in [21].

The wideband ellipse model and its geometrical relationships for massive MIMO systems

Here, for modeling massive MIMO channels, a novel wideband ellipse two-dimensional channel model is investigated [15]. Referring to Fig. 8.9 (which is from Fig. 8.1 in [15]), the major assumptions and key parameter definitions are set as follows for illustrating the geometrical relationships of the model:

(1) Near-field effects of spherical wavefront are considered on both line-of-sight (LOS) component and NLOS components.
(2) The antenna arrays on Tx and Rx are with uniform linear distribution. There are numbers of M_T and M_R omnidirectional antenna elements on Tx and Rx arrays, respectively.

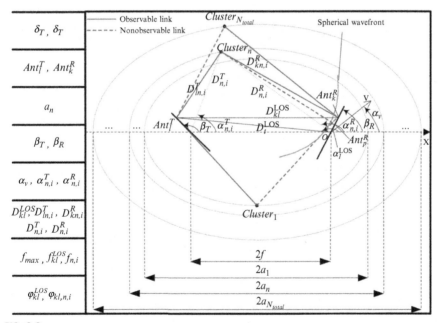

FIG. 8.9

The geometrical relationships of the proposed wideband multiconfocal ellipse model with key parameter definitions.

(3) Initial phase of the signal at Tx is φ_0; LOS component is always linked to each antenna element. Rician factor is K.

(4) Mean power of the nth cluster is P_n and there are totally S rays emitted within one ellipse of the cluster. The theoretical model generally set parameter S incline to infinite.

(5) δ_T (δ_R) is the separated space for adjacent antenna elements of the array on Tx (Rx). The array centers of Tx and Rx are located at the relative focal points of the ellipses with $2f$ distance.

(6) Ant_l^T (Ant_k^R) represent the lth (kth) antenna elements of array at Tx (Rx). Note the relative relationship: $1 \leq l \leq M_T$ ($1 \leq k \leq M_k$).

(7) The nth cluster is on the nth ellipse with major axis $2\,a_n$, that is, a_n is the semimajor axis of nth ellipse. $2f$ is the distance between the Tx and Rx (or focal length).

(8) The angle β_T (β_R) depicts the tilt angle of Tx (Rx) antenna array.

(9) α_v, $\alpha_{n,i}^T$, and $\alpha_{n,i}^R$ denote the parameters for angular dependence. α_v denotes the direction of the movement of the Rx (e.g., MS), which is the angle compared to the x-axis. $\alpha_{n,i}^T$ is the angle of departure (AOD) of the ith multicomponent of nth cluster to the Tx array center ($i = 1, 2, ..., S$); $\alpha_{n,i}^R$ is the AOA of the ith multicomponent of nth cluster to the Rx array center ($i = 1, 2, ..., S$). Note that the α_l^{LOS} refers to the LOS component from Ant_l^T to the Rx array center.

(10) D_{kl}^{LOS}, D_l^{LOS}, $D_{ln,i}^T$, $D_{kn,i}^R$, $D_{n,i}^T$, and $D_{n,i}^R$ denote the distances in space. D_{kl}^{LOS} is the LOS distance between relative lth and kth antenna elements on Tx and Rx arrays; the D_l^{LOS} refers to the LOS component from Ant_l^T to the Rx array center; $D_{ln,i}^T$ ($D_{kn,i}^R$) is the distance between nth cluster and Ant_l^T (Ant_k^R) antenna element on Tx (Rx) via ith ray; $D_{n,i}^T$ ($D_{n,i}^R$) is the distance between nth cluster and the Tx (Rx) via ith ray.

(11) f_{max}, f_{kl}^{LOS}, and $f_{n,i}$ denote the parameters for Doppler frequency. f_{max} is the maximum Doppler frequency; f_{kl}^{LOS} is the Doppler frequency of the LOS component between relative lth and kth antenna elements on Tx and Rx arrays; $f_{n,i}$ is the Doppler frequency of the ith ray within nth cluster.

(12) φ_{kl}^{LOS} and $\varphi_{kl,n,i}$ are the phase of the LOS component between relative lth and kth antenna elements on Tx and Rx arrays and the phase of the ith ray within nth cluster between relative lth and kth antenna elements on Tx and Rx arrays.

For massive MIMO channel modeling, the geometrical relationships should be investigated in detail (refer to Fig. 8.9). Furthermore, the nonstationary behavior of the channel modeling will depend on the nonstationary behavior of clusters on the array axis [18,19]; that is, different clusters may be observable to different antenna elements. The examples are illustrated in Fig. 8.9. The *cluster*$_1$ is observable to Ant_l^T while it is nonobservable to Ant_k^R; *cluster*$_{N_{total}}$ is nonobservable to Ant_l^T while it is observable to Ant_k^R; *cluster*$_n$ is observable to both Ant_l^T and Ant_k^R while it is not observable to Ant_p^R.

Let C_l^T (C_k^R) denote the cluster set in which clusters are observable to Ant_l^T (Ant_k^R). These parameters are of great importance and will be modeled as a birth-death process in the following sections. The total number of clusters N_{total} can be defined as [22]

$$N_{total} = card\left(\bigcup_{l=1}^{M_T}\bigcup_{k=1}^{M_R}\left(C_l^T(t)\cap C_k^R(t)\right)\right)$$
(8.3)

where $card(\cdot)$ denotes the cardinality of a set. Eq. (8.3) depicts that the total number of clusters will be observable to at least one antenna element on Tx or Rx. Moreover, an arbitrary cluster is observable to both antenna elements on Tx and Rx if and only if this cluster is a subset of set $\{C_l^T\cap C_k^R\}$.

Based on the previous definitions and analyses, the final massive MIMO channel can be expressed as a complex matrix $\mathbf{H}(t,\tau) = [h_{kl}(t,\tau)]_{M_R\times M_T}$, where $k = 1,2,\dots,M_R$ and $l = 1,2,\dots,M_T$. The multipath complex gains $h_{kl}(t,\tau)$ between relative lth and kth antenna elements of the array on Tx and Rx can be calculated with a delay τ as

$$h_{kl}(t,\tau) = \sum_{n=1}^{N_{total}} h_{kl,n}(t)\delta(\tau-\tau_n)$$
(8.4)

where $h_{kl,n}(t)$ is the complex gain, which is contributed by nth cluster ($Cluster_n$):

$$h_{kl,n}(t,\tau) = \delta(n-1)\sqrt{\frac{K}{K+1}}e^{j\left(2\pi f_{kl}^{LOS}\cdot t+\varphi_{kl}^{LOS}\right)} + \sqrt{\frac{P_n}{K+1}}\lim_{S\to\infty}\frac{1}{\sqrt{S}}\sum_{i=1}^{S}e^{j\left(2\pi f_{n,i}\cdot t+\varphi_{kl,n,i}\right)}$$
(8.5)

where the forward represent the LOS component (when n is equal to 1) while the backward represent NLOS components. Eq. (8.5) can only be established when $Cluster_n \in \{C_l^T\cap C_k^R\}$. In contrast, if $Cluster_n \notin \{C_l^T\cap C_k^R\}$, the $h_{kl,n}(t) = 0$.

As an assumption of spherical wavefront and relative geometrical relationships shown in Fig. 8.9, the phase of the LOS component φ_{kl}^{LOS} can be counted as

$$\varphi_{kl}^{LOS} = \varphi_0 + \frac{2\pi}{\lambda}\left[\left(D_l^{LOS}\right)^2 + \left(\frac{M_R-2k+1}{2}\delta_R\right)^2 - (M_R-2k+1)\delta_R D_l^{LOS}\cos\left(\alpha_l^{LOS}-\beta_R\right)\right]^{1/2}$$
(8.6)

where the λ is the wavelength and D_l^{LOS}:

$$D_l^{LOS} = \left[(2f)^2 + \left(\frac{M_T-2l+1}{2}\delta_T\right)^2 - 2f(M_T-2l+1)\delta_T\cos\left(\beta_T\right)\right]^{1/2}$$
(8.7)

The AOA of the LOS component α_l^{LOS} can be computed as

$$\alpha_l^{LOS} = \pi - \arcsin\left[\frac{\sin\beta_T}{D_l^{LOS}}\left(\frac{M_T-2l+1}{2}\delta_T\right)\right]$$
(8.8)

Simultaneously, the Doppler frequency of the LOS component f_{kl}^{LOS} can be computed as

$$f_{kl}^{LOS} = f_{max}\cos\left(\pi - \arcsin\left[\frac{D_l^{LOS}}{D_{kl}^{LOS}}\sin\left(\alpha_l^{LOS}-\beta_R\right)\right]\right)$$
(8.9)

where

$$D_{kl}^{LOS} = \left[\left(D_l^{LOS}\right)^2 + \left(\frac{M_R - 2k + 1}{2}\delta_R\right)^2 - (M_R - 2k + 1)\delta_R D_l^{LOS} \cos\left(\alpha_l^{LOS} - \beta_R\right)\right]^{1/2} \quad (8.10)$$

For the NLOS components, the relationships for AOAs $\alpha_{n,i}^R$ and AODs $\alpha_{n,i}^T$ are computed within an ellipse model:

$$\alpha_{n,i}^T = \begin{cases} \Phi\left(\alpha_{n,i}^R\right) & \text{if } 0 < \alpha_{n,i}^R \leq \alpha_0 \\ \Phi\left(\alpha_{n,i}^R\right) + \pi & \text{if } \alpha_0 < \alpha_{n,i}^R \leq 2\pi - \alpha_0 \\ \Phi\left(\alpha_{n,i}^R\right) + 2\pi & \text{if } 2\pi - \alpha_0 < \alpha_{n,i}^R \leq 2\pi \end{cases} \quad (8.11)$$

where

$$\Phi\left(\alpha_{n,i}^R\right) = \arctan\left[\frac{\left(k_0^2 - 1\right)\sin\left(\alpha_{n,i}^R\right)}{2k_0 + \left(k_0^2 + 1\right)\cos\left(\alpha_{n,i}^R\right)}\right] \quad (8.12)$$

$$\alpha_0 = \pi - \arctan\left(\frac{k_0^2 - 1}{2k_0}\right) \quad (8.13)$$

$$k_0 = a_n \Big/ f \quad (8.14)$$

The a_n of nth cluster can be determined by excess delay τ_n relative to the first cluster semimajor axis a_1 as

$$a_n = c\tau_n + a_1 \quad (8.15)$$

where c is the speed of light.

According to the geometrical relationships shown in Fig. 8.9, the parameters $D_{n,i}^T$ and $D_{n,i}^R$ can be expressed accordingly as well as parameters $D_{ln,i}^T$ and $D_{kn,i}^R$:

$$D_{n,i}^T = \frac{2a_n \sin\left(\pi - \alpha_{n,i}^R\right)}{\sin\alpha_{n,i}^T + \sin\left(\pi - \alpha_{n,i}^R\right)} \quad (8.16)$$

$$D_{n,i}^R = \frac{2a_n \sin\alpha_{n,i}^T}{\sin\alpha_{n,i}^T + \sin\left(\pi - \alpha_{n,i}^R\right)} \quad (8.17)$$

$$D_{ln,i}^T = \left[\left(D_{n,i}^T\right)^2 + \left(\frac{M_T - 2l + 1}{2}\delta_T\right)^2 - (M_T - 2l + 1)\delta_T D_{n,i}^T \cos\left(\beta_T - \alpha_{n,i}^T\right)\right]^{1/2} \quad (8.18)$$

$$D_{kn,i}^R = \left[\left(D_{n,i}^R\right)^2 + \left(\frac{M_R - 2k + 1}{2}\delta_R\right)^2 - (M_R - 2k + 1)\delta_R D_{n,i}^R \cos\left(\alpha_{n,i}^R - \beta_R\right)\right]^{1/2} \quad (8.19)$$

The received phase of Ant_k^R from Ant_l^T via ith ray from the nth cluster can be computed as

$$\varphi_{kl,n,i} = \varphi_0 + \frac{2\pi}{\lambda}\left(D^T_{ln,i} + D^R_{kn,i}\right) \tag{8.20}$$

Finally, the Doppler frequency of the ith ray within the nth cluster is

$$f_{n,i} = f_{max} \cos\left(\alpha^R_{n,i} - \alpha_v\right) \tag{8.21}$$

So far, Eqs. (8.6)–(8.21) list all the geometrical relationships that will be used in the massive MIMO channel model. In the next section, the dynamic properties of clusters, C^T_l and C^R_k, will be studied by a birth-death process on the antenna array axis; that is, the appearance or disappearance of one cluster for an antenna element will follow statistical law.

Cluster evolution on array axis for the wideband ellipse model

For the massive MIMO channel modeling, Refs. [15,22] adapt the birth-death process into an antenna array axis while the process is originally used on a time axis [21].

First, the key parameter definitions are set as follows for generation of cluster sets C^T_l an C^R_k:

1. λ_G (per meter) and λ_R (per meter) represent the generation rate and the recombination rate of cluster, respectively.
2. D^a_c is defined as a scenario-dependent correlation factor on the array axis [15].
3. $\xrightarrow{E}$ is the operator that represents the evolution on both array and time axis. When the cluster set is going to evolve from previous antenna element Ant^T_{l-1} to Ant^T_l in the Tx or Ant^R_{k-1} to Ant^R_k in the Rx, this process can be described by using the operator

$$C^T_{l-1} \xrightarrow{\quad E \quad} C^T_l \quad \text{or} \quad C^R_{k-1} \xrightarrow{\quad E \quad} C^R_k$$

4. For each time the cluster set evolves, the survival probabilities of the cluster in the cluster set determine the evolution of the cluster set. The survival probabilities of the clusters at Tx and Rx are defined as $P^T_{survival}$ and $P^R_{survival}$, respectively [23]:

$$P^T_{survival} = e^{-\lambda_R \frac{\delta_T}{D^a_c}} \tag{8.22}$$

$$P^R_{survival} = e^{-\lambda_R \frac{\delta_R}{D^a_c}} \tag{8.23}$$

5. The duration between two cluster appearances and the duration between two cluster disappearances follow the exponential distribution as the birth-death process [23].
6. The number of newly generated clusters is modeled as a Poisson process with the expectation that can be expressed as [23]

$$E\left[N_{new}^T\right] = \frac{\lambda_G}{\lambda_R}\left(1 - e^{-\frac{\delta_L}{D_c^a}}\right) \tag{8.24}$$

$$E\left[N_{new}^R\right] = \frac{\lambda_G}{\lambda_R}\left(1 - e^{-\frac{\delta_R}{D_c^a}}\right) \tag{8.25}$$

where the $E[\cdot]$ denotes the expectation evaluating.

Then, Fig. 8.10 (which is from Fig. 8.2 in [15]) gives the flowchart of the cluster generation algorithm. Owing to the process of cluster generation for Tx equivalent to those of Rx, we are concerned with the process in Tx only. First, N is the initial number of the cluster for Ant_1^T and the $C_1^T = \{c_1^T, c_2^T, c_x^T, ..., c_N^T\}$, where c_x^T denotes the cluster member of the cluster set for Ant_1^T. Accordingly, the cluster sets of the rest of the antenna elements in the Tx array will be generated from the evolution of C_1^T; see Fig. 8.10. In the process of cluster evolution from C_{l-1}^T to C_l^T, the members of C_{l-1}^T will exist with survival probability. Meanwhile, C_l^T will be supplemented with newly generated cluster members based on a Poisson process. The antenna correlations that have been embedded into the cluster generation process (see Eqs. 8.22 and 8.23) will issue that one cluster will not appear again after its disappearance to ensure the gradual evolution of the cluster.

Then, the work of randomly shuffling and pairing clusters will mimic the arbitrariness of the propagation environment after the generation of the cluster sets for each antenna element. $\cup_{l=1}^{M_T} C_l^T$ and $\cup_{l=1}^{M_R} C_k^R$ are the full collections of clusters that are observable to the Tx and Rx array, respectively. The cluster indexes in $\cup_{l=1}^{M_T} C_l^T$ and $\cup_{l=1}^{M_R} C_k^R$ will be shuffled and arranged in a random order; see Fig. 8.11. Then, the cluster indexes of Tx and Rx are paired for determining that each cluster will be observable for at least one antenna element in Tx and one antenna element in Rx. For example, as illustrated in Fig. 8.11, a common cluster will be shared between the Tx antenna element that can observe cluster index 6 (c_6^T) and the Rx antenna element that can observe cluster index 5 (c_5^R). Every cluster will be determined in this way.

Generally, it is inevitable for cluster pairing that clusters are not observable to either Tx array or Rx array; that is, the cardinalities of $\cup_{l=1}^{M_T} C_l^T$ and $\cup_{l=1}^{M_R} C_k^R$ are not equal. In this case, the unnecessary clusters will be eliminated from the full Rx cluster set after pairing (example is in Fig. 8.11). Then, the remaining clusters will be reassigned by new cluster indexes from 1 to N_{total}.

The finishing touches for a complete wideband ellipse model are the generations of ray properties from clusters, of which the delays and mean power randomly obey the exponential distribution [24] and AOAs follow von Mises distribution [25]. The delay is prior property to be obtained, as the semimajor axis of the ellipse will be computed via Eq. (8.15) accordingly. Ref. [24] indicates the relationship between mean power and delay of the ray. Then, the mean power is related to the size of its corresponding ellipse directly.

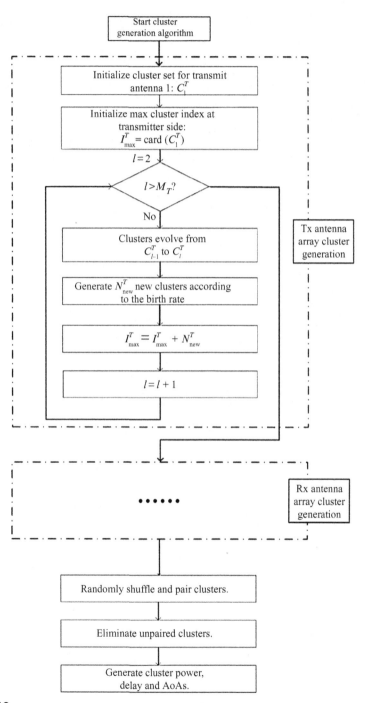

FIG. 8.10

Flowchart for cluster generation.

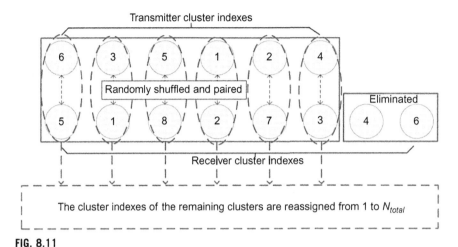

FIG. 8.11

An example of random shuffling and pairing for Tx and Rx cluster indexes.

To model the process of cluster evolution on both array and time axis, the recommended model can be further enhanced based on the birth-death process used on the time axis [15,21].

Cluster evolution on time axis for array-time evolution model

Basically, there are two issues that should be noticed for the moving channel modeling. One is that the clusters and the Rx are moving; the other is that the clusters, Tx and Rx are all moving. Considering that the latter can be equivalent to the former with the principles of relative motion, the former will be studied in detail [15].

The time-variant channel modeling will lie on the changing of the clusters and Rx. Accordingly, the previous defined parameters, α_n^T, α_n^R, D_n^T, D_n^R, D_{ln}^T, D_{kn}^R, a_n, f, α_v, and τ_n will further evolve into $\alpha_n^T(t)$, $\alpha_n^R(t)$, $D_n^T(t)$, $D_n^R(t)$, $D_{ln}^T(t)$, $D_{kn}^R(t)$, $a_n(t)$, $f(t)$, $\alpha_v(t)$, and $\tau_n(t)$ from initial time instance $t=t_m$ to $t=t_{m+1}$ $(m=1,2,...)$. The v is the speed of moving Rx at $t=t_m$. Moreover, the $Cluster_n$ $(n=1,2,...,N)$ moves in the direction $\varphi_{c,n}$ with speed of v_c at $t=t_m$. Referring to Fig. 8.12 (which is from Fig. 8.4 in [15]), in the next time instance $t=t_{m+1}$, the distance between two focal points of ellipse $2f(t_{m+1})$ can be computed as

$$2f(t_{m+1}) = \left[[2f(t_m)]^2 + [v(t_{m+1}-t_m)]^2 + 4f(t_m)v(t_{m+1}-t_m)\cos[\alpha_v(t_m)] \right]^{1/2} \tag{8.26}$$

Simultaneously, for the relationships between Tx and $Cluster_n$, the distance between the Tx and $Cluster_n$ $D_n^T(t_m)$ needs to be updated to $D_n^T(t_{m+1})$ [15]

$$D_n^T(t_{m+1}) = \left[[D_n^T(t_m)]^2 + v_c^2(t_{m+1}-t_m)^2 + 2D_n^T(t_m)v_c(t_{m+1}-t_m)\cos\left(\alpha_n^T(t_m)-\varphi_{c,n}\right) \right]^{1/2} \tag{8.27}$$

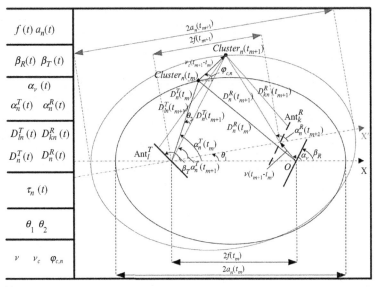

FIG. 8.12

The geometrical relationships of the array-time evolution model with evolved parameters.

Following the relationships in the instant ellipse of $Cluster_n$ at $t = t_{m+1}$, the $\alpha_n^T(t_{m+1})$ and $D_{ln}^T(t_{m+1})$ can be computed as

$$\alpha_n^T(t_{m+1}) = \alpha_n^T(t_m) - \theta_1 - \theta_2 \tag{8.28}$$

$$\theta_1 = \arcsin\left[\frac{v(t_{m+1} - t_m)\sin(\pi - \alpha_v(t_m))}{2f(t_{m+1})}\right] \tag{8.29}$$

$$\theta_2 = \arcsin\left[\frac{v_{c,n}(t_{m+1} - t_m)\sin(\pi - \alpha_n^T(t_m) + \varphi_{c,n})}{D_n^T(t_{m+1})}\right] \tag{8.30}$$

$$D_{ln}^T(t_{m+1}) = \left[D_n^T(t_{m+1})^2 + \left(\frac{M_T - 2l + 1}{2}\delta_T\right)^2 - D_n^T(t_{m+1})(M_T - 2l + 1)\delta_T \cos\left(\beta_T - \alpha_n^T(t_{m+1})\right)\right]^{1/2} \tag{8.31}$$

Similarly, the parameters related to Rx and $Cluster_n$ can be computed as [15]

$$D_n^R(t_{m+1}) = \left[[2f(t_{m+1})]^2 + [D_n^T(t_{m+1})]^2 - 4f(t_{m+1})D_n^T(t_{m+1})\cos\left(\alpha_n^T(t_{m+1})\right)\right]^{1/2} \tag{8.32}$$

$$\alpha_n^R(t_{m+1}) = \arcsin\left[\frac{D_n^T(t_{m+1})\sin\left[\alpha_n^T(t_{m+1})\right]}{D_n^R(t_{m+1})}\right] \tag{8.33}$$

$$D_{kn}^R(t_{m+1}) = \left[D_n^R(t_{m+1})^2 + \left(\frac{M_R - 2k + 1}{2} \delta_R \right)^2 \right.$$
$$\left. - D_n^R(t_{m+1})(M_R - 2k + 1)\delta_R \cos \left(\alpha_n^R(t_{m+1}) - \beta_R + \theta_1 \right) \right]^{1/2} \qquad (8.34)$$

Furthermore, other necessary parameters for modeling, for example, Doppler frequency, AOA, and so on, will be computed as follows:

$$2a_n(t_{m+1}) = D_n^T(t_{m+1}) + D_n^R(t_{m+1}) \qquad (8.35)$$

$$\tau_n(t_{m+1}) = \tau_n(t_m) + \frac{2a_n(t_{m+1}) - 2a_n(t_m)}{c} \qquad (8.36)$$

$$\alpha_v(t_{m+1}) = \alpha(t_m) - \theta_1 \qquad (8.37)$$

$$f_n\left(\alpha_n^R(t_{m+1})\right) = f_{\max} \cos \left(\alpha_n^R(t_{m+1}) - \alpha_v(t_{m+1}) \right) \qquad (8.38)$$

These relationships cover all the geometrical relationships in the array-time evolution model. The remaining question is how clusters evolve on the time axis. In Ref. [21], the evolutions of time-variant $C_l^T(t)$, $C_k^R(t)$, and $N_{new}(t)$ follow the birth-death process on the time axis. From time instance $t = t_m$ to $t = t_{m+1}$, the cluster sets will reserve the surviving cluster member and eliminate the disappeared cluster member. Certain cluster sets will be supplemented with newly generated cluster members [15,21].

8.3.2 MILLIMETER-WAVE MASSIVE MIMO CHANNEL SOUNDING

In this section, several efficient sounding techniques will be discussed based on recently arisen sounding systems.

Channel sounding techniques are developed for studying wireless channel properties. The channel sounder is the instrument that is dedicated to wireless channel sounding. Taking radar as an example, the detective radio suffering different attenuations will help detect potential targets. Obviously, the major difference between channel sounding and radar is whether bistatic or monostatic properties of the transceiver are used in the system. The final outputs of channel sounding can be channel impulse response (CIR) or channel properties (e.g., angular properties).

The previous works in mmWave channel sounding, especially for MIMO channel sounding, may be limited to measurement environments (indoor or outdoor, static or dynamic) or system requirements (bandwidth, single-input single-output (SISO) or MIMO, signal processing, etc.). It has reached a consensus that different channel sounders are dedicated to limited measurements. For example, Refs. [26,27] present the mmWave channel sounder based on a sliding correlation approach and rotation of horn antenna to detect properties of angular distribution and propagation loss. But the sounder can only measure static channels or very low dynamic channels. Hence, the requirements of measurement and technical challenges of hardware or software are the prior issues to determine.

First, as an appropriate channel sounder should be determined depending on the measurement requirements, the following key factors of sounding approach and system architecture should be considered before developing a channel sounder:

1. Frequency and bandwidth of the channel measurement
2. SISO channel detecting or MIMO channel detecting
3. Static channel or dynamic channel, that is, the measurement speed for the sounding system
4. Approach for sounding signal processing
5. System cost and its extension flexibility

Second, the stabilized performances for mmWave radio frequency (RF) components are the challenges for massive MIMO channel sounding, so the system should give consideration to mmWave frequency band (up to several tens of GHz), ultra-broad wideband (up to several GHz), and massive MIMO (even up to a hundred antennas in both Tx and Rx). The related components of a high-performance RF should include an up-and-down-convertor, suitable antenna array, power amplifier, low-noise amplifier, band-pass filter, and RF switch for mmWave massive MIMO channel sounding.

Third, signal generation, signal acquisition and effective data storage are the key challenges for integrated channel measurement. The processes of signal generation and acquisition make a request for high performance with deep bit depth wideband digital-to-analog convertors (DACs) and analog-to-digital convertors (ADCs). Then, efficient data storage and streaming are of great importance for long time measurement, especially for mmWave massive MIMO measurement. A simple example is that 1 GHz bandwidth signal transmitting in a SISO system will demand 1 G sample rates. When converting sample rate into data rate, the data rate will be multiplied proportional to the employment of a massive MIMO system and a deeper bit depth of DAC and ADC.

Fourth, the sounding system for an mmWave massive MIMO system needs accurate synchronization and calibration to meet the growing requirements of higher path delay resolution, more sensitivity in interchannel delay measurement, and nonideal amplitude/phase.

Efficient sounding techniques for mmWave massive MIMO channel sounding aim to answer the aforementioned requirements and challenges. In Ref. [28], a wideband signal correlation approach is employed instead of the commonly used sliding correlator and frequency swept approaches because the wideband signal correlation approach provides much faster processing speed that can possibly support extracting for CIR from raw data simultaneously [28]. Furthermore, the wideband signal correlation approach receives a complex signal (including amplitude and phase information) that can speculate angular distribution in a MIMO system with an advanced angle estimation algorithm (e.g., space-alternative generalized expectation-maximization [29,30]). These advances make the wideband signal correlation approach seductive in mmWave massive MIMO sounding.

Then, to minimize cross-interference among massive MIMO array elements and keep a faster measurement speed, a method using a switch at Tx and parallel acquisition at Rx is adopted [28]. Moreover, this method avoids the extra time consumption in using a set of orthogonal codes and interference cancellation techniques. The switch is controlled by high-speed electrical hardware which allows the switch shifting rapidly.

It is noteworthy that in Ref. [28], with the huge amount of raw data in Rx, the author presents a new approach that can efficiently store data in embedded memory or transfer the data stream to a disk. The key ideal to resolve the problem is utilizing an embedded field-programmable gate array in the Rx to directly process real-time correlation to extract effective CIR information from raw data with a far smaller data size. This process helps embedded memory store CIR data for a longer time or in real time to a disk for data streaming.

Meanwhile, the sounding system proposed in Ref. [28] uses high-performance rubidium clocks for system synchronization and calibration.

Furthermore, Ref. [31] develops a dual-polarized ultra-wideband multichannel-sounder that is competent for recording dynamic effects while providing parallel fully polarimetric measurements via multiantennas at mmWave frequencies. The system combines fast signal acquisition speed with a high dynamic range of multiparallel channels. The maximum signal bandwidth is up to 7 GHz and the M-sequence principle is employed for signal generation.

8.4 CONCLUSION

In this chapter, we first point out the specific features of the mmWave massive channels in terms of the propagation mechanisms, static channel, and dynamic channel, respectively. For the propagation mechanisms, the high path loss and near-LOS propagation, frequency dispersion, and distance-dependent transmission are the most distinguished features. For the static channel, frequency selective fading, mutual orthogonality, and spherical frontwave assumption are the key issues that the mmWave massive MIMO channel modeling should address. For the dynamic channel, until now there has been no widely acknowledged measurement data available. But, many efforts have been made toward the characterization of mobile mmWave massive MIMO channels in dynamic environments.

In the latter half of this section, based on the latest, most representative studies, the state-of-the-art of mmWave massive MIMO channel modeling and sounding are discussed in detail, respectively. To sum up, until now there has been no official or mature mmWave massive MIMO channel model; however, this is one of the most interesting topics on which many researchers have been making efforts toward B5G. Very different propagations and channels will result in very distinguished design considerations on the mmWave massive MIMO systems, such as antenna and RF transceiver architecture, channel estimation, interference mitigation, precoding, and multicell cooperation.

REFERENCES

[1] A.L. Swindlehurst, E. Ayanoglu, P. Heydari, F. Capolino, Millimeter-wave massive MIMO: the next wireless revolution? IEEE Commun. Mag. 52 (9) (2014) 56–62.

[2] S.L.H. Nguyen, K. Haneda, J. Jarvelainen, A. Karttunen, J. Putkonen, On the mutual orthogonality of millimeter-wave massive MIMO channels, in: 2015 IEEE 81st Vehicular Technology Conference (VTC Spring), 11–14 May 2015, 2015, pp. 1–5.

[3] T.L. Marzetta, Noncooperative cellular wireless with unlimited numbers of base station antennas, IEEE Trans. Wirel. Commun. 9 (11) (2010) 3590–3600.

[4] I. Akyildiz, J. Jornet, C. Han, TeraNets: ultra-broadband communication networks in the terahertz band, IEEE Wirel. Commun. 21 (4) (2014) 130–135.

[5] D. Cassioli, L.A. Annoni, S. Piersanti, Characterization of path loss and delay spread of 60-GHz UWB channels vs. frequency, in: 2013 IEEE International Conference on Communications (ICC), 2013, pp. 5153–5157.

[6] J.M. Jornet, I.F. Akyildiz, Channel modeling and capacity analysis for electromagnetic wireless nanonetworks in the terahertz band, IEEE Trans. Wirel. Commun. 10 (10) (2011) 3211–3221.

[7] T. Kürner, S. Priebe, Towards THz communications — status in research, standardization and regulation, J. Infrared Millim. Te. 35 (1) (2013) 53–62, http://dx.doi.org/10.1007/s10762-013-0014-3. Electronic publication.

[8] R. Piesiewicz, C. Jansen, D. Mittleman, T. Kleine-Ostmann, M. Koch, T. Kürner, Scattering analysis for the modeling of THz communication systems, IEEE Trans. Antennas Propag. 55 (11) (2007) 3002–3009.

[9] C. Jansen, S. Priebe, C. Möller, M. Jacob, H. Dierke, M. Koch, T. Kürner, Diffuse scattering from rough surfaces in THz communication channels, IEEE Trans. Terahertz Sci. Technol. 1 (2) (2011) 462–472.

[10] S. Priebe, M. Jacob, T. Kürner, AoA, AoD and ToA characteristics of scattered multipath clusters for THz indoor channel modeling, in: 17th European Wireless Conference (EW), 9 pages (electronic), Vienna, April 2011, 2011.

[11] M. Jacob, S. Priebe, R. Dickhoff, T. Kleine-Ostmann, T. Schrader, T. Kürner, Diffraction in mm and sub-mm wave indoor propagation channels, IEEE Trans. Microw. Theory Tech. 60 (3) (2012) 833–844.

[12] S. Priebe, M. Jacob, T. Kürner, Calibrated broadband ray tracing for the simulation of wave propagation in mm and sub-mm wave indoor communication channels, in: 18th European Wireless Conference (EW), 10 pages (electronic), Poznan, April 2012, 2012.

[13] K. Zheng, S. Ou, X. Yin, Massive MIMO channel models: a survey. Int. J. Antennas Propag. 2014 (2014) 10, http://dx.doi.org/10.1155/2014/848071, Article ID 848071.

[14] T.S. Rappaport, S. Sun, R. Mayzus, H. Zhao, Y. Azar, K. Wang, G.N. Wong, J.K. Schulz, M. Samimi, F. Gutierrez, Millimeter wave mobile communications for 5G cellular: it will work! IEEE Access 1 (2013) 335–349.

[15] S. Wu, C.X. Wang, H. Haas, E.H.M. Aggoune, M.M. Alwakeel, B. Ai, A non-stationary wideband channel model for massive MIMO communication systems, IEEE Trans. Wirel. Commun. 14 (3) (2015) 1434–1446.

[16] C.A. Balanis, Antenna Theory Analysis and Design, third ed., Wiley, Hoboken, NJ, 2005.

[17] S.R. Saunders, A. Aragon-Zavala, Antennas and Propagation for Wireless Communication Systems, second ed., Wiley, West Sussex, 2007.

[18] S. Payami, F. Tufvesson, Channel measurements and analysis for very large array systems at 2.6 GHz, in: Proceedings of 6th European Conference Antennas Propagation, Prague, Czech Republic, 2012, pp. 433–437.

[19] X. Gao, F. Tufvesson, O. Edfors, F. Rusek, Measured propagation characteristics for very-large MIMO at 2.6 GHz, in: Proceedings of 46th Annual Asilomar Conference on Signals, Systems, and Computers, Pacific Grove, CA, USA, 2012, pp. 295–299.

[20] F. Bohagen, P. Orten, G.E. Oien, Design of Capacity-Optimal HighRank Line-Of-Sight MIMO Channels, Res. Rep., University of Oslo, Oslo, Norway, 2007.

[21] T. Zwick, C. Fischer, D. Didascalou, W. Wiesbeck, A stochastic spatial channel model based on wave-propagation modeling, IEEE J. Sel. Areas Commun. 18 (1) (2000) 6–15.

[22] S. Wu, C.-X. Wang, El-H.M. Aggoune, M.M. Alwakeel, Y. He, A non-stationary 3-D wideband twin-cluster model for 5G massive MIMO channels, IEEE J. Sel. Areas Commun. 32 (6) (2014) 1207–1218.

[23] A. Papoulis, S.U. Pillai, Probability, Random Variables, Stochastic Processes, fourth ed., McGraw-Hill, New York, 2002.

[24] P. Kyosti, et al., WINNER II Channel Models, Winner, Munich, 2007. D1.1.2 V1.1, Sep. 2007.

[25] A. Abdi, J.A. Barger, M. Kaveh, A parametric model for the distribution of the angle of arrival and the associated correlation function and power spectrum at the mobile station, IEEE Trans. Veh. Technol. 51 (3) (2002) 425–434.

[26] T.S. Rappaport, S. Sun, et al., Millimeter wave mobile communications for 5G cellular: it will work! in: Global Communications Conference (GLOBECOM), 2013 IEEE, 2013, pp. 3948–3953.

[27] H. Zhao, T.S. Rappaport, et al., 28 GHz millimeter wave cellular communication measurements for reflection and penetration loss in and around buildings in New York city, in: 2013 IEEE International Conference on Communications (ICC), 2013, pp. 5163–5167.

[28] Z. Wen, H. Kong, mmWave MIMO channel sounding for 5G, in: 2014 1st International Conference on 5G for Ubiquitous Connectivity (5GU), 2014, pp. 192–197.

[29] J.A. Fessler, A.O. Hero, Space-alternating generalized expectationmaximization algorithm, IEEE Trans. Sig. Process. 42 (10) (1994) 2664–2677.

[30] B.H. Fleury, M. Tschudin, R. Heddergott, D. Dahlhaus, K.L. Pedersen, Channel parameter estimation in mobile radio environments using the SAGE algorithm, IEEE J. Sel. Areas Commun. 17 (3) (1999) 434–450.

[31] R. Muller, D.A. Dupleich, C. Schneider, R. Herrmann, R.S. Thoma, Ultrawideband 3D mmWave channel sounding for 5G, in: 2014 XXXIth URSI General Assembly and Scientific Symposium (URSI GASS), 2014, pp. 1–4.

mmWave communication enabling techniques for 5G wireless systems: A link level perspective

T.E. Bogale*, X. Wang*, L.B. Le†
Western University, London, ON, Canada* University of Quebec, Montreal, QC, Canada†

CHAPTER OUTLINE

9.1 INTRODUCTION

The exponential growth of capacity hungry multimedia services and the desire for ubiquitous data access by mobile users obliged the telecommunication operators to rethink the way cellular networks are designed. To meet the demand, wireless industries have outlined a new fifth generation (5G) standard. Research on next-generation 5G wireless systems, which aims to resolve several unprecedented technical requirements and challenges, has attracted growing interests from both academia and industry over the last few years. To meet the 5G requirements, one may think of improving the spectral and energy efficiency of the current fourth generation (4G) network, which is mainly overcrowded from 600 MHz to 3 GHz. However, studies show that such improvement alone will not meet the 5G network capacity.

mmWave Massive MIMO. http://dx.doi.org/10.1016/B978-0-12-804418-6.00009-1

For this reason, mmWave frequency bands ranging from 30 to 300 GHz have attracted significant interest regarding meeting the capacity requirements of 5G network where there is an enormous amount of available bandwidth [1, 2]. Although the available bandwidth of mmWave frequencies is very large, the propagation characteristics are significantly different from that of the microwave frequency bands, which are briefly summarized as follows:

1. **Path loss**: From Friis's law, the isotropic path loss increases with the carrier frequency. As an example, the free-space path loss decays with the square of carrier frequency. Thus, in a point-to-point communication, one may expect significant path loss when we move from 3 to 60 GHz carrier frequency [3].

2. **Diffraction and blockage**: Diffraction leads to wave propagation in the geometrical shadow region behind obstacles. Diffraction may cause a nonnegligible multipath propagation under both line-of-sight (LOS) and nonline-of-sight (NLOS) conditions [4]. From electromagnetic theory, it is well understood that electromagnetic waves experience a difficulty to diffract when they propagate at obstacles with physical dimensions significantly larger than the wavelength [5]. Furthermore, signals at microwave frequencies can penetrate more easily through solid materials and buildings than mmWave. For these reasons, mmWave signals are influenced by the effect of shadowing and diffraction to a much greater extent than microwave signals [6]. For instance, one can observe more than 35 dB blockage losses due in bricks, concretes, etc., and around 35 dB due to human body, where these losses are negligible at microwave frequency bands [7].

3. **Rain attenuation**: In general, the losses due to a rain attenuation at mmWave frequency bands are much larger than those of microwave bands. If we consider a typical mmWave frequency of 73 GHz, one can observe a rain attenuation of roughly 10 dB/km, which is quite large [8] (see Fig. 2 of [8]).

4. **Atmospheric absorption**: Field measurement results have shown that mmWave signals are more susceptible to oxygen absorption than that of microwave signals. For instance, one can observe roughly 20 dB loss around 60 GHz mmWave signal [9] (see Fig. 1 of [9]).

5. **Foliage loss**: The attenuation of radio signals caused due to the presence of trees obstructing the radio link is termed foliage loss [10]. Foliage losses for mmWaves are significant and can be a limiting factor for some propagation environments. Empirical results demonstrate that at 10 m foliage penetration, the loss at 80 GHz mmWave frequency reaches around 23.5 dB, which is about 15 dB higher compared to that of the 3 GHz microwave frequency [3].

All these evidences validate that the overall losses of mmWave systems are larger than those of microwave systems. Fortunately, the small wavelengths of mmWave signals enable a large number of antenna elements to be placed in the same physical antenna area thereby providing high spatial processing gains that can theoretically compensate for at least the corresponding isotropic path loss. However, since the mmWave systems are equipped with several antennas, a number of computation and implementation challenges arise to maintain the expected performance gain. Toward this end, this chapter discusses key enabling techniques of the mmWave communication for 5G networks from the link level perspective. The link level performance of the mmWave wireless system depends on a number of enabling approaches including the transmission scheme (i.e., whether we employ beamforming, multiplexing, or both), the approach to identifying the channel, how to design the transmitted signal waveform structure and access strategies, which are briefly summarized as follows.

Multiple input multiple output (MIMO) is one of the promising techniques for improving the link level performance of wireless channels. To exploit the full potential of MIMO, one can leverage conventional digital beamforming (DB) where one radio frequency (RF) chain is dedicated to each antenna. In fact, the DB is a realistic design approach for the previous and current wireless systems as these systems are equipped with few antennas (around 10) [11, 12]. However, as mmWave systems are likely equipped with large antenna arrays, the dedication of separate RF chains for each antenna is not feasible. This motivates researchers to create a mmWave system where the number of RF chains is much smaller than that of antennas. One approach to realizing such a system is by deploying hybrid beamforming (HB) where it is realized both in analog and digital domains (i.e., hybrid analog-digital beamforming) [13–17]. The MIMO communications are also well known for their ability to transmit independent information over different antenna elements (i.e., spatial multiplexing). Specifically, mmWave systems are suitable to deploy a large number of antennas (in the order of 100–1000). Furthermore, they are operated at very high frequency bands where the available bandwidths are significantly large (around 7 GHz) at 60 GHz. This encourages consideration of spatial multiplexed transmissions for mmWave channels to strongly boost the link level performance [18, 19]. Spatial multiplexing provides multiplexing gain that increases transmission throughput by subdividing an outgoing signal stream into multiple pieces, where each piece is transmitted simultaneously and in parallel on the same RF channel through different antennas. Furthermore, for the spatial multiplexed transmission to work well, the channel must provide sufficient decorrelation between different closely spaced antennas. Spatial multiplexing can be enabled for a single user system by exploiting the spatial antenna separation, and for a multiuser system by utilizing the user equipments (UEs) directional information.

The performance of beamforming and spatial multiplexing depends on the availability of channel state information (CSI) where better capacity is achieved when CSI acquisition is performed efficiently. In general, the quality of the channel estimator can be improved by increasing the training period of a MIMO channel [20, 21]. However, as practical channels have a limited coherence time where the channel is treated as almost constant, the training period cannot be chosen as desired. Furthermore, as mmWave systems are likely equipped with a smaller number of RF chains than antennas, the CSI may need to be identified with reduced spatial dimensions, which consequently worsens the channel estimation quality and the link level performance [22]. The transmitted signal waveform is another important factor affecting the link level performance of a wireless channel. In fact, different waveform designs lead to different system performance in terms of channel estimation and beamforming [23, 24]. For example, if one utilizes the well-known orthogonal frequency division multiplexing (OFDM) waveform, such a design requires high dynamic range power amplifier (PA) for each antenna element. This will create another price overhead for mmWave systems. On the contrary, if one uses low cost PAs, some of the transmitted signals will be clipped, which will in turn lose the orthogonality of each subcarrier. Furthermore, this design yields a very high peak to average power ratio (PAPR) for the transmitted signals, which seriously degrades the system performance in terms of bit error rate (BER) [25]. Another important enabler of the link level performance of mmWave systems is the wireless channel access strategy. In this regard, the idea of simultaneously connecting the UE to both microwave and mmWave bands has been promoted for overlaid small cell mmWave systems (i.e., a small cell is overlaid with the macro base station (BS)). In doing so, the control and data planes can be split so that critical control data are transmitted over reliable microwave links between UEs and macro BSs while high-speed data communications between UEs and small-cell BSs are realized over mmWave frequency bands. Consequently, reliable and stable communications can be maintained while reaping the benefits of mmWave bands [2, 26, 27]. In the following sections, we provide a detailed treatment of the aforementioned enabling techniques for the future 5G network with emphasis on mmWave systems and the resulting link level performances (Table 9.1).

Table 9.1 Acronyms Used in This Chapter

Acronym	Meaning
2G	Second generation
3G	Third generation
4G	Fourth generation
5G	Fifth generation
ADC	Analog to digital converter
AoA	Angle of arrival
AoD	Angle of departure
BER	Bit error rate
BS	Base station
CDMA	Code division multiple access
CSI	Channel state information
DAC	Digital to analog converter
DB	Digital beamforming
DVB2	Digital video broadcasting—Second generation
FDD	Frequency division duplex
FDMA	Frequency division multiple access
FFT	Fast Fourier transform

(Continued)

Table 9.1 Acronyms Used in This Chapter —cont'd

Acronym	Meaning
FTN	Faster than Nyquist
HB	Hybrid beamforming
IFFT	Inverse fast Fourier transform
i.i.d.	Independent and identically distributed
LOS	Line of sight
LTE	Long-term evolution
LTE-A	Long-term evolution advanced
LS	Least squares
MIMO	Multiple input multiple output
MMSE	Minimum mean square error
MP	Matching pursuit
MSE	Mean square error
mmWave	Millimeter wave
NLOS	Nonline of sight
OFDM	Orthogonal frequency division multiplexing
OFDMA	Orthogonal frequency division multiple access
PAPR	Peak to average power ratio
PA	Power amplifier
PS	Phase shifter
RF	Radio frequency
SINR	Signal to interference plus noise ratio
SNR	Signal to noise ratio
TDD	Time division duplex
TDMA	Time division multiple access
ULA	Uniform linear array
UE	User equipment
VGA	Variable gain amplifier
WiFi	Wireless fidelity
WiMax	Worldwide interoperability for microwave access
ZF	Zero forcing
ZMCSCG	Zero mean circularly symmetric complex Gaussian

9.2 BEAMFORMING

The deployment of multiple antennas at the transmitter and/or receiver improves the overall performance of the system (see [23, 28, 29] for the detailed discussion about different aspects of MIMO communication for microwave frequency bands). This performance improvement is enabled by utilizing beamforming where it is realized by exploiting CSI. The performance of a beamformer achieves the maximum benefit when the channel coefficients corresponding to different transmit-receive antenna pairs experience independent fading. For a given carrier frequency, such an independent fading channel is exhibited when the distance between two antenna elements is at least 0.5λ, where λ is the wavelength [23, 26, 30]. Thus, for a fixed spatial dimension, the number of deployed antennas increases as the carrier frequency increases, which consequently helps to pack large antenna arrays at mmWave frequencies compared to those of microwaves [31]. To exemplify, in a 1 m antenna form factor, one can deploy around 214 and 14 antenna elements by operating at 30 and 2 GHz carrier frequencies, respectively.

9.2.1 DIGITAL BEAMFORMING

Beamforming is the process of controlling the transmitted and/or received signal amplitude and phase according to the desired application and channel environment. In a MIMO system, beamformed transmission is the most widely adopted approach where it can be realized both for single user and multiuser systems. A number of beamforming approaches are available in several research works and standard MIMO communication books (see, e.g., [23, 28, 29]). In a conventional DB, one RF chain is required for each antenna element at the BS and UE where an RF chain includes low-noise amplifier, down-converter, digital to analog converter (DAC), analog to digital converter (ADC), and so on [32, 33]. To show the basics of the DB, we examine a scenario where a BS with N antennas serves a K decentralized single antenna UEs with the parameters as summarized in Table 9.2. In each symbol duration, the data symbol of all UEs are first concatenated in vector form $\mathbf{d} = [d_1, d_2, ..., d_K]$, where d_k is the data symbol of the kth UE. Then, the kth UE symbol will be multiplied by its corresponding beamforming vector $\mathbf{b}_k$ to form the transmitted vector corresponding to the kth UE as $\mathbf{x}_k = \mathbf{b}_k d_k$. Finally, all transmitted vectors will be summed up as $\bar{\mathbf{d}} = \sum_{i=1}^{K} \mathbf{x}_i$, passed through the RF chains and transmitted to the air via all antennas. Now if we would like to transmit S_t symbols for each UE, the transmitter broadcasts $\bar{\mathbf{d}}_1, \bar{\mathbf{d}}_2, ..., \bar{\mathbf{d}}_{S_t}$ sequences of information. This process is pictorially depicted in Fig. 9.1. In general, the beamforming matrix $\mathbf{B}$ is the channel and noise variance dependent design parameter, which may need to be optimized

Table 9.2 Digital Beamforming (Precoding) Operations

Variables	Definition
d_k	Transmitted symbol for the kth UE per symbol period
$\mathbf{d} = [d_1, d_2, ..., d_k]$	Transmitted symbol of all UEs per symbol period
$\mathbf{b}_k \in C^{N \times 1}$	Beamforming vector for the kth UE
$\mathbf{B} = [\mathbf{b}_1, \mathbf{b}_2, ..., \mathbf{b}_k]$	Beamforming matrix for all UEs
$\mathbf{x}_k = \mathbf{b}_k d_k$	Beamformed sample of the kth UE per symbol period
$\bar{\mathbf{d}} = \sum_{i=1}^{K} \mathbf{x}_i = \mathbf{Bd}$	Transmitted samples per symbol period

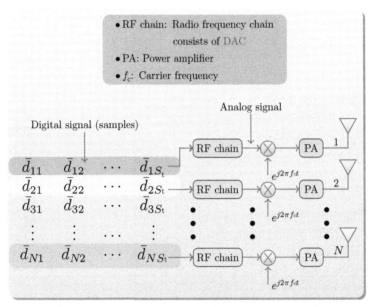

FIG. 9.1

Transmitter architecture with digital beamforming (precoding).

based on different design criteria (see, e.g., [12, 34, 35] and their references for more details about the design of **B**).

As can be seen from this figure, DB requires that each antenna is connected to one RF chain. The DB is commonly adopted for microwave frequency applications as the number of deployed antennas is in the order of 10 (e.g., the current LTE network). However, since mmWave systems are likely be equipped with several antennas and the cost and power consumptions of ADC and DAC is very high, the dedication of a separate RF chain for each antenna is practically infeasible [36–41]. For this reason, beamforming design with a limited number of RF chains has recently received significant attention.

9.2.2 HYBRID BEAMFORMING

The dedication of a separate RF chain for each antenna is not economical for large antenna array applications, such as mmWave. This drives researchers to think of a system where the number of RF chains is much smaller than that of antennas. One approach of realizing such a system is by deploying HB using both analog and digital domains. In the digital domain, beamforming can be realized using microprocessors (supercomputers), whereas in the analog domain, beamforming is implemented by using phase shifters (PSs) [26, 42, 43] and possibly variable gain amplifiers (VGAs). Fig. 9.2 shows a typical transmitter architecture for a system where a BS having N antennas and $N_a < N$ RF chains incorporating a hybrid analog-digital beamforming

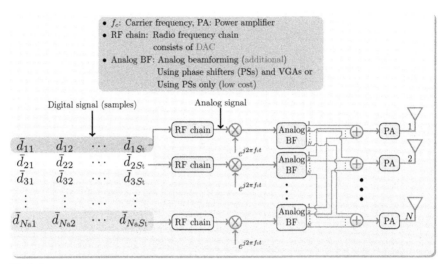

FIG. 9.2

Transmitter architecture with hybrid analog-digital beamforming (precoding).

serves a K decentralized single antenna UEs and all the other settings are as shown in Table 9.2. With this hybrid architecture, the beamforming operation is summarized in Table 9.3.

Since the price of RF chains is much higher than that of the analog electronic components when realizing the analog beamforming matrix **A** of the HB, one can control the cost of the HB design by controlling the number of RF chains N_a. However, as the hybrid analog-digital architecture will loose $N - N_a$ degrees of freedom compared to that of the digital architecture, the performance of the HB cannot be better than that of the DB. Hence, the tradeoff between hybrid and digital beamforming is cost versus performance. The hybrid analog-digital architecture at the receiver side can also be designed similar to Fig. 9.2 and its details are omitted for

Table 9.3 Hybrid Beamforming (Precoding) Operations

Variables	Definition
d_k	Transmitted symbol for the kth UE per symbol period
$\mathbf{d} = [d_1, d_2, ..., d_k]$	Transmitted symbol of all UEs per symbol period
$\tilde{\mathbf{b}}_k \in \mathcal{C}^{N_a \times 1}$	Beamforming vector for the kth UE ($N_a < N$)
$\tilde{\mathbf{B}} = [\tilde{\mathbf{b}}_1, \tilde{\mathbf{b}}_2, ..., \tilde{\mathbf{b}}_k]$	Beamforming matrix for all UEs
$\tilde{\mathbf{x}}_k = \tilde{\mathbf{b}}_k d_k$	Beamformed sample of the kth UE per symbol period
$\mathbf{A} \in \mathcal{C}^{N \times N_a}$	Analog beamforming matrix of all UEs
$\bar{\mathbf{d}} = \mathbf{A} \sum_{i=1}^{K} \tilde{\mathbf{x}}_i = \mathbf{A} \tilde{\mathbf{B}} \mathbf{d}$	Transmitted samples per symbol period (**A** is common for all UEs)

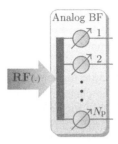

FIG. 9.3

Single PS-based hybrid beamforming architecture.

conciseness. Different HB implementation approaches can be found in Zhang et al. [44] and Gholam et al. [45].

The analog beamforming part of the HB architecture can be realized by utilizing VGAs and PSs. However, as the price of VGAs is much higher than PSs, it is the design of analog beamforming using PSs only that has received significant attention [14, 32]. For these reasons, several HB architectures have been proposed over the past few years by assuming that the DB part is realized using microprocessors, whereas the analog beamforming part is realized using PSs only [26, 42–45]. To this end, one of the fundamental issues of the hybrid analog-digital architecture is how to employ the smallest number of RF chains while maintaining almost the same system performance as digital. The main challenge arises due to the fact that $N_a < N$ and the elements of the analog beamforming matrix $\mathbf{A}$ are designed by employing PSs alone having constant modulus.[1] In this respect, a number of research works have been undertaken to jointly design the beamforming matrices $\tilde{\mathbf{B}}$ and $\mathbf{A}$ for several design criteria and system setups. For the fixed $\mathbf{A}$, one can design $\tilde{\mathbf{B}}$ by applying the techniques used in the conventional DB [14, 17, 32]. The remaining question will therefore be regarding how to design $\mathbf{A}$. In general, two design approaches have been suggested by researchers to design $\mathbf{A}$: single PS and paired PSs.

Single PS approach

The first approach tries to design $\mathbf{A}$ by constraining each of its elements to have a constant amplitude as shown in Fig. 9.3 (i.e., single PS design approach). As we can see from this figure, the signal coming from each RF chain will pass through N_p PSs. When $N_p = N$, the signal transmitted from each RF chain will contribute to each antenna. For this reason, the HB design is termed as a fully connected single PS hybrid analog-digital architecture [14, 16, 17]. And when $N_p < N$, the transmitted signal of each RF chain will contribute only to some of the antennas and hence it is termed as a partially connected HB architecture [46].

---

[1]In this chapter, we assume a unity modulus as in several works without loss of generality [26, 42–45].

With such an approach, almost all the HB algorithms utilize the codebook method to optimize **A** where its main idea is to choose the columns of **A** from a set of predefined vectors. In this regard, matching pursuit (MP) technique has been commonly adopted in several pieces of literature (see, e.g., [14–17]). In general, the MP-based HB algorithm achieves very good performance both for single user and multiuser massive MIMO systems when the channel has very few scatterers. However, the MP-based HB approach is not able to achieve the same performance as that of the BD one for general channel matrices. In particular, the performance gap between the HB and DB is very large when the number of scatterers is much larger than the number of RF chains (this will be justified in the next subsection). Also, in Liang et al. [47], a single PS hybrid precoding scheme for multiuser massive MIMO systems is considered. The paper employs the zero forcing (ZF) hybrid precoding where it is designed to maximize the sum rate of all users. Neither of such an HB approach is able to maintain the same performance as that of the DB (see also [48–50] for other HB designs with single PS). In none of these works is it clear how one can analytically quantify the relation between the HB and DB performances. The potential reason for this is that such an approach will modify the phase of each element of **A** only. This motivates researchers to design the analog beamforming matrix of the HB design with PSs only while ensuring that both the amplitude and phase of **A** can be controlled, which leads to a paired PS design approach.

Paired PS approach

We understand that the single PS design approach helps to modify the phase of each element of **A**. In consequence, its performance is degraded and such an approach is not able to analytically quantify the relation between the HB and DB performances. With the paired PS, as the elements of **A** can have different amplitude and phase, the objective becomes obtaining some relationship between HB and DB designs. This can be addressed by quantifying the number of RF chains and paired PSs such that the HB and DB designs achieve the same performance. One approach of determining these numbers is by considering a particular design criteria and channel model (e.g., sum rate maximization with uniform linear array (ULA) channel model). However, such a design approach is not flexible as it may not be extended for other design criteria and channel models (e.g., for weighted sum rate maximization, max-min signal to interference plus noise ratio (SINR) problems and so on). This motivates researchers to address the above issue by considering arbitrary design criteria.

In fact, the performance of HB will be the same as that of DB when the transmitted and received signals in both cases are the same. Consequently, the optimal $\mathbf{A}\tilde{\mathbf{B}}$ may need to be selected as:

$$\mathbf{A}\tilde{\mathbf{B}} = \mathbf{B} = \mathbf{U}\mathbf{Q} \tag{9.1}$$

where **B** and $\{\mathbf{A}, \tilde{\mathbf{B}}\}$ are as defined in Tables 9.2 and 9.3, respectively, and $\mathbf{U}\mathbf{Q} = \mathrm{QR}(\mathbf{B})$ is the QR decomposition of **B** with $\mathbf{U}^H\mathbf{U} = \mathbf{I}$. By using matrix theory, one can prove that the equalities in Eq. (9.1) cannot be satisfied when $N_\mathrm{a} \leq \mathrm{rank}(\mathbf{B}) \leq K$ for arbitrary $\mathbf{B}$, which justifies that $N_a \geq K$. Now if we have $N_a = K$, one may think of directly set $\mathbf{A} = \mathbf{U}$ and $\widetilde{\mathbf{B}} = \mathbf{Q}$. However, since the amplitude of each of the elements of $\mathbf{U}$ is not necessarily the same, it is not clear how one can realize this matrix using PSs only. Hence, such a direct plug-in will not help to design the HB architecture. For this reason, Zhang et al. [33] first came up with a novel and clever method to represent "any" vector $\mathbf{v} \in \mathcal{C}^{N \times 1}$ as (see Theorem 1 of Zhang et al. [33]):

$$\mathbf{v} = \mathbf{W}\mathbf{z} \tag{9.2}$$

where $\mathbf{W} \in \mathcal{C}^{N \times 2}$ and $\mathbf{z} \in \mathcal{C}^{2 \times 1}$ with $|\mathbf{W}_{(i,j)}|^2 = 1, \forall i, j$, which leads Zhang et al. [33] to conclude that the performance of any DB can be achieved with the HB if the number of RF chains is at least two times the number of data streams (i.e., two times the rank of $\mathbf{B}$) [32, 33]. Now if we utilize this technique to the HB architecture, a maximum of $N_a = 2K$ RF chains and $2KN$ PSs is needed to achieve the same performance as that of the DB design.

It is shown in Zhang et al. [33] that $\mathbf{W}$ of Eq. (9.2) is not unique. This prompts [13] the further reduction of the number of RF chains and PSs by utilizing the degree of freedom in $\mathbf{W}$. The main idea of Bogale et al. [13] was to exploit the following relation (see Theorem 1 of Bogale et al. [13]): for any real number x with $-2 \leq x \leq 2$, it is shown in Bogale et al. [13] that:

$$x = e^{j\cos^{-1}\left(\frac{x}{2}\right)} + e^{-j\cos^{-1}\left(\frac{x}{2}\right)} \tag{9.3}$$

$$jx = e^{j\sin^{-1}\left(\frac{x}{2}\right)} + e^{j\left(\pi - \sin^{-1}\left(\frac{x}{2}\right)\right)} \tag{9.4}$$

where $j = \sqrt{-1}$. These equations have also been proven in parallel work [51] by using different methods. By applying Eq. (9.3), one can express the (m, n)th element of $\mathbf{U}$ as:

$$\begin{aligned} \mathbf{U}_{(m,n)} &\triangleq a_{mn}e^{j\phi_{mn}} \\ &= e^{j\left(\cos^{-1}\left(\frac{a_{mn}}{2}\right) + \phi_{mn}\right)} + e^{-j\left(\cos^{-1}\left(\frac{a_{mn}}{2}\right) - \phi_{mn}\right)} \end{aligned} \tag{9.5}$$

Consequently, each element of $\mathbf{U}$ can be equivalently expressed as a sum of two PSs. Furthermore, as the maximum number of nonzero elements of $\mathbf{U}$ is KN, the solution obtained in DB can be achieved by employing $2NK$ PSs and $N_a \leq K$ RF chains (i.e., reduced by half compared to the work of Zhang et al. [33]), which leads the paired PS architecture as shown in Fig. 9.4. The number of PSs can be reduced slightly and different customized versions of the paired PSs can be found in Bogale et al. [13]. Furthermore, such a paired PS architecture can also be either fully connected or partially connected by selecting N_p of Fig. 9.4 appropriately.

9.2.3 LINK LEVEL PERFORMANCE

In this subsection, simulation results are presented to assess the link level performance of DB and HB with single PS architecture designs. For the simulation, we consider a multiuser system where the BS is equipped with N ULA antennas and

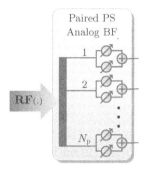

FIG. 9.4

Paired PS-based hybrid beamforming architecture.

K UEs where each UE is equipped with M_k ULA antennas. The transmitter and receiver number for the RF chains is set to P_t and P_k, respectively. Furthermore, the number of scatterers between the BS and kth UE is set to L_k. Under such settings, the channel between the kth UE and BS can be expressed as [14, 17, 29]:

$$\mathbf{H}_k = \sqrt{\frac{NM_k}{L_k\rho_k}}\sum_{i=1}^{L_k} g_{ki}\mathbf{a}_{tk}(\theta_{tk}(i))\mathbf{a}_{rk}^H(\theta_{rk}(i)) \tag{9.6}$$

where g_{ki} is the complex gain of the kth UE ith path with $\mathrm{E}\{|g_{ki}|^2\} = 1$, ρ_k is the path-loss between the transmitter and kth receiver, $\theta_t(i) \in [0, 2\pi]$, $\theta_{rk}(i) \in [0, 2\pi]$, $\forall i$, and $\mathbf{a}_t(.)$ and $\mathbf{a}_r(.)$ are the antenna array response vectors at the transmitter and receiver, respectively. Here, we adopt ULA antenna elements, where $\mathbf{a}_{tk}(.)$ and $\mathbf{a}_{rk}(.)$ are modeled as [37]:

$$\mathbf{a}_{tk}(\theta) = \frac{1}{\sqrt{N}}[1, \exp^{j\frac{2\pi}{\lambda}d\sin(\theta)}, \dots, \exp^{j(N-1)\frac{2\pi}{\lambda}d\sin(\theta)}]^T$$

$$\mathbf{a}_{rk}(\theta) = \frac{1}{\sqrt{M_k}}[1, \exp^{j\frac{2\pi}{\lambda}d\sin(\theta)}, \dots, \exp^{j(M_k-1)\frac{2\pi}{\lambda}d\sin(\theta)}]^T$$

where $j = \sqrt{-1}$, λ is the transmission wave length and d is the antenna spacing.

We have used $N = 128$, $M_k = 32$, $d = 0.5\lambda$, $K = 4$, $L_k = 16$, and $P_t = KP_{rk}$. The SNR, which is defined as $\mathrm{SNR} = \dfrac{P_{av}}{\sigma^2}$ is controlled by varying σ^2 while keeping the total transmitted power $P_{max} = KS_k\mathrm{mW}$ and $P_{av} = \dfrac{P_{max}}{S_k}$. The channel parameters are $\rho = [0.2338\ 0.2333\ 0.0402\ 0.5290]$, and $\theta_{tk}(i)(\theta_{rk}(i))$ are taken randomly from uniform random variables in $[0, 2\pi]$ (see Eq. 9.7). The total sum rate is computed as $R_t = \sum_{k=1}^{K}\sum_{i=1}^{S_k} R_{ki}$, where the rate of each symbol is calculated as $R_{ki} = \log_2(1 + \gamma_{ki})$ with γ_{ki} as the achieved SINR of the kth UE ith symbol. All of the plots are generated by averaging 1000 realizations of g_k, $\forall k$:

$$\boldsymbol{\theta}_t(\text{in }\pi) = \begin{bmatrix} 0.71 & 0.66 & 0.20 & 0.70 \\ 1.29 & 0.16 & 0.81 & 1.88 \\ 0.65 & 0.91 & 0.22 & 0.52 \\ 1.70 & 0.64 & 0.94 & 1.61 \\ 0.80 & 1.85 & 1.42 & 1.33 \\ 0.43 & 1.11 & 1.99 & 1.38 \\ 1.24 & 0.61 & 1.53 & 1.95 \\ 0.80 & 0.10 & 0.77 & 1.22 \\ 0.47 & 0.07 & 0.38 & 0.39 \\ 0.43 & 1.50 & 0.83 & 1.83 \\ 0.04 & 0.71 & 1.27 & 1.24 \\ 0.60 & 0.53 & 0.16 & 0.32 \\ 1.50 & 1.76 & 0.08 & 0.71 \\ 0.77 & 1.48 & 0.05 & 1.23 \\ 0.74 & 1.89 & 1.70 & 0.20 \\ 0.02 & 0.83 & 0.36 & 1.71 \end{bmatrix} , \quad \boldsymbol{\theta}_r(\text{in }\pi) = \begin{bmatrix} 0.25 & 1.87 & 0.73 & 0.45 \\ 1.04 & 0.73 & 0.73 & 0.16 \\ 0.39 & 0.74 & 1.51 & 1.30 \\ 0.23 & 1.52 & 1.69 & 1.40 \\ 0.77 & 0.16 & 0.55 & 1.04 \\ 1.92 & 1.63 & 1.81 & 1.71 \\ 1.80 & 1.58 & 0.07 & 1.86 \\ 1.37 & 0.44 & 0.93 & 0.16 \\ 1.85 & 1.05 & 1.19 & 0.75 \\ 0.49 & 0.64 & 1.31 & 0.35 \\ 0.20 & 1.09 & 1.28 & 0.24 \\ 0.85 & 1.39 & 0.98 & 1.20 \\ 1.13 & 0.10 & 1.13 & 1.46 \\ 0.88 & 1.92 & 1.59 & 1.08 \\ 0.84 & 1.93 & 1.68 & 1.47 \\ 1.56 & 0.25 & 1.44 & 1.96 \end{bmatrix} \qquad (9.7)$$

In the first simulation, we compare the performance of the DB and HB designs. To this end, we set $S_k = 8$ and $P_{rk} = 16$, $\forall k$. Fig. 9.5 shows the sum rate achieved by the

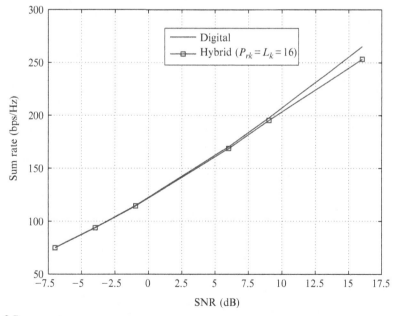

FIG. 9.5

Comparison of digital and hybrid beamformings [17] when $L_k = 16$, $S_k = 8$, and $P_{rk} = 16$.

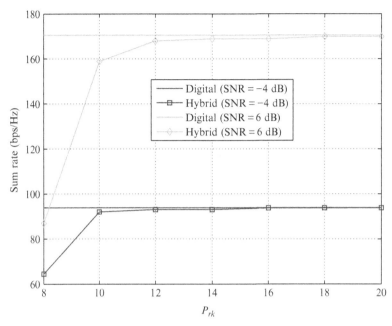

FIG. 9.6

Effect of P_{rk} on the hybrid beamforming.

From T.E. Bogale, L.B. Le, Beamforming for multiuser massive MIMO systems: digital versus hybrid analog-digital, in: Proceedings of IEEE Global Telecommunications Conference (GLOBECOM), Austin, TX, 2014, pp. 4066–4071.

digital and hybrid beamformings. As can be seen from this figure, the performance of the HB is almost the same as that of the digital one from the low to moderate SNR regions. And little performance gap is observed at high SNR regions. Next, we examine the effect of P_{rk} on the performance of the HB design. Fig. 9.6 illustrates the performance of HB design for different settings of P_{rk} at SNR=$\{-4\,dB, 6\,dB\}$ and $S_k = 8$. From this figure, one can observe that the performance of the HB design improves as the number of RF chains and ADCs (i.e., P_{rk}) increase, which is expected. Finally, we analyze the joint effects of S_k and P_{rk} on the performance of the digital and hybrid beamformings. To this end, Fig. 9.7 shows the performance of the digital and HBs when SNR=$-$ 4 dB. As can be seen from this figure, the achieved sum rates of the digital and hybrid beamformings increase as S_k increases. This is because as S_k increases, $P_{max} = KS_k$ also increases. Also the performance gap between hybrid and digital beamformings decreases as S_k decreases. Hence, for a limited number of RF chains and ADCs (i.e., P_{rk}), the performance gap between hybrid and digital beamformings can be decreased by reducing the number of multiplexed symbols (S_k).

One can observe from all these figures that, for the given number of RF chains and ADCs, the performance gap between the digital and hybrid beamformings can be

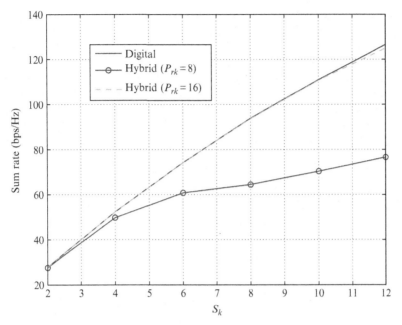

FIG. 9.7

Effect of S_k and P_{rk} on the digital and hybrid beamforming.

From T.E. Bogale, L.B. Le, Beamforming for multiuser massive MIMO systems: digital versus hybrid analog-digital, in: Proceedings of IEEE Global Telecommunications Conference (GLOBECOM), Austin, TX, 2014, pp. 4066–4071.

decreased by decreasing the number of multiplexed symbols. Moreover, for the given number of multiplexed symbols, increasing the number of RF chains and ADCs increases the achievable sum rate of the HB, which is expected. The detailed comparison of the single PS and paired PSs HB architecture for different system setups can be found in Bogale et al. [13].

9.3 SPATIAL MULTIPLEXING

As stated in the introduction section, low frequency bands are less susceptible to blockage. Furthermore, they can easily be diffracted in many physical materials and bodies such as bricks, buildings, glass, human, etc. Such behavior means that the received signals of the microwave frequency bands incorporate several signals reflected from different scatterers, which are often independent. For this reason, the channel between any transmit-receive antenna pairs often tend to be independent in a rich scattering environment [52, 53]. In such an environment, the capacity of a MIMO wireless channel can be expressed as:

$$C_{\text{MIMO}} = \alpha \min(M, N) \log 2(1 + \gamma(M, N)) \tag{9.8}$$

where $N(M)$ is the number of transmit (receive) antennas, $\gamma(M, N)$ is the SINR achieved by the MIMO link which depends on the number of antennas, and α depends on the transmission coding scheme.[2] As we can see from this expression, C_{MIMO} increases linearly with $\min(M,N)$. It is shown in Zheng and Tse [53] that different coding schemes yield different α and $\gamma(M, N)$. Specifically, any coding scheme increasing one decreases the other and vice verse. Now if we consider uncoded transmission, which is the focus of this chapter, the gain achieved by transmitting $\min(M,N)$ symbols determines the multiplexing gain whereas, $\gamma(M, N)$ determines the diversity gain (i.e., link reliability) of the MIMO channel. One can observe from C_{MIMO} that when the transmission medium has acceptable link reliability $\gamma(M, N)$, it is advantageous to utilize all the available degrees of freedom to transmit $\min(M,N)$ symbols per symbol period to get better C_{MIMO}.

We have discussed in the above sections that mmWave systems are suitable for deploying a number of antennas (in the order of 100–1000). Furthermore, they are operated at very high frequency bands where the available bandwidths are significantly large. Potential mmWave spectrum bands are: the 28 GHz (27.5–29.5 GHz), 38 GHz (36–40 GHz), 60 GHz (57–64 GHz), E-band (71–76 and 81–86 GHz), and W-bands (92–95 GHz) [6, 54]. This motivates researchers to consider spatial multiplexed transmissions for mmWave channels to strongly boost the system capacity as the $N(M)$ is extremely large. However, mmWave channels are expected to be specular and have low rank, and their statistical characteristics may not help them to exploit all the available spatial degrees of freedom. For this reason, mmWave systems experience a difficulty in transmitting multiple streams on the same time-frequency resource, which consequently limits the multiplaxing gains [39, 55, 56]. Spatial multiplexing provides multiplexing gain that increases transmission throughput by subdividing an outgoing signal stream into multiple pieces, where each piece is transmitted simultaneously and in parallel on the same channel through different antennas [39]. Furthermore, for the spatial multiplexed transmission to work well, the channel must provide sufficient decorrelation between different closely spaced antennas, which may not hold true in most practical systems such as mmWave indoor.

One approach of exploiting the spatial multiplexed transmission is by enabling multiuser mmWave system applications, where the transmitter transmits different data streams to different UEs that are nonoverlapped geographically [57]. However, such an approach cannot be employed when two UEs are spaced closer to each other. A number of other works have tried to exploit spatial multiplexing for mmWave channels for single user system. In Torkildson et al. [55], the possibility of enabling spatial multiplexing has been discussed for indoor mmWave channels by modeling indoor channels as three rays; direct line of sight and, left and right reflected waves. It is validated through numerical simulation that the multiplexing ability of such a system depends on the relative positions of the transmitter and receiver node.

---

[2] In the uncoded transmission $\alpha = 1$.

Furthermore, even if we have only an LOS channel, the number of dominant singular values of the channel above a given threshold can be maximized by appropriately selecting the antenna positions of the transmitter and receiver nodes, which consequently helps to increase the channel capacity [39, 55, 58].

For the LOS channel, the number of spatially multiplexed channels depends on the antenna array structure. For instance, the maximum number of spatially multiplexed channels varies as the inverse of λ and λ^2 for ULA and rectangular antenna arrays, respectively [19]. To enable LOS spatial multiplexing, the principles of diffraction limited optics has been exploited [18, 19]. And, for the given $N \times N$ ($N \times 1$) size antenna arrays and link range R, the minimum distance between two antenna elements may need to be selected as [55]:

$$D = \sqrt{\frac{R\lambda}{N}} \tag{9.9}$$

This distance ensures that the angular separation of the transmitter antenna elements is greater than or equal to the angular resolution of the receiver array. This has the same significance as increasing the rank of the LOS channel (i.e., increasing spatial multiplexing gain) [19]. The LOS spatial multiplexing has been demonstrated in Sheldon et al. [19] by employing an array of size 4 at 60 GHz carrier frequency.

9.4 CHANNEL ESTIMATION

Beamforming and spatial multiplexing are key enabling techniques to scale up the link capacity of mmWave systems. It is well known that the performance of beamforming and spatial multiplexing highly depends on the availability of CSI and better capacity is achieved when CSI acquisition is performed reliably. In general, channel estimation can be performed either using frequency division duplex (FDD) or time division duplex (TDD) approaches. In the FDD approach, first, the transmitter sends pilot sequences to the receiver. Then, the receiver estimates its own channel. Finally, the estimated channel is fed back to the transmitter via a feedback channel. In the TDD system, first, the receiver transmits its own pilot symbols to the transmitter. Then, the transmitter will estimate the channel and assign the conjugate of the estimated channel as the channel between the transmitter and receiver. The TDD system is employed when the CSI between the transmitter and receiver has reciprocity characteristics with its reverse channel, which exhibits when both links employ the same frequency band. However, in the FDD system, the downlink and uplink channels generally employ sufficiently separated frequency bands. There are a number of TDD- and FDD-based CSI acquisition approaches for MIMO systems. An extensive survey of different CSI acquisition approaches for different system setups can be found in Elijah et al. [59] and Liu et al. [60] (see also [21, 61]). Of these methods, the least square (LS) and minimum mean square error (MMSE) are commonly used as they are general and simple and are discussed in the following section.

To elaborate on the LS and MMSE channel estimation methods, we consider a TDD-based channel estimation for a multiuser system where the BS equipped with N antennas is serving a K decentralized single antenna UEs, and the CSI is estimated at the BS. For this system, one can employ K training periods to estimate the CSI of all UEs. The received signal at the jth training duration is given by [20–22]:

$$\mathbf{y}_{tj} = \sum_{i=1}^{K} \mathbf{h}_i p_{ij}^* + \mathbf{n}_{tj} \tag{9.10}$$

where K is the number of UEs, p_{ij}^* is the pilot symbol of the ith UE at symbol time j, $\mathbf{n}_j \in \mathcal{C}^{N \times 1}$ is the received noise vector at symbol time j, and $\mathbf{h}_i \in \mathcal{C}^{N \times 1}$ is the channel between the ith UE and BS, which can be modeled as:

$$\mathbf{h}_k = \mathbf{R}_k^{1/2} \widetilde{\mathbf{h}}_k$$

where the entries of $\widetilde{\mathbf{h}}_k \in \mathcal{C}^{N \times 1}$ are modeled as independent and identically distributed (i.i.d.) zero mean circularly symmetric complex Gaussian (ZMCSCG) random variables, each with unit variance, and $\mathbf{R}_k \in \mathcal{C}^{N \times N}$ is a positive semi-definite channel covariance matrix. By spending K training symbols, the BS receives the following signal:

$$\mathbf{Y}_t = \mathbf{H}\mathbf{P}^H + \mathbf{N}_t$$

where $\mathbf{H} = [\mathbf{h}_1, \mathbf{h}_2, ..., \mathbf{h}_k]$, $\mathbf{Y}_t = [\mathbf{y}_{t1}, \mathbf{y}_{t2}, ..., \mathbf{y}_{tk}]$, $\mathbf{P} = [\mathbf{p}_1, \mathbf{p}_2, ..., \mathbf{p}_k]$, $\mathbf{N}_t = [\mathbf{n}_{t1}, \mathbf{n}_{t2}, ..., \mathbf{n}_{tk}]$ is the additive noise, which is modeled as $\mathbf{N}_t \sim \mathcal{CN}(\mathbf{0}, \sigma_t^2 \mathbf{I})$ and $\mathbf{p}_i = [p_{i1}, p_{i2}, ..., p_{ik}]^T$. For simplicity, the entries of $\mathbf{P}$ are selected from a constellation satisfying $\mathbf{P}^H \mathbf{P} = \mathbf{I}$. For instance, for the LTE system, Zadoff Chu sequences are employed. The BS can decouple the channels of each UE by multiplying the overall received signal $\mathbf{Y}$ with $\mathbf{P}$, i.e.,

$$\mathbf{E} \triangleq \mathbf{Y}_t \mathbf{P} = \mathbf{H} + \mathbf{N}_t \mathbf{P}, \Rightarrow \mathbf{e}_k = \mathbf{h}_k + \widetilde{\mathbf{n}}_k \tag{9.11}$$

where $\widetilde{\mathbf{n}}_k = \mathbf{N}_t \mathbf{p}_k$. As we can see from Eq. (9.11), $\mathbf{e}_k$ does not contain $\mathbf{h}_i$, $i \neq k$.

The LS channel estimation approach treats $\mathbf{e}_k$ as the estimate of $\mathbf{h}_k$. However, the MMSE method incorporates additional steps to estimate the kth channel from $\mathbf{e}_k$. This is performed by introducing an MMSE matrix $\mathbf{W}_k^H$ for the kth UE and expressing the estimated channel $\hat{\mathbf{h}}_k$ as:

$$\hat{\mathbf{h}}_k = \mathbf{W}_k^H \mathbf{e}_k \tag{9.12}$$

where $\mathbf{W}_k$ is designed such that the mean square error (MSE) between $\hat{\mathbf{h}}_k$ and $\mathbf{h}_k$ is minimized as follows:

$$\begin{aligned} \xi_k &= \mathrm{tr}\{\mathrm{E}\{|\hat{\mathbf{h}}_k - \mathbf{h}_k|^2\}\} = \mathrm{tr}\{\mathrm{E}\{|\mathbf{W}_k^H(\mathbf{h}_k + \widetilde{\mathbf{n}}_k) - \mathbf{h}_k|^2\}\}, \\ &= \mathrm{tr}\{(\mathbf{W}_k^H - \mathbf{I}_N)\mathbf{R}_k(\mathbf{W}_k^H - \mathbf{I}_N)^H + \sigma_t^2 \mathbf{W}_k^H \mathbf{W}_k\} \end{aligned} \tag{9.13}$$

where tr(.) and E{.} are the trace and expectation operators, respectively. The optimal $\mathbf{W}_k$ is obtained from the gradient of ξ_k as:

$$\frac{\partial \xi_k}{\partial \mathbf{W}_k^H} = 0, \Rightarrow \mathbf{W}_k^\star = (\mathbf{R}_k + \sigma_i^2 \mathbf{I})^{-1} \mathbf{R}_k \tag{9.14}$$

As can be seen from the expressions (9.11) and (9.12), the difference between the LS and MMSE channel estimators is that the former does not employ the statistical information of the noise and channel covariance matrix, whereas the latter requires the statistics of noise and channel covariance matrix. From these estimators, one can also observe that the CSI information of all UEs cannot be decoupled if we employ less than K training periods, which is equal to the overall number of transmitter antennas of the uplink channel. From the above discussions, one can also notice that N pilot symbols at least are needed for efficient CSI acquisition if the training is sent from the BS. In a microwave communication system as K and N are comparable, sending pilot symbols to the BS or UEs will not make a significant difference. Furthermore, providing the CSI feedback either to UEs or BS is not expensive. For this reason, one can apply either the TDD or FDD system. However, when $N \gg K$ (i.e., massive MIMO system), sending pilots from the UEs is economical, which consequently leads to the promotion of the TDD-based CSI acquisition approach [21, 22, 62].

Although there are extensive surveys on different CSI acquisition methods of a MIMO system, most of the methods cannot be applied directly to mmWave systems. The main reasons are:

- The Doppler shift scales linearly with frequency, thus, the coherence time of mmWave channels is roughly an order of magnitude smaller than that of comparable microwave bands. In a typical setting, for example, one may experience channel coherence times of 500 and 35 μs when the system is deployed at the carrier frequencies 2 and 28 GHz, respectively [63, 64]. For this reason, the overall challenges of the CSI learning overhead of mmWave channels are expected to be significantly higher than that of microwave massive MIMO systems for a comparable mobility environment [56]. Furthermore, mmWave channels experience severe blockage where the CSI acquisition becomes challenging as the probability of link outage is much larger due to beamforming (link) misalignment [7].
- Almost all the conventional MIMO systems assume that the number of deployed RF chains is the same as that of antennas, which may not be feasible for mmWave applications, as such systems will be equipped with large antenna arrays. Furthermore, when the number of RF chains at the BS is much less than the number of BS antennas, an orthogonal channel estimation approach cannot be used. This is due to the fact that the receiver will lose a certain degree of freedom to reliably estimate the CSI coefficients.

For the above two reasons, a number of research efforts have been tried to come up with different techniques to enable efficient channel estimation. In general, the quality of the channel estimator can be improved by increasing the training period. However, as practical channels have a limited coherence time where the channel is treated as almost constant, a large training period cannot be chosen. Furthermore, as the acquired CSI is used for data transmission, more CSI acquisition time will lead to a reduced network throughput for the given coherence time. For this reason, the CSI acquisition may need to be performed efficiently while taking into account both accuracy and training duration [65].

As mentioned in Section 9.2.2, one can employ either the single PS or paired PSs hybrid analog-digital architecture for mmWave systems equipped with less RF chains than that of antennas. The channel estimation problem for mmWave systems have been formulated as a compressive sensing in Bajwa et al., Alkhateeb et al., and Malloy and Nowak [66–68] by employing a single PS hybrid analog-digital architecture. The work of Alkhateeb et al. [67] employs adaptive codebook approach to design its trainings for ULA antenna array models. In Ramasamy et al. [69], a compressed channel sensing approach with analog beamforming architecture was proposed where the approach tries to estimate a predefined number of paths using a grid of spatial frequencies followed by further fine tuning. Along the same lines, compressed channel sensing with random training sequences has been provided with hybrid architecture in Lee et al. [70].

Bogale et al. [22] recently considered joint channel estimation and beamforming while taking into account the channel coherence time by employing the paired PS hybrid analog-digital architecture. In Akdeniz et al. [7], extensive outdoor experiments were conducted to characterize mmWave channels. This paper discusses some of the large scale characteristics of mmWave channels, such as the angle of arrival (AoA), the angle of departure (AoD), the angular beam spread around the center, and the power and group delay profile of each cluster, where all these are varying slowly compared to the small scale fading. This result validates that mmWave channel covariance matrices would be almost constant for a considerable symbol duration [71, 72]. On the other hand, Eliasi et al. [73] recently proposed different channel covariance matrix estimation methods for millimeter wave channels by utilizing hybrid analog-digital architecture. Furthermore, as the mmWave systems are equipped with large antenna arrays, TDD approach is most reasonable. These ideas motivated Bogale et al. [22] to employ a periodic frame structure of the TDD-based channel estimation and data transmission by exploiting the covariance information of mmWave channels as depicted in Fig. 9.8. With this frame structure, the latter paper studies the training-throughput tradeoff for the multiuser mmWave MIMO systems with limited number of RF chains. The work of Bogale et al. [22] exploits the fact that the optimal training duration that maximizes the overall network throughput depends on the operating SNR, available number of RF chains, channel coherence time (T_c) and covariance matrices of all UEs. Specifically, it is shown that the training time optimization problem can be formulated as a concave maximization problem where its global optimal solution can be obtained efficiently using of-the-shelf convex optimization tools such as CVX [74].

In the following, we provide numerical results for the link level performance of the mmWave channel estimation, which utilizes a paired PS hybrid analog-digital

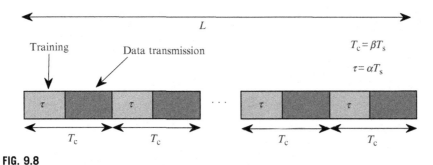

FIG. 9.8

Channel estimation and data transmission frame structure.

From T.E. Bogale, L.B. Le, X. Wang, Hybrid analog-digital channel estimation and beamforming:
training-throughput tradeoff, IEEE Trans. Commun. 63 (12) (2015) 5235–5249.

architecture [22]. We consider the achieved link level throughput for fixed channel coherence time. In this regard, we consider a single user wireless network where the BS equipped with N antennas is serving one single antenna UE. The number of RF chains at the BS is set to 1. We consider a frame structure as shown in Fig. 9.8 with T_c as the channel coherence time, τ is the training time and $T_c - \tau$ is the data transmission duration, and $\beta = 2T_o$, $\alpha = \theta$ and T_o is introduced for convenience. The channel covariance matrix of the UE **R** is taken from a widely used exponential correlation model as $\mathbf{R}_{(i,j)} = \rho^{|i-j|}$, which is assumed to be almost constant for sufficiently large number of frames L, where $0 \le \rho < 1$ [75–77]. We have chosen this model as it captures the characteristics of different propagation environments. For instance, mmWave channels are often represented by low rank channel covariance matrices that can be modeled by selecting large ρ.

The normalized channel estimation MSE for $N = 64$ and $T_o = 64$, $\rho = \rho_1$ and different θ and SNR is plotted in Fig. 9.9A. As can be seen from this figure, increasing θ improves the quality of the channel estimator by reducing the MSE, which is expected. On the other hand, for a given operating SNR, the required training duration to achieve a given target MSE decreases as ρ increases. This result is expected because as ρ increases, $\dfrac{\lambda_i}{\lambda_1}$ with $\lambda_1 \ge \lambda_2, \ldots, \lambda_N$ decreases with i, which consequently helps to reduce the training duration.

Fig. 9.9B shows the achievable throughput versus training duration for different SNR and ρ. As can be seen from this figure, despite the channel estimation approach used in the microwave systems, the optimal training duration of the hybrid analog-digital channel estimator is not necessarily the same for all ρ. In particular, the optimal training duration decreases as the rank of the channel covariance matrix decreases (i.e., as ρ increases). As an example, when SNR=10dB, the optimal training durations are around $4T_s$, $12T_s$, and $20T_s$ for $\rho = 0.9$, 0.5, and 0 (i.e., i.i.d. Rayleigh fading channel), respectively. On the other hand, a shorter training period is employed when the UE's channel is highly correlated, which consequently results in a higher normalized throughput. These results confirm that the training period

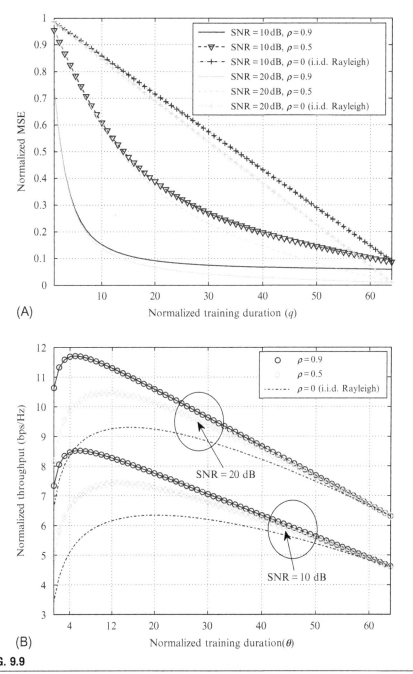

FIG. 9.9

Hybrid analog-digital channel estimation and beamforming for single user system.
(A) Normalized MSE versus training duration. (B) Normalized throughput versus training
duration.

From T.E. Bogale, L.B. Le, X. Wang, Hybrid analog-digital channel estimation and beamforming: training-throughput tradeoff, IEEE Trans. Commun. 63 (12) (2015) 5235–5249.

should be selected carefully when the BS has fewer RF chains than antennas, which is often the case for mmWave systems.

As can be seen from Fig. 9.9, the channel estimation procedure for mmWave systems is different from that of the microwave systems. Furthermore, there is a tradeoff between the training duration and achievable throughput when the number of RF chains is limited, and hybrid analog-digital channel estimation and beamforming is applied.

9.5 **WAVEFORM DESIGN**

The link level performance of a wireless channel is also determined by how we design the transmitted signal waveforms. In fact, different waveform designs lead to different system performances in terms of channel estimation and beamforming, which in turn yields different link level capacity [23, 24]. In this regard, the performance of any waveform is determined by its ability to efficiently realize signaling and multiplexing. A number of waveform design approaches exist in the existing wireless standards: time division multiple access (TDMA) and frequency division multiple access (FDMA) have been employed in 2G systems for their ability to multiplex information symbols. And the spread spectrum (CDMA) becomes more dominant for the 3G system. OFDM has been chosen by a number of recent wireless standards as it has several advantages compared to its TDMA, FDMA, and CDMA counterparts [24]. All these techniques employ a Nyquist transmission approach where only one item of information content is incorporated per symbol duration. Since the future 5G network requires significant capacity, recent literature try to suggest a transmission strategy that can handle more information content per symbol period. In this respect, the faster than Nyquist (FTN) transmission is considered [78]. The first work on FTN transmission was proposed by Mazo in 1975, where it is claimed that it is possible to enable a transmission rate beyond what is defined by the Nyquist criterion without imposing additional bandwidth [79]. The magnitude of error probability of a transmitted symbol plays a fundamental role in understanding the communication channel, which can be directly related to the minimum Euclidean distance d_{min} between two distinct symbols. The key idea of Mazo [79] is that for the scenario with Nyquist symbol duration T_{Ny} and sinc pulse shaping filter, d_{min} will be the same when the symbol duration is set as τT_{Ny} with $\tau \in [0.802, 1]$. This result justifies that one can still maintain the same error probability as that of the Nyquist transmission by selecting $\tau \approx 0.82$, which in turn helps to improve the bandwidth efficiency by $\frac{1}{0.802} \approx 25\%$. It is shown that different pulse-shaping filters have different Mazo limits (i.e., minimum τ). For instance, the minimum τ with a root raised cosine filter with excess bandwidth $\beta = 0.3$ yields a 42% increase in spectrum efficiency (i.e., $\tau_{min} = 0.703$) [80].

Although, the FTN transmission achieves better channel capacity than that of the Nyquist one, it requires a nonlinear decoding operation, which is not desirable in practice. For this reason, OFDM enabled Nyquist transmission is still popular for several wireless standards LTE, LTE-A, WiFi, WiMax, DVB2, etc. [23, 81, 82]. The main advantages of OFDM waveforms are:

1. To achieve very high data rate transmission, large bandwidth is required, which in turn experiences a frequency selective channel. The OFDM waveform design approach helps to transform the frequency selective channels to a parallel flat fading channel. This will consequently help to examine signal components experiencing flat fading channels (i.e., subcarriers) independently.

2. The OFDM waveform can be generated with an inverse fast Fourier transform (IFFT) matrix operation, and the transmitted signal can be recovered with the fast Fourier transform (FFT) operation. As the FFT and IFFT operations can be realized efficiently, OFDM is attractive from the implementation aspect.

3. Since OFDM allows each subcarrier to be handled separately, different subcarriers can be allocated for different users, which consequently enables low complexity multiplexing ability. Furthermore, it can be extended to multiuser MIMO OFDM per subcarrier basis [54].

The OFDM transmission, however, has one major disadvantage. Since the transmitter employs the IFFT operation on the transmitted symbols (i.e., precoding), such an operation will significantly increase the PARP of the transmitted waveform, especially when the IFFT size is large. This will consequently require a high-quality PA, which is often expensive, especially for terminals with low transmission power capability. Furthermore, if one uses low cost PA, some of the transmitted signals will be clipped, which will in turn result in the lose of the orthogonality of each subcarrier. Furthermore, high PAPR issue results in serious degradation of BER performance [25]. To circumvent this, different modified OFDM transmissions have been proposed from which single carrier transmission is the most commonly employed approach (e.g., in the up-link channel transmission) [24, 83].

This will raise a question as to whether OFDM can still be chosen as a viable solution for mmWave applications. As stated in the above sections, a significant bandwidth is available at mmWave frequency bands (e.g., around 12.9 GHz bandwidth is available in the E-band [26]), which implies that high capacity mmWave communications can be achieved. On the other hand, since mmWave frequency bands have very small channel coherence times, the slot duration of each transmission block will be very low compared to that of microwave frequency bands. For instance, the slot duration of the current LTE system is around 0.5 ms, which is capable of handling 6–7 OFDM symbols each having 66.7 µs duration. Furthermore, microwave frequency PAs are less expensive than that of mmWave application PAs. Since mmWave bands employ massive antenna arrays, the deployment of high-quality PAs for mmWave application will make OFDM transmission not preferable. In addition, since the slot duration is very small, and bandwidth is high, a simple TDMA (FDMA) scheme could be practically feasible for multiuser systems while incorporating a single carrier system to enable low complexity receiver design. All these reasons favor the use of single-carrier-enabled TDMA (OFDM) waveform design approaches for mmWave systems [54].

9.6 **ACCESS STRATEGY**

The link level performance of mmWave systems also relies on the access strategy employed in the network. In this regard, a dual wave connectivity has been proposed by researchers to achieve the benefits of both microwave and mmWave bands for overlaid small cell mmWave systems (i.e., a small cell is overlaid with the macro BS). The main idea of dual wave connectivity is that each UE is operated at microwave and mmWave frequency bands simultaneously. In doing so, the control and data planes can be split so that critical control data are transmitted over reliable microwave links between UEs and macro BSs, while high-speed data communications between UEs and small-cell BSs are realized over mmWave frequency bands. Consequently, a reliable and stable communications can be maintained while reaping the benefits of mmWave bands [2, 26, 27]. For instance, a dual band small-cell with prioritized and layered cell association strategy, as shown in Fig. 9.10, can exploit the advantages of these different frequency bands. In this design, the coverage area is divided into three regions where the UEs in the inner region (i.e., radius a)/middle region (i.e., radius b) are served by the small cell at mmWave/microwave frequency bands. And the UEs in the outer region (i.e., radius $> b$) are served by the macrocell at microwave frequency bands. Here, the outer region operates at microwave frequency bands since these bands allow long-range communications to support mobility. The parameters a and b can be chosen or optimized adaptively based on different criteria. This dual-band small-cell design isolates the mmWave and microwave frequency UEs (i.e., the mmWave UEs inside the radius a do not experience any interference from the macro BS which is desirable to scale up the link level capacity).

In Akdeniz et al. [84], the capacity of mmWave picocellular systems has been analyzed. It is shown that mmW networks will operate in an extremely power

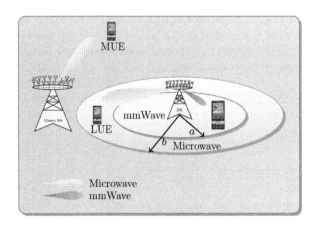

FIG. 9.10

Dual wave connectivity for small cell mmWave systems [2]: MUE (LUE) denotes UE served by the macro-BS (small cell).

(noise) limited regimes where, the full spatial and bandwidth degrees of freedom are not fully utilized. For this reason, the beamforming pattern of one UE would not strongly interfere those of other UEs [85]. This has promoted the use of mmWave frequency for both access and backhaul networks. By doing so, more deployment flexibility can be achieved [26]. In this regard, a number of research works have been conducted. In-band wireless backhaul solution where the backhaul and access are multiplexed on the same frequency band via the TDD-based scheduling is considered in Taori and Sridharan [86] (see also Niu et al. [87]). Resource allocation problems for joint design of access and backhaul networks has been examined in Bernardos et al. [88]. The design and implementation challenges associated with the joint mmWave access and backhaul communications has also been discussed in Dehos et al. [89]. All these factors help improve the mmWave link level performance.

9.7 CONCLUSIONS

This chapter discusses key enabling techniques affecting the link level performance of the future mmWave-based 5G wireless systems. In particular, beamforming, spatial multiplexing, channel estimation, waveform design, and access strategies have been discussed extensively. It is shown that the link level capacity of mmWave systems depends not only on the number of antennas deployed, but also on the number of RF chains. In fact, the best performance is achieved when the mmWave systems are engineered such that the number of RF chains is the same as that of antennas. However, such a design strategy will be very expensive, as mmWave systems are likely equipped with large antenna arrays. In general, the cost of mmWave system can be reduced to an acceptable level by utilizing a hybrid analog-digital architecture at the expense of some loss of performance. It is shown in the above sections that the deployment of hybrid analog-digital architecture changes the beamforming, spatial multiplexing, channel estimation, and transmission waveform performances significantly. It is also discussed briefly that spatial multiplexed communication is feasible for LOS mmWave channels, which facilitates ultrahigh data rate indoor communications. On the other hand, employing an architecture that permits dual wave connectivity is also the unique aspect of mmWave communications, which yields different link level capacity.

REFERENCES

[1] F. Boccardi, R.W. Heath, A. Lozano, T.L. Marzetta, P. Popovski, Five disruptive technology directions for 5G, IEEE Commun. Mag. 52 (2) (2014) 74–80.
[2] T. Bogale, L. Le, Massive MIMO and mmWave for 5G wireless HetNet: potential benefits and challenges, IEEE Veh. Technol. Mag. 11 (1) (2016) 64–75.
[3] F. Khan, Z. Pi, mmWave mobile broadband (MMB): unleashing the 3–300 GHz spectrum, in: Proceedings of IEEE Sarnoff Symposium, 2011, pp. 1–6.

[4] M. Jacob, S. Priebe, R. Dickhoff, T. Kleine-Ostmann, T. Schrader, T. Kurner, Diffraction in mm and sub-mm wave indoor propagation channels, IEEE Trans. Microw. Theory Tech. 60 (3) (2012) 833–844.

[5] Z. Shi, R. Rongxing Lu, J. Chen, X.S. Shen, Three-dimensional spatial multiplexing for directional millimeter-wave communications in multi-cubicle office environments, in: Proceedings of IEEE Global Telecommunications Conference, 2013, pp. 4384–4389.

[6] C. Kourogiorgas, S. Sagkriotis, A.D. Panagopoulos, Coverage and outage capacity evaluation in 5G millimeter wave cellular systems: impact of rain attenuation, in: Proceedings of European Conference on Antennas and Propagation (EuCAP), 2015, pp. 1–5.

[7] M.R. Akdeniz, Y. Liu, M.K. Samimi, S. Sun, S. Rangan, T.S. Rappaport, E. Erkip, Millimeter wave channel modeling and cellular capacity evaluation, IEEE J. Sel. Areas Commun. 32 (6) (2014) 1164–1179.

[8] T.S. Rappaport, J.N. Murdock, F. Gutierrez, State of the art in 60 GHz integrated circuits & systems for wireless communications, Proc. IEEE 99 (8) (2011) 1390–1436.

[9] Z. Qingling, Rain attenuation in millimeter wave ranges, in: Proceedings of International Symposium on Antennas, Propagation & EM Theory (ISAPE), 2006, pp. 1–4.

[10] S. Joshi, S. Sancheti, Foliage loss measurements of tropical trees at 35 GHz, in: Proceedings of IEEE International Conference on Recent Advances in Microwave Theory and Applications, 2008, pp. 531–532.

[11] D.P. Palomar, M.A. Lagunas, J.M. Cioffi, Optimum linear joint transmit-receive processing for MIMO channels with QoS constraints, IEEE Trans. Signal Process. 52 (5) (2004) 1179–1197.

[12] T.E. Bogale, L. Vandendorpe, Weighted sum rate optimization for downlink multiuser MIMO coordinated base station systems: centralized and distributed algorithms, IEEE Trans. Signal Process. 60 (4) (2011) 1876–1889.

[13] T.E. Bogale, L.B. Le, A. Haghighat, L. Vandendorpe, On the number of RF chains and phase shifters, and scheduling design with hybrid analog-digital beamforming, IEEE Trans. Wirel. Commun. 15 (5) (2016) 3311–3326.

[14] O.E. Ayach, S. Rajagopal, S. Abu-Surra, Z. Pi, R.W. Heath, Spatially sparse precoding in millimeter wave MIMO systems, IEEE Trans. Wirel. Commun. 13 (3) (2014) 1499–1513.

[15] G. Kwon, Y. Shim, H. Park, H.M. Kwon, Design of millimeter wave hybrid beamforming systems, in: Proceedings of IEEE Vehicular Technology Conference (VTC-Fall), 2014, pp. 1–5.

[16] Y.-Y. Lee, C.-H. Wang, Y.-H. Huang, A hybrid RF/baseband precoding processor based on parallel-index-selection matrix-inversion-bypass simultaneous orthogonal matching pursuit for millimeter wave MIMO systems, IEEE Trans. Signal Process. 63 (2) (2015) 305–317.

[17] T.E. Bogale, L.B. Le, Beamforming for multiuser massive MIMO systems: digital versus hybrid analog-digital, in: Proceedings of IEEE Global Telecommunications Conference (GLOBECOM), Austin, TX, 2014, pp. 4066–4071.

[18] D. Gesbert, H. Bolcskei, D. Gore, A. Paulraj, Outdoor MIMO wireless channels: models and performance prediction, IEEE Trans. Commun. 50 (12) (2002) 1926–1934.

[19] C. Sheldon, M. Seo, E. Torkildson, M. Rodwell, U. Madhow, Four-channel spatial multiplexing over a millimeter-wave line-of-sight link, in: Proceedings of IEEE International Microwave Symposium Digest (MTT-S), 2009, pp. 389–392.

[20] T.L. Marzetta, How much training is required for multiuser MIMO? in: Proceedings of the 40th Asilomar Conference on Signals, Systems and Computers, Pacific Grove, CA, 2006, pp. 359–363.

[21] T.E. Bogale, L.B. Le, Pilot optimization and channel estimation for multiuser massive MIMO systems, in: Proceedings of IEEE Conference on Information Sciences and Systems (CISS), 2014, pp. 1–5.

[22] T.E. Bogale, L.B. Le, X. Wang, Hybrid analog-digital channel estimation and beamforming: training-throughput tradeoff, IEEE Trans. Commun. 63 (12) (2015) 5235–5249.

[23] D. Tse, P. Viswanath, Fundamentals of Wireless Communication, Cambridge University Press, Cambridge, UK, 2005.

[24] J.G. Andrews, S. Buzzi, W. Choi, S.V. Hanly, A. Lozano, A.C.K. Soong, J.C. Zhang, What will 5G be? IEEE Sel. Areas Commun. 32 (6) (2014) 1065–1082.

[25] J.C. Chen, C.P. Li, Tone reservation using near-optimal peak reduction tone set selection algorithm for PAPR reduction in OFDM systems, IEEE Signal Process. Lett. 17 (11) (2010) 933–936.

[26] Z. Pi, F. Khan, An introduction to millimeter-wave mobile broadband systems, IEEE Commun. Mag. 49 (6) (2011) 101–107.

[27] K. Zheng, L. Zhao, J. Mei, M. Dohler, W. Xiang, Y. Peng, 10 Gb/s HetSNets with millimeter-wave communications: access and networking challenges and protocols, IEEE Commun. Mag. 53 (1) (2015) 86–92.

[28] D.P. Palomar, A unified framework for communications through MIMO channels, PhD thesis, Technical University of Catalonia (UPC), Barcelona, Spain, 2003.

[29] T.E. Bogale, Transceiver design for single-cell and multi-cell downlink multiuser MIMO systems, PhD thesis, University Catholique de Louvain (UCL), Louvain la Neuve, Belgium, 2013.

[30] A.L. Swindlehurst, E. Ayanoglu, P. Heydari, F. Capolino, Millimeter-wave massive MIMO: the next wireless revolution? IEEE Commun. Mag. 52 (9) (2014) 56–62.

[31] T. Bai, A. Alkhateeb, R.W. Heath, Coverage and capacity of millimeter-wave cellular networks, IEEE Commun. Mag. 52 (9) (2014) 70–78.

[32] V. Venkateswaran, A.-J.V. Veen, Analog beamforming in MIMO communications with phase shift networks and online channel estimation, IEEE Trans. Signal Process. 58 (8) (2010) 4131–4143.

[33] X. Zhang, A.F. Molisch, S.-Y. Kung, Variable-phase-shift-based RF-baseband codesign for MIMO antenna selection, IEEE Trans. Signal Process. 53 (11) (2005) 4091–4103.

[34] A. Wiesel, Y.C. Eldar, S. Shamai, Linear precoding via conic optimization for fixed MIMO receivers, IEEE Trans. Signal Proc. 54 (1) (2006) 161–176.

[35] T. Endeshaw, L. Vandendorpe, Sum rate optimization for coordinated multi-antenna base station systems, in: Proceedings of IEEE International Conference on Communications (ICC), Kyoto, Japan, 2011, pp. 1–5.

[36] L. Chen, Y. Yang, X. Chen, W. Wang, Multi-stage beamforming codebook for 60 GHz WPAN, in: Proceedings of the 16th International ICST Conference on Communications and Networking, China, 2011, pp. 361–365.

[37] S. Hur, T. Kim, D.J. Love, J.V. Krogmeier, T.A. Thomas, A. Ghosh, Millimeter wave beamforming for wireless backhaul and access in small cell networks, IEEE Trans. Commun. 61 (10) (2013) 4391–4403.

[38] Y. Tsang, A. Poon, S.S. Addepalli, Coding the beams: improving beamforming training in mmWave communication system, in: Proceedings of IEEE Global Telecommunications Conference (GLOBECOM), 2011, pp. 1–6.

[39] S. Sun, T.S. Rappaport, R.W. Heath, A. Nix, S. Rangan, MIMO for millimeter-wave wireless communications: beamforming, spatial multiplexing, or both? IEEE Commun. Mag. 52 (12) (2014) 110–121.

[40] S. Han, I. Chih-Lin, Z. Xu, C. Rowell, Large-scale antenna systems with hybrid analog and digital beamforming for millimeter wave 5G, IEEE Commun. Mag. 53 (1) (2015) 186–194.

[41] A. Alkhateeb, J. Mo, N. Gonzalez-Prelcic, R.W. Heath, MIMO precoding and combining solutions for millimeter-wave systems, IEEE Commun. Mag. 52 (2014) 186–195.

[42] L. Wei, R.Q. Hu, Y. Qian, G. Wu, Key elements to enable millimeter wave communications for 5G wireless systems, IEEE Wirel. Commun. Mag. 52 (2014) 136–143.

[43] J. Wang, Z. Lan, C. Pyo, T. Baykas, C. Sum, M. Rahman, J. Gao, R. Funada, F. Kojima, H. Harada, Beam codebook based beamforming protocol for multi-Gbps millimeter-wave WPAN systems, IEEE J. Sel. Areas Commun. 27 (8) (2009) 1390–1399.

[44] J.A. Zhang, X. Huang, V. Dyadyuk, Y.J. Guo, Massive hybrid antenna array for millimeter-wave cellular communications, IEEE Wirel. Commun. Mag. 22 (1) (2015) 79–87.

[45] F. Gholam, J. Via, I. Santamaria, Beamforming design for simplified analog antenna combining architectures, IEEE Trans. Veh. Technol. 60 (5) (2011) 2373–2378.

[46] J. Singh, S. Ramakrishna, On the feasibility of codebook-based beamforming in millimeter wave systems with multiple antenna arrays, IEEE Trans. Wirel. Commun. 14 (5) (2015) 2670–2683.

[47] L. Liang, W. Xu, X. Dong, Low-complexity hybrid precoding in massive multiuser MIMO systems, IEEE Wirel. Commun. Lett. 3 (6) (2014) 653–656.

[48] S. Han, I. Chih-Lin, Z. Xu, S. Wang, Reference signals design for hybrid analog and digital beamforming, IEEE Commun. Lett. 18 (7) (2014) 1191–1193.

[49] S.-H. Wu, K.-Y. Lin, L.-K. Chiu, Hybrid beamforming using convex optimization for SDMA in millimeter wave radio, in: Proceedings of IEEE Personal, Indoor and Mobile Radio Communications (PIMRC), 2009, pp. 823–827.

[50] B. Li, Z. Zhou, H. Zhang, A. Nallanathan, Efficient beamforming training for 60-GHz millimeter-wave communications: a novel numerical optimization framework, IEEE Trans. Veh. Technol. 63 (2) (2014) 703–717.

[51] E. Zhang, C. Huang, On achieving optimal rate of digital precoder by RF-baseband codesign for MIMO systems, in: Proceedings of IEEE Vehicular Technology Conference (VTC Fall), 2014, pp. 1–5.

[52] G.J. Foschini, K. Karakayali, R.A. Valenzuela, Coordinating multiple antenna cellular networks to achieve enormous spectral efficiency, IEE Proc. Commun. 153 (4) (2006) 548–555.

[53] L. Zheng, D.N.C. Tse, Diversity and multiplexing: a fundamental tradeoff in multiple-antenna channels, Trans. Commun. 49 (5) (2003) 1073–1096.

[54] A. Ghosh, T.A. Thomas, M.C. Cudak, R. Ratasuk, P. Moorut, F.W. Frederick, W. Vook, T.S. Rappaport, G.R. MacCartney, S. Sun, S. Nie, Millimeter-wave enhanced local area systems: a high-data-rate approach for future wireless networks, IEEE J. Sel. Areas Commun. 32 (6) (2014) 1152–1163.

[55] E. Torkildson, C. Sheldon, U. Madhow, M. Mark Rodwell, Millimeter-wave spatial multiplexing in an indoor environment, in: Proceedings of IEEE Global Telecommunications Conference, 2009, pp. 1–6.

[56] A. Adhikary, E.A. Safadi, M.K. Samimi, R. Wang, G. Caire, T.S. Rappaport, A. F. Molisch, Joint spatial division and multiplexing for mm-wave channels, IEEE J. Sel. Topics Commun. 32 (6) (2014) 1239–1255.

[57] A. Arvanitis, G. Anagnostou, N. Moraitis, P. Constantinou, Capacity study of a multiple element antenna configuration in an indoor wireless channel at 60 GHz, in: Proceedings of IEEE Vehicular Technology Conference (VTC-Spring), 2007, pp. 609–613.

[58] F. Bohagen, P. Orten, G. Oien, Design of optimal high-rank Line-of-sight MIMO channels, IEEE Trans. Wirel. Commun. 6 (4) (2007) 1420–1425.

[59] O. Elijah, C.Y. Leow, T.A. Rahman, S. Nunoo, S.Z. Iliya, A comprehensive survey of pilot contamination in massive MIMO-5G system, IEEE Commun. Surveys Tuts. 18 (2) (2015) 905–923.

[60] Y. Liu, Z. Tan, H. Hu, L.J. Cimini, G.Y. Li, Channel estimation for OFDM, IEEE Commun. Surveys Tuts. 16 (4) (2014) 1891–1908.

[61] H. Bolcskei, R.W. Heath, A.J. Paulraj, Blind channel identification and equalization in OFDM-based multiantenna systems, IEEE Trans. Signal Process. 50 (1) (2002) 96–109.

[62] T.L. Marzetta, Noncooperative cellular wireless with unlimited numbers of base station antennas, IEEE Trans. Wirel. Commun. 9 (11) (2010) 3590–3600.

[63] Z. Pi, F. Khan, A millimeter-wave massive MIMO system for next generation mobile broadband, in: Proceedings of Asilomar Conference on Signals, Systems, and Computers, 2012, pp. 693–698.

[64] R.W. Heath, Comparing massive MIMO and mmWave MIMO, 2014. https://www.ieee-ctw.org/2014/slides/session3/Heath-CTW_v6.pdf.

[65] M. Biguesh, A.B. Gershman, Training-based MIMO channel estimation: a study of estimator tradeoffs and optimal training signals, IEEE Trans. Signal Process. 54 (3) (2006) 884–893.

[66] W. Bajwa, J. Haupt, A. Sayeed, R. Nowak, Compressed channel sensing: a new approach to estimating sparse multipath channels, Proc. IEEE 98 (6) (2010) 1058–1076.

[67] A. Alkhateeb, O.E. Ayach, G. Leusz, R.W. Heath, Channel estimation and hybrid precoding for millimeter wave cellular systems, IEEE J. Sel. Topics Signal Process. 8 (2014) 831–846.

[68] M. Malloy, R. Nowak, Near-optimal adaptive compressed sensing, IEEE Trans. Inform. Theory 60 (7) (2014) 4001–4012.

[69] D. Ramasamy, S. Venkateswaran, U. Madhow, Compressive adaptation of large steerable arrays, in: Proceedings of Information Theory and Applications Workshop (ITA), 2012, pp. 234–239.

[70] J. Lee, G.-T. Gil, Y.H. Lee, Exploiting spatial sparsity for estimating channels of hybrid MIMO systems in millimeter wave communications, in: Proceedings of IEEE Global Telecommunications Conference, 2014, pp. 3326–3331.

[71] T.S. Rappaport, S. Sun, R. Mayzus, H. Zhao, Y. Azar, K. Wang, G.N. Wong, J.K. Schulz, M. Samami, F. Gutierrez, Millimeter wave mobile communications for 5G cellular: it will work! IEEE Access 1 (2013) 335–349.

[72] M. Samimi, K. Wang, Y. Azar, G.N. Wong, R. Mayzus, H. Zhao, J.K. Schulz, S. Sun, F. Gutierrez, T.S. Rappaport, 28 GHz angle of arrival and angle of departure analysis for outdoor cellular communications using steerable beam antennas in New York city, in: Proceedings of IEEE VTC-Fall, 2013, pp. 1–6.

[73] P.A. Eliasi, S. Rangan, T.S. Rappaport, Low-rank spatial channel estimation for millimeter wave cellular systems, 2014. CoRR abs/1410.4831.

[74] S. Boyd, L. Vandenberghe, Convex optimization, Cambridge University Press, Cambridge, UK, 2004, pp. 1–716. http://www.ams.org/mathscinet-getitem?mr=2061575.

[75] S.L. Loyka, Channel capacity of MIMO architecture using the exponential correlation matrix, IEEE Commun. Lett. 5 (9) (2001) 369–371.

[76] T.E. Bogale, L.B. Le, X. Wang, Hybrid analog-digital channel estimation and beamforming: training-throughput tradeoff (draft with more results and details), 2015. http://arxiv.org/abs/1509.05091.

[77] M. Ding, Multiple-input multiple-output wireless system designs with imperfect channel knowledge, PhD thesis, Queens University Kingston, Ontario, Canada, 2008.

[78] J.B. Anderson, F. Rusek, V. Owall, Faster-than-Nyquist signaling, Proc. IEEE 101 (8) (2013) 1817–1830.

[79] J.E. Mazo, Faster-than-Nyquist signaling, Bell Syst. Technical J. 54 (8) (1975) 1451–1462.

[80] A.D. Liveris, C.N. Georghiades, Exploiting faster-than-Nyquist signaling, IEEE Trans. Commun. 51 (9) (2003) 1502–1511.

[81] A. Morello, V. Mignone, DVB-S2: the second generation standard for satellite broadband services, Proc. IEEE 94 (1) (2006) 210–227.

[82] J.G. Andrews, A. Ghosh, R. Muhamed, Fundamentals of WiMAX: understanding broadband wireless networking, Pearson Education, Upper Saddle River, NJ, 2007, pp. 1–478.

[83] N. Benvenuto, R. Dinis, D. Falconer, S. Tomasin, Single carrier modulation with nonlinear frequency domain equalization: An idea whose time has come again, Proc. IEEE 98 (1) (2010) 69–96.

[84] M.R. Akdeniz, Y. Liu, S. Rangan, E. Erkip, Millimeter wave picocellular system evaluation for urban deployments, in: Proceedings of IEEE Global Telecommunications Conference, 2013, pp. 105–110.

[85] M. Abouelseoud, G. Charlton, System level performance of millimeter-wave access link for outdoor coverage, in: Proceedings of IEEE Wireless Communications and Networking Conference, 2013, pp. 4146–4151.

[86] R. Taori, A. Sridharan, Point-to-multipoint in-band mmWave backhaul for 5G networks, IEEE Commun. Mag. 53 (1) (2015) 195–201.

[87] Y. Niu, C. Gao, Y. Li, L. Su, D. Jin, A.V. Vasilakos, Exploiting device-to-device transmissions in joint scheduling of access and backhaul for small cells in 60 GHz band, IEEE J. Sel. Commun. 33 (10) (2015) 2052–2069.

[88] C.J. Bernardos, A.D. Domenico, J. Ortin, P. Rost, D. Wubben, Challenges of designing jointly the backhaul and radio access network in a cloud-based mobile network, in: Proceedings of Future Network and Mobile Summit, 2013, pp. 1–10.

[89] C. Dehos, J.L. Gonzalez, A.D. Domenico, D. Ktenas, L. Dussopt, Millimeter-wave access and backhauling: the solution to the exponential data traffic increase in 5G mobile communications systems? IEEE Commun. Mag. 52 (9) (2014) 88–95.

MAC layer design for mmWave massive MIMO

10

G. Lee, Y. Sung

Korea Advanced Institute of Science and Technology, Daejeon, South Korea

CHAPTER OUTLINE

10.1 INTRODUCTION

Medium access control or multiple access control (MAC) has been one of the main topics under discussion in relation to cellular networks in which a base station (BS) communicates to multiple users in downlink and multiple users communicate to a BS in uplink in each cell. In classical circuit-switched networks with voice communication as their main application, such as 1G and 2G cellular networks, the channel resource between the BS and multiple users is divided into a fixed number of orthogonal channels, and these orthogonal channels are assigned to users who request voice communication. When all channels are occupied by users there are no more channels left for additional users and additional users requiring communication service experience communication resource shortage called blocking. In such schemes, the

mmWave Massive MIMO. http://dx.doi.org/10.1016/B978-0-12-804418-6.00010-8

number of orthogonal channels is determined to maintain a certain probability of blocking with consideration of the average number of users in the cell and their activity factor. Examples of such schemes include frequency-division multiple access (FDMA) in 1G and time-division multiple access (TDMA) and code-division multiple access (CDMA) in 2G.

With the pervasive use of the Internet in late 1990s, efficient wireless packet data communication was incorporated into cellular networks. Since then, wireless packet data communication has been crucial in cellular networks. The design of wireless packet data cellular networks is inherently different from that of circuit-switched cellular networks. First, the arrival of wireless packets, e.g., emails and image files, for each user is bursty. Second, most applications for packet communication require much less stringent delay constraints. These two aspects of wireless packet data combined with the properties of wireless channels enabled a new paradigm for MAC design in cellular networks. Traditionally, wireless channel fading or fluctuation due to mobility was considered to be a detrimental effect that should be compensated for by power control to provide a constant quality communication link for each voice user. In wireless packet data communication, however, channel fading is exploited rather than compensated for to yield good overall system performance [1, 2]. Suppose that the communication resource such as time is divided into blocks in time domain and each resource block is assigned to one of many packet data users, as in early single user (SU) selection scheduling methods [3–6], and that each user experiences independent channel fading over time blocks. Then, there is a high probability that the wireless channel of one of the users is in good condition when there are many users in the cell. By assigning each communication resource block to the user with the best channel condition at that time block, it is possible to achieve good overall system performance. In this case, the channel fading is not compensated for but exploited for efficient system performance. The diversity gain provided by the selection of a user or users with a good channel condition among the many users with independent channel fading is called *multiuser (MU) diversity* [1, 2] and the process of dynamically selecting some users for each communication resource block among the many active users in the cell based on their channel conditions is called *user scheduling*. MU diversity and user scheduling are the key aspects of MAC design for cellular packet networks [7].

MU diversity gain varies with channel environments and scheduling methods, and thus various scheduling algorithms have been proposed under different channel and system environments over the last two decades. In the early packet data cellular networks, user scheduling was performed in one dimension, for example, the TDMA-based downlink of the IS-856 cellular standard [2, 6]. In this one-dimensional case, the communication resource such as time is divided into one-dimensional blocks and a single user is scheduled to each resource block. Thus, the design of a scheduler is a rather simple problem. For example, the proportional fair (PF) scheduling algorithm applied to the one-dimensional case optimizes the system data rate and the fairness among the users, and it was adopted in IS-856. However, as orthogonal frequency-division multiplexing/multiple access (OFDM/OFDMA) was adopted as the physical-layer transmission technology in IEEE 802.16 (WiMAX), IEEE 802.20 (MBWA), and 3GPP LTE, more flexible resource allocation was possible over the two-dimensional communication resource space of time and frequency, but optimal resource allocation of two-dimensional blocks with power allocation to users in the cell is more complicated than the one-dimensional case. Various scheduling and resource allocation methods have been proposed for OFDM/OFDMA downlink and uplink networks under various criteria [8–13]. Furthermore, the MIMO technology was incorporated to OFDM/OFDMA for 4G cellular networks. With the MIMO technology, multiple users can simultaneously be supported at the same time-frequency resource block based on spatial-division multiple access (SDMA). Thus, in 4G cellular networks adopting MIMO-OFDM user scheduling can be performed in the three-dimensional resource space of time, frequency, and space, and optimal user scheduling in the three-dimensional resource allocation is a far more difficult problem than that of the one- or two-dimensional systems. Thus, many practical scheduling algorithms simplify the problem by using the capability of multiple data stream transmission provided by MIMO to transmit multiple data streams to one user. In this case, for each resource block SU MIMO is implemented between the BS and the selected user for the resource block and scheduling in MIMO-OFDM networks reduces to the previous two-dimensional problem [14–16]. However, it was shown that the sum rate of SU-MIMO-based MIMO downlink scheduling achieves only a small fraction of the sum rate capacity of a MIMO broadcast channel (BC) when the number of transmit antennas at the BS is larger than that of the receive antennas at each user [17]. This is mainly because the extra number of transmit antennas at the BS is used to increase the power gain rather than the multiplexing gain associated with multiple antennas. On the contrary, scheduling based on MU-MIMO, where multiple data streams provided by MIMO are assigned to multiple users at the same time-frequency resource block based on beamforming, can achieve a large fraction of the sum rate capacity of a MIMO BC by attaining the full multiplexing gain [18, 19]. It has been shown that when the number of active users in the cell grows without bound, the optimal sum rate capacity of a MIMO BC obtained by dirty paper coding (DPC) can be achieved asymptotically by some smart MU scheduling. Specifically, both the sum rate capacity of a MIMO BC and the sum rates of smart MU-MIMO scheduling algorithms scale as $M \log \log K$ for fixed M under independent Rayleigh fading assumption, where M and K are the numbers of transmit antennas and active users in the cell, respectively [18, 19].

Recently, the MIMO technology has evolved to massive MIMO, which is considered to be one of the key technologies for 5G. In massive MIMO, large-scale antenna arrays are used at BSs to harness the gain of MIMO beyond the small-scale MIMO of four or eight antennas at the BS used in 4G. With massive MIMO it is possible to attain significant capacity increases and energy efficiency [20]. In rich scattering environments, such as in sub-6 GHz radio bands, the use of a large number of antennas results in asymptotic orthogonality among users' channel vectors and thus simple linear processing, such as maximum ratio transmission (MRT) in downlink and maximum ratio combining (MRC) in uplink, can be used to remove inter-user interference. Furthermore, the use of a large number of antennas also results in channel hardening, which means that small-scale channel fading vanishes due to the law of large numbers. This channel hardening effect simplifies resource allocation and power control. Specifically, frequency-domain user scheduling may not be necessary because all subcarriers of OFDM/OFDMA have almost the same channel gain due to the channel hardening. Thus, the whole bandwidth can be assigned to each scheduled user separated by MIMO in the spatial domain. In this case, the user scheduling problem focuses on how to select multiple users simultaneously supported in the spatial domain at each scheduling time interval. Massive MIMO generally requires a large amount of feedback from each user due to the large dimension of channel vectors and thus it is important to devise scheduling algorithms having reasonable performance with a relatively small amount of feedback based on partial channel state information (CSI) feedback from each user. Several scheduling algorithms have been proposed and analyzed for massive MIMO in this respect [21–25].

Another key technology for 5G is the use of mmWave bands for commercial cellular networks. The use of mmWave bands provides wide bandwidth for ultra-high data rates. However, the radio propagation characteristics in mmWave bands are quite different from those in sub-6 GHz bands mostly characterized as rich scattering with the popular Rayleigh or Rician fading channel model. In mmWave bands, the radio propagation experiences severe path loss and there exist very few multiple paths [26, 27]. Such large path losses can be compensated for by very sharp beamforming based on large antenna arrays at the BS and user terminals. Consequently, it is natural to incorporate massive MIMO into wireless communication in mmWave bands. From the perspective of user scheduling, it is important to analyze the performance of conventional MU-MIMO scheduling algorithms in the new mmWave channel environment and devise MU-MIMO scheduling algorithms suitable for the mmWave channel environment [24, 28–31].

This chapter introduces several key MU-MIMO scheduling algorithms for MIMO and massive MIMO, and some theoretical results related to the introduced scheduling algorithms in sub-6 GHz and mmWave bands, mainly focusing on downlink scheduling. The remainder of this chapter is organized as follows. Section 10.2 describes basic user scheduling methods in single-input single-output (SISO) networks as background. Sections 10.3 and 10.4 provide several key scheduling methods in MIMO and some results on their scheduling gain in massive MIMO under rich scattering environments, respectively. Efficient user scheduling methods for mmWave massive MIMO are discussed in Section 10.5, followed by a summary in Section 10.6.

10.2 BASIC SCHEDULING ALGORITHMS

As background of later sections, this section explains three basic scheduling methods in a TDMA network to provide a basic understanding of user scheduling. Consider the situation where multiple active downlink users exist in a cell, the channel gain from the BS to each user varies due to long- and/or short-term channel fading, as illustrated in Fig. 10.1, and the BS selects one user for each scheduling time interval for downlink data transmission based on channel gain feedback from each user. The first method is the *maximum rate scheduler* with the data rate as the optimization goal. This scheduler selects the user that has the maximum instantaneous rate for the given scheduling interval among all active users in the cell. That is, the index $k*$ of the scheduled user is given by:

$$k^* = \arg\max{}_{k\in\mathcal{K}}R_k,\tag{10.1}$$

where R_k is the instantaneous data rate for user k depending on the channel gain of user k, and $\mathcal{K}=\{1,2,...,K\}$ is the set of active user indices. Since the maximum rate scheduler aims to select the best rate user at each scheduling interval, it is advantageous in terms of the system rate. Indeed, it is optimal for maximizing the system rate of a fast fading channel model under an average transmit power constraint when combined with downlink power control [2]. However, this scheduling algorithm is not fair from the perspective of an individual user's rate when the average channel gain of each user is different due to long-term fading resulting from shadowing effects and the different distance from the BS to the user. In such situations, the users close to the BS with good channel gains are mostly served and the users far from the BS with bad channel gains are not served. The second scheduler to fix this problem of inequality is the *round-robin scheduler*. This scheduler serves all active users in the cell one by one for each scheduling interval so that users with bad channel gains are also served with equal opportunities. However, with the round-robin scheduler the opportunity of attaining larger system data rate by assigning the communication resource to users with good channel gains is lost by equally distributing the communication resource to all users regardless of their channel gains and thus discarding the MU diversity gain. The third scheduler, named the *proportionally fair (PF) scheduler*, tries to compromise the MU diversity gain and the fairness among users [2, 6]. In the PF scheduling policy, the BS finds a user k that has the ratio of the maximum instantaneous rate at the scheduling interval to the average served rate until the previous scheduling interval, i.e.:

$$k^*[t] = \arg\max{}_k \frac{R_k[t]}{\mu_k[t]},\tag{10.2}$$

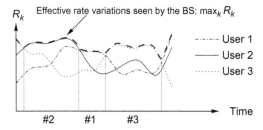

FIG. 10.1

Maximum rate scheduler.

E. Dahlman, S. Parkvall, J. Skold, 4G:LTE/LTE-Advanced for Mobile Broadband, second edition, Academic Press, New York, 2013.

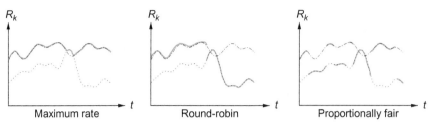

FIG. 10.2

Comparison of basic schedulers in a two-user case.

E. Dahlman, S. Parkvall, J. Skold, 4G:LTE/LTE-Advanced for Mobile Broadband, second edition, Academic Press, New York, 2013.

where $R_k[t]$ is the instantaneous rate of user k at scheduling interval t and $\mu_k(t)$ is the average served rate of user k until scheduling interval $t - 1$ updated as:

$$\mu_k[t+1] = \begin{cases} (1-\delta)\mu_k[t] + \delta R_k[t], & \text{if } k = k^*[t], \\ (1-\delta)\mu_k[t], & \text{otherwise.} \end{cases} \quad (10.3)$$

Here, δ is the parameter of the first-order autoregressive (AR) averaging filter. Since the PF scheduler selects a user that has the ratio of the maximum instantaneous rate at the scheduling interval to the average served rate until the previous scheduling interval, a user with a bad average channel gain can also be scheduled by the PF scheduler due to the low average served rate for such a user. Fig. 10.2 illustrates the operation of the three basic schedulers in a two-user case.

10.3 USER SCHEDULING IN MU-MIMO

In this section, we mainly consider the MU multiple-input single-output (MU-MISO) downlink situation where the BS has M transmit antennas and there exists $K (> M)$ single-receive antenna users in the cell, as illustrated in Fig. 10.3.

The capacity of a communication link can be expressed as:

$$C = D\log\left(1 + \frac{P}{N}\right), \quad (10.4)$$

where D is the degrees-of-freedom (DoF) for communication, P is the received signal power, and N is the noise power. The multiple-antenna transmission technology provides two major gains: the *power gain*, which increases the received signal power P and the *multiplexing gain*, which provides more DoF D for communication. Under the considered MU-MISO downlink, if only one user is selected at each scheduling interval, a point-to-point MISO channel is formed from the BS to the selected user and thus only the power gain by beamforming is available in this case. However, if M users out of all K users are selected, then a MIMO channel is formed from the BS to the selected M users. If all M channel vectors of the selected users are linearly independent, the DoF of the $M \times M$ MIMO channel from the BS to the selected M

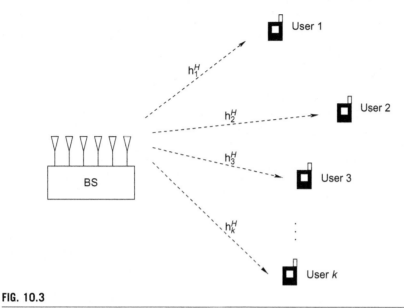

FIG. 10.3

Multiuser MISO downlink.

users is M. Since the DoF changes the slope of the capacity increase with respect to signal-to-noise ratio (SNR), whereas the power gain only increases the SNR value itself, the multiplexing gain increasing the DoF is far more beneficial to improve the system rate in the medium and high SNR range. The situation is similar in the case that each user has multiple receive antennas, but the BS has more transmit antennas than each user [17]. Thus, it is preferable to apply MU-MIMO and schedule multiple users simultaneously for multiple antennas at each scheduling interval rather than to select only one user based on SU-MIMO at each scheduling interval, although the latter can simply be implemented by applying the criteria discussed in Section 10.2 with feedback of the rate R_k of each user k given by:

$$R_k = \log \det \left(\mathbf{I} + \mathbf{H}_k \mathbf{Q} \mathbf{H}_k^H \right), \tag{10.5}$$

where $\mathbf{H}_k$ is the channel matrix from the BS to user k and $\mathbf{Q}$ is the covariance matrix of the transmit signal at BS.

The MU-MISO cellular downlink described in Fig. 10.3 can be modeled as a MISO BC, where the data model is given by:

$$\begin{bmatrix} y_1 \\ y_2 \\ \vdots \\ y_K \end{bmatrix} = \begin{bmatrix} \mathbf{h}_1^H \\ \mathbf{h}_2^H \\ \vdots \\ \mathbf{h}_K^H \end{bmatrix} \begin{bmatrix} x_1 \\ \vdots \\ x_M \end{bmatrix} + \begin{bmatrix} w_1 \\ w_2 \\ \vdots \\ w_K \end{bmatrix} = \begin{bmatrix} \mathbf{h}_1^H \\ \mathbf{h}_2^H \\ \vdots \\ \mathbf{h}_K^H \end{bmatrix} \mathbf{x} + \mathbf{w} = \mathbf{H}_{\text{overall}} \mathbf{x} + \mathbf{w}. \tag{10.6}$$

Here, x_i is the transmitted signal at transmit antenna i, y_j is the received signal at receiver j, $\mathbf{h}_j^H$ is the $1 \times M$ channel vector from the BS to receiver j, and w_j is the zero-mean white Gaussian thermal noise at receiver j. Typically, the BS has a transmit power constraint given by $\mathbb{E}\{\text{tr}(\mathbf{xx}^H)\} = \mathbb{E}\{\mathbf{x}^H\mathbf{x}\} \leq P_t$. The main point of the MISO BC is that the intended message for user j is decoded solely based on its own received signal y_j without help from other users. It is known that the capacity of the MISO BC (10.6) is achieved by the DPC scheme [32]. Suppose that the scheduling of M users out of all $K(> M)$ users is required. Then, under the assumption that all channel vectors are fed back to the BS, one can compute the capacity of the M-user MISO BC composed of the BS and the selected M users achievable again by DPC for each selection of M users, and one can then select the best M users out of all $\binom{K}{M}$ possible combinations. However, such a brute-force method requires heavy computational complexity for large K and M. Hence, more efficient algorithms have been developed. Among many scheduling methods based on DPC, a well-known method is the greedy user-selection method based on zero-forcing (ZF) beamforming and DPC (gZF-DP), proposed by Tu and Blum [33] and analyzed later by Dimic and Sidiropoulos [34]. However, DPC is still difficult to implement in a practical way [35]. To circumvent the difficulty of DPC implementation, several low-complexity scheduling schemes have been proposed based on linear beamforming for the last decade [18, 19, 36]. Among them two representative scheduling methods from which many other scheduling methods are derived are the *semiorthogonal user selection (SUS) with ZF beamforming* proposed by Yoo and Goldsmith [19] and the *random beamforming (RBF)* proposed by Sharif and Hassibi [18]. The amount of feedback for the two methods is quite different since the former requires full vector CSI feedback but the latter only requires partial CSI feedback; however, both methods are asymptotically optimal in the sense that the achievable rates of the two methods scale at the same capacity rate of MU-MISO BC (10.6) as the number K of users grows without bound. In this section we introduce several MU-MISO scheduling methods, including RBF and SUS, and the related results before considering user scheduling in mmWave massive MIMO.

10.3.1 RBF AND SUS

First, consider the RBF method [18] in the MU-MISO downlink network illustrated in Fig. 10.3 described by the model (10.6). The specific procedure of RBF is as follows:

1. The BS generates an arbitrary set of random orthonormal beam vectors $\{\mathbf{u}_b\}_{b=1}^M$ that are an orthonormal basis of the M (complex) dimensional space.
2. During the training period, the BS transmits each of the generated beam vectors sequentially in time. At the end of training, each user k computes the signal-to-interference-plus-noise ratio (SINR) for each beam b, and then feeds the maximum SINR value (i.e., $\max_b \text{SINR}_{k,b}$) and the corresponding beam index

(i.e., $\arg\max_b \text{SINR}_{k,b}$) back to the BS. Here, under the assumption that one user is assigned to each and every transmit beam, the SINR of user k for beam b is given by:

$$\text{SINR}_{k,b} = \frac{\rho|\mathbf{h}_k^H\mathbf{u}_b|^2}{1 + \rho\sum_{b'\neq b}|\mathbf{h}_k^H\mathbf{u}_{b'}|^2}, \tag{10.7}$$

where $\mathbf{h}_k$ is the channel vector of user k, $\rho = \dfrac{P_t}{M}$, and P_t is the total transmit power.

3. After the feedback is over, for each beam b, the BS schedules the user that has the maximum SINR among the users reporting the beam index b as their maximum SINR index.
4. After scheduling, the BS transmits M data streams to the M scheduled users with the beam vectors $\{\mathbf{u}_b\}_{b=1}^M$.

The main advantage of the RBF method is that only one real number (the maximum SINR value) and one integer (the maximum SINR beam index) value are required for RBF, whereas scheduling based on full CSI feedback requires $2M$ real numbers for each user [18]. Surprisingly, when each channel vector follows the independent and identically distributed (i.i.d.) Rayleigh fading channel model, i.e.:

$$\mathbf{h}_k \overset{\text{i.i.d.}}{\sim} \mathcal{CN}(\mathbf{0}, \mathbf{I}), \tag{10.8}$$

the sum rate of RBF has the same scaling as the capacity of the MU-MISO BC (10.6) for fixed M, as the number K of active users goes to infinity [18]. That is:

$$\lim_{K\to\infty} \frac{\mathcal{R}_{\text{RBF}}}{M\log\log K} = 1 \tag{10.9}$$

for fixed finite M, where $\mathcal{R}_{\text{RBF}} = \mathbb{E}\left[\sum_{b=1}^M \log\left(1 + \max_{1\leq k\leq K}\text{SINR}_{k,b}\right)\right]$ is the sum rate of RBF.

Whereas RBF is based on partial CSI feedback as seen in the above, the SUS method requires full CSI feedback. In the SUS method [19], the BS collects full CSI $\{\mathbf{h}_k\}_{k=1}^k$ from all users, and schedules users in a greedy sequential fashion so that the selected channel vectors are almost orthogonal and have large magnitudes. Specifically, the BS first selects the user k^* that has the largest channel magnitude among all K users. Then, the BS constructs a hyperslab based on the selected channel vector $\mathbf{h}_{k^*}$ defined as [19]:

$$\mathcal{H}_1 = \left\{\mathbf{h} : \frac{|\mathbf{h}_{k^*}^H\mathbf{h}|}{\|\mathbf{h}_{k^*}\| \cdot \|\mathbf{h}\|} \leq \xi\right\}, \tag{10.10}$$

where $\xi \in (0, 1)$ is a parameter that controls the thickness of the hyperslab. Note that when ξ is properly small, the channel vectors contained in the hyperslab $\mathcal{H}_1$ are semi-orthogonal to the firstly selected channel vector $\mathbf{h}_{k^*}$. Now, the BS considers only the users whose channel vectors are in $\mathcal{H}_1$, and selects the user among them that has the largest projection magnitude of its channel vector onto the orthogonal complement of $\mathbf{h}_{k^*}$. After the second user is selected, the BS constructs another hyperslab $\mathcal{H}_2$ contained in $\mathcal{H}_1$ based on the second user's channel vector. The BS now considers only the users whose channel vectors are in the second hyperslab $\mathcal{H}_2$ (note that the channel vectors in $\mathcal{H}_2$ are semiorthogonal to both the channel vectors of the two selected users), and selects the user that has the largest projection magnitude of its channel vector onto the orthogonal complement of the column space of the selected two users' channel vectors. In this way, the BS sequentially schedules semiorthogonal users with large channel magnitudes until M users are selected or there is no user whose channel vector is in the newly constructed hyperslab. After scheduling, the BS finally transmits data streams to the selected users with ZF beamforming (ZFBF) or minimum-mean-square-error (MMSE) beamforming. The purpose of the later ZFBF is to remove inter-user interference completely since the channel vectors of the selected users are not completely orthogonal and thus there still exists inter-user interference when the channel vectors themselves are used as the transmit beam vectors. It is known that this method is also asymptotically optimal under the i.i.d. Rayleigh fading channel model [19], but has superior performance to RBF in finite-user cases due to the full CSI feedback. As hinted by the two scheduling methods, optimal scheduling for MU-MISO downlink with linear beamforming involves selecting users that have roughly orthogonal channel vectors with large magnitudes.

10.3.2 LIMITED FEEDBACK, QUANTIZATION, AND TWO-PHASE USER SCHEDULING

Even though exploiting full CSI enables better user selection and more elaborate beamforming, it imposes a heavy feedback burden on the network, especially when the number of active users or the number of antennas is large. One approach to reduce the feedback overhead is to quantize the CSI fed back for user scheduling. There have been abundant studies on user scheduling with finite-rate feedback (FRF) and beamforming with FRF in MIMO BCs [37–40]. Jindal [37] showed that full multiplexing gain can be achieved when the feedback rate increases with the logarithm of SNR. Caire et al. [39] analyzed the achievable rate by considering not only the FRF, but also the channel estimation error via downlink training. Yoo et al. [38] considered MU-MISO, combined scheduling and FRF, proposed a ZFBF-based scheduling method with FRF, and analyzed its performance. Under the scheduling method in Yoo et al. [38], each user feeds channel direction information (CDI) and a scalar SINR value back to the BS for user selection and ZFBF design. Trivellato et al. [40] extended the work of Yoo et al. [38] to the case of multiantenna receivers. However, the scheduling methods based on full CSI feedback still requires heavy feedback overhead even with FRF, especially when the number of active users in the cell is large.

Another effort to reduce the feedback overhead is to apply two-phase feedback to user scheduling and beam design. Several scheduling algorithms were proposed by applying two-phase feedback to SUS [41, 42]. In the proposed method of Zakhour and Gesbert [41], each user feeds back a coarse quantized version of his/her channel vector and the BS selects users by the SUS algorithm based on the coarse channel vector. After user selection the BS receives refinements of the selected users' channel vectors and transmits to the selected users with ZFBF based on the refined channel information. Xu and Zhao [42] proposed another SUS-based two-phase scheduling method. In their method, each user feeds back the magnitude of its channel vector and the BS selects users whose channel magnitudes are larger than a predetermined threshold. Then, the BS requests CSI feedback from the selected users and chooses users for data transmission by the SUS algorithm based on the CSI feedback from the selected users. In this method, the number of selected users in the first phase still needs to be large to have a set of roughly orthogonal M users for data transmission by applying the SUS algorithm in the later stage.

Although SUS-based two-phase feedback scheduling methods reduce the feedback overhead compared to the original SUS algorithm, they still impose heavy feedback overhead on the network. The idea of two-phase feedback can also be applied to RBF-type scheduling methods based on partial CSI feedback to further reduce its overhead and enhance the performance over RBF. In this direction of two-phase feedback scheduling, the BS schedules users based on minimum feedback of partial CSI from each user as in RBF in the first phase, then requires the feedback of CDI or CSI only from the scheduled users in the second phase, and finally transmits multiple data streams to the selected users with elaborate beamforming based on the received CDI or CSI from the selected users. Kountouris et al. [43] proposed a RBF-based two-phase scheduling method in which the RBF method is used to select users in the first phase and after scheduling the BS controls the power allocation over the random beams without changing the direction of the initial beams based on the additional feedback from the selected users in the second phase. This method basically intends to overcome the performance degradation of the RBF method resulting from the predetermined training beams also used for data transmission. Recently, a new two-phase method that maintains the advantage of SUS, overcomes the disadvantage of RBF, and requires an amount of feedback comparable to RBF was proposed under the name of REference-based Distributed semiorthogonal user Selection with Postselection Beam Refinement (ReDOS-PBR) by Lee and Sung [25]. In this method, as in RBF, the BS first transmits a set of M orthonormal reference beam vectors $\{\mathbf{u}_b\}_{b=1}^{M}$ sequentially in time. Around each reference beam direction i, a nonoverlapping double cone $\mathcal{C}_i$ is defined with a predetermined $\xi' \in (0, 1)$ as:

$$\mathcal{C}_i = \left\{ \mathbf{h} : \frac{|\mathbf{h}^H \mathbf{u}_i|}{\|\mathbf{h}\|} \geq \xi' \right\}, \qquad i = 1, 2, \dots, M, \qquad (10.11)$$

as shown in Fig. 10.4. At the end of the first phase transmission, each user k knows $\{\mathbf{h}_k^H \mathbf{u}_1, \mathbf{h}_k^H \mathbf{u}_2, \dots, \mathbf{h}_k^H \mathbf{u}_K\}$ from its received signals. Based on $\{\mathbf{h}_k^H \mathbf{u}_1, \mathbf{h}_k^H \mathbf{u}_2, \dots, \mathbf{h}_k^H \mathbf{u}_K\}$, each user k computes its channel norm $\|\mathbf{h}_k\| = \sqrt{\sum_{i=1}^{M} |\mathbf{h}_k^H \mathbf{u}_i|^2}$ and check if its channel vector $\mathbf{h}_k$ belongs to one of the M cones. If its channel vector belongs to one of the M cones, each user k feeds back the *cone index* and its *channel norm* $\|\mathbf{h}_k\|$ instead of SINR reported by RBF to the BS. After the first feedback of the partial CSI, the BS

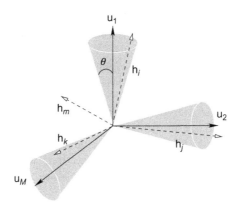

FIG. 10.4

User selection double cones (the other half of each double cone is not shown).

selects one user for each cone that has the maximum channel vector length in the cone. The selected M users in this way have roughly orthogonal channel vectors with large magnitudes since the M cones are roughly orthogonal by construction. Indeed, this follows the principle of SUS by selecting roughly orthogonal users with large channel magnitudes, but such selection is done in a distributed way based only on partial CSI feedback in the case of ReDOS-PBR. Since the channel vectors of the selected users are not perfectly orthogonal, the use of the reference beams as the data transmission beam vectors would yield inter-user interference. Hence, the actual data transmission beams are designed by applying ZFBF based on the feedback of the channel vectors of only the selected M users in the second feedback phase to eliminate inter-user interference. Note that this scheme requires the feedback of K or less real numbers (in the first feedback phase) and M^2 complex numbers (in the second feedback phase) in total. Thus, the reduction in the amount of feedback is significant for large K with small M as in small-scale MIMO systems. It is shown that this method is also asymptotically optimal in MU-MISO downlink and almost achieves the performance of SUS that requires full CSI feedback from all users, with significantly reduced feedback overhead, as shown in Fig. 10.5.

10.4 USER SCHEDULING IN MASSIVE MIMO

As seen in the previous section, the sum rate of RBF and SUS scales with the same rate as the capacity with scaling $O(M \log \log K)$ in MU-MISO downlink networks for fixed M under the i.i.d. Rayleigh fading assumption, as K increases [18, 19]. The considered situation with fixed M and large K models MU-MISO networks with small-scale BS antenna arrays of four or eight transmit antennas at the BS and several hundred or thousand active users in a cell, as in a 4G network. In massive MIMO, the

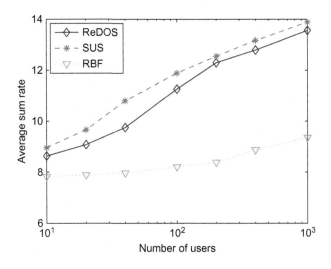

FIG. 10.5

Performance of ReDOS-PBR in MU-MISO downlink: $M = 4$, SNR $= 15$ dB, channel covariance matrix $[\mathbf{R}]_{i,j} = \nu^{|i-j|}$ with $\nu = 0.2$.

number M of transmit antennas at the BS is very large, in the order of hundreds. To analyze the performance of user scheduling in massive MIMO, it is necessary to consider the scenario of large M and the performance behavior as M increases. One of the important findings in this regard is that the scheduling gain obtained from smart user selection is insignificant under rich scattering environments, commonly modeled as i.i.d. Rayleigh fading channel elements (10.8), when M is very large [21–23]. This is mainly due to the "channel hardening effect" [44] and the difficulty of finding a subset of semiorthogonal users in the massive MIMO case [21, 30]:

-   *Channel hardening effect:* This refers to the phenomenon that in massive MIMO the norm of each channel vector from the same Rayleigh fading distribution converges to the same constant by the law of large numbers. Assume the channel vector of user k $\mathbf{h}_k \overset{\text{i.i.d.}}{\sim} \mathcal{CN}(\mathbf{0}, \mathbf{I})$. Then, $\|\mathbf{h}_k\|^2$ is a chi-square random variable with $2M$ degrees-of-freedom and one can show [21]:

$$\mathbb{P}\left(\|\mathbf{h}_k\|^2 \leq x\right) \geq 1 - e^{-M(x-1-\log x)}, \tag{10.12}$$

for $x > 1$, using large deviations principle. Then:

$$\mathbb{P}\left(\max_k \|\mathbf{h}_k\|^2 \leq x\right) = \mathbb{P}\left(\|\mathbf{h}_k\|^2 \leq x\right)^K \tag{10.13}$$

$$\geq \left(1 - e^{-M(x-1-\log x)}\right)^K. \tag{10.14}$$

Since $\eta \triangleq x - 1 - \log x > 0$ for $x > 1$, the right-hand side (RHS) of (10.14) converges to:

$$(1 - e^{-\eta M})^K \to 1, \qquad (10.15)$$

if $\lim_{M,K \to \infty} \frac{\log K}{M} = 0$. This implies that $\mathbb{P}(\max_k \|\mathbf{h}_k\|^2 \le x) \uparrow 1$ for all $x > 1$, when $\log K$ grows slower than M such as $K = \mu M$, $K = M^3$, and $K = e^{M^{0.9}}$. It is obvious that $\mathbb{P}(\max_k \|\mathbf{h}_k\|^2 \le x) \downarrow 0$ for all $x < 1$ due to $\mathbb{E}[\|\mathbf{h}_k\|^2] = 1$ by the law of large numbers. Consequently, $\max_k \|\mathbf{h}_k\|^2$ converges to the constant 1 in probability. This observation shows that there is no user that has a significantly large channel vector magnitude when $\lim_{M,K \to \infty} \frac{\log K}{M} = 0$, i.e., $K = o(e^{c'M})$ with some constant $c' > 0$.

- *The negligible probability of finding a subset of semiorthogonal users:* Let us consider the probability of finding a set of M roughly orthogonal users from K users with $\mathbf{h}_k \sim^{\text{i.i.d.}} \mathcal{CN}(\mathbf{0}, \mathbf{I})$, $k = 1, 2, ..., K$. This probability can be computed by considering a set of M double cones constructed around each axis of the M-dimensional space:

$$\mathcal{C}_i' = \left\{ \mathbf{h} : \frac{|\mathbf{h}^H \mathbf{e}_i|}{\|\mathbf{h}\|} \ge \xi' \right\}, \qquad i = 1, 2, ..., M, \qquad (10.16)$$

where $\mathbf{e}_i$ is the ith column of the $M \times M$ identity matrix. Note that a set of M channel vectors each of which is contained in one distinct cone of the cone set $\{\mathcal{C}_1', \mathcal{C}_2', ..., \mathcal{C}_M'\}$ is a set of M roughly orthogonal channel vectors. The probability that the channel vector $\mathbf{h}_k$ falls in the cone $\mathcal{C}_i'$ is given by [30]:

$$\mathbb{P}\{\mathbf{h}_k \in \mathcal{C}_i'\} = \mathbb{P}\{|h_{k,i}| \ge \xi' \|\mathbf{h}_k\|\}$$
$$\approx e^{-(\xi')^2 M}, \qquad (10.17)$$

where the second step results from the fact that $\|\mathbf{h}_k\| \to M$ for large M by the law of large numbers. Therefore, the probability that the cone $\mathcal{C}_i'$ contains at least one out of the K channel vectors is given by:

$$\mathbb{P}\{\mathcal{C}_i' \text{ contains at least one user}\}$$
$$= 1 - \mathbb{P}\{\mathbf{h}_k \notin \mathcal{C}_i'\}^k \qquad (10.18)$$
$$\approx 1 - \left(1 - e^{-(\xi')^2 M}\right)^k \to 0$$

if $\lim_{M,K \to \infty} \frac{\log K}{M} = 0$. This observation also shows that there are no subset of roughly orthogonal M users for very large M if $\lim_{M,K \to \infty} \frac{\log K}{M} = 0$, i.e., $K = o(e^{c'M})$ with some constant $c' > 0$.

The above two facts show that if K grows slower than the exponential rate with respect to M, there is neither significant MU diversity gain nor a user subset of size M with roughly orthogonal channel vectors. Hence, as the number M of transmit antennas at the BS increases, K should increase exponentially with respect to M

to obtain a nontrivial scheduling gain. However, in massive MIMO with M in the order of hundreds, the required number K of active users in the cell is too large, thus the scheduling gain is trivial due to the insufficient number of active users in practical networks under rich scattering environments.

The channel hardening effect leads to simplified resource allocation because fast-fading channel variation in the frequency domain also disappears due to the channel hardening effect. Thus, scheduling over the frequency domain may not be required and the whole spectrum can be assigned simultaneously to all users scheduled in the antenna domain.

Consistent with the above observations, the sum rate of RBF is shown to scale as [18]:

$$\mathcal{R}_{\text{RBF}} \sim cM, \tag{10.19}$$

for $M = O(\log K)$ where $x \sim y$ indicates $\lim_{M,K\to\infty} x/y = 1$ and c is a positive constant. Furthermore, when $\lim_{M,K\to\infty} \frac{\log K}{M} = 0$, [18]:

$$\lim_{M,K\to\infty} \frac{\mathcal{R}_{\text{RBF}}}{M} = 0. \tag{10.20}$$

In other words, in order to achieve linear sum rate scaling by RBF with respect to the number M of antennas, the number K of active users in the cell should increase exponentially with respect to the number M of antennas. This result presents a pessimistic prospect as to the scheduling methods based on partial CSI feedback, and suggests that it is necessary to obtain accurate CSI in the massive MIMO regime, since the sum capacity of an arbitrary subset of users with cardinality M among K users scales with M. The observation agrees with the result in Ravindran and Jindral [45]: given a constraint on the total amount of feedback, it is preferable to acquire accurate CSI feedback from a small number of users than to get coarse channel feedback from a large number of users. From negligible scheduling gains in massive MIMO under rich scattering environments, efficient user scheduling is less important. Thus, in massive MIMO under rich scattering environments simple scheduling methods that preselect users according to a certain probability distribution and exploit the CSI of only the selected users were considered [22, 23].

10.5 USER SCHEDULING IN mmWave MASSIVE MIMO

In the previous section, the results on user scheduling in massive MIMO under the Rayleigh fading channel model suitable for rich-scattering environments at lower frequency bands are provided. However, the results provided are limited to the Rayleigh fading channel model suitable for rich-scattering environments and are not valid in mmWave massive MIMO networks, since the scheduling gain depends on the radio channel environment and the radio propagation and channel environment in the mmWave band are different from those in lower frequency bands.

In the mmWave band, the nature of radio propagation is quasi-optical: there exist very few multiple paths with large path losses [27, 46]. To analyze the scheduling and MU diversity gain in mmWave massive MIMO, relevant channel models capturing the radio propagation in the mmWave band are necessary. One widely considered model for mmWave MISO channels with uniform linear arrays (ULAs) at BSs is the *uniform-random single-path (UR-SP) channel model*, which captures the highly directional radio propagation in the mmWave band and is analytically tractable [24, 30, 47]. In the UR-SP channel model, the channel vector $\mathbf{h}_k$ from the BS equipped with a ULA of M transmit antennas to user k is expressed as:

$$\mathbf{h}_k = \alpha_k \sqrt{M} \boldsymbol{\eta}(\theta_k), \tag{10.21}$$

where $\alpha_k \overset{\text{i.i.d.}}{\sim} \mathcal{CN}(0,1)$ is the path gain, $\theta_k \overset{\text{i.i.d.}}{\sim} \text{Unif}[-1,1]$ is the normalized angle of departure (AoD) for user k, and $\boldsymbol{\eta}(\theta)$ is the ULA steering vector given by:

$$\boldsymbol{\eta}(\theta) = \frac{1}{\sqrt{M}} \left[1, e^{-\imath \pi \theta}, \ldots, e^{-\imath \pi (M-1)\theta} \right]^T, \qquad \imath = \sqrt{-1}. \tag{10.22}$$

The normalized AoD θ_k is related to the actual AoD ϕ_k as $\theta_k = \dfrac{2d \sin(\phi_k)}{\lambda}$, where d is the antenna element spacing and λ is the wavelength. With $d = \dfrac{\lambda}{2}$, $\phi_k \in [-\pi/2, \pi/2]$ translates to $\theta_k \in [-1, 1]$. Under the UR-SP channel model (10.21), the channel vector of each user has a single path component with a random AoD θ_k and a random path gain α_k. Since there is only one path in each user's channel, the UR-SP channel model is a simplified channel model capturing the dominant propagation path in highly directional propagation environments, whereas the Rayleigh fading channel model captures rich scattering environments.

In this section, we first introduce some results on RBF in massive MIMO captured by the asymptotic situation of $M \to \infty$ under the UR-SP channel model, and then provide additional results on RBF under a generalized multipath channel model relevant to sparse mmWave channels. Finally, we introduce efficient user scheduling methods for sparse mmWave channels under the generalized channel model.[1]

10.5.1 RANDOMLY DIRECTIONAL BEAMFORMING UNDER THE UR-SP CHANNEL MODEL

We first investigate the MU diversity gain in the K-user MU-MISO network with $P_t = 1$ described by Eq. (10.6) under the UR-SP channel model by considering a simple SU randomly directional beamforming (RDB) scheme. In this scheme, during the training period the BS randomly chooses an angle ϑ according to the uniform distribution over $[-1, 1]$, and transmits a beam $\mathbf{x}$ to the chosen angle direction. Thus, the transmit beam vector in this case is given by

---

[1]The contents here mainly follow from Refs. [30, 31].

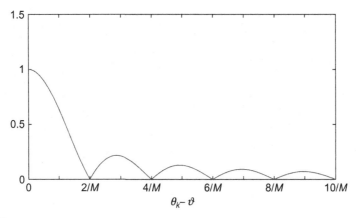

FIG. 10.6

$B_M(\theta_k - \vartheta)$ with $M = 100$.

$$\mathbf{x} = \boldsymbol{\eta}(\vartheta), \qquad (10.23)$$

where $\vartheta \sim \text{Unif}[-1, 1]$. Then, each user k computes its received signal power $|\mathbf{h}_k^H \mathbf{x}|^2$ from y_k and feeds it back to the BS. After the feedback is over, the BS selects the user that has the maximum received signal power and transmits a data stream to the selected user by using the beam $\mathbf{x}$. The expected rate $\mathcal{R}_1$ in this case is given by:

$$\mathcal{R}_1 = \mathbb{E}\left[\log \left(1 + \max_{1 \le k \le K} |\alpha_k|^2 M |\boldsymbol{\eta}(\theta_k)^H \boldsymbol{\eta}(\vartheta)|^2 \right) \right]. \qquad (10.24)$$

Here, the term $|\boldsymbol{\eta}(\theta_k)^H \boldsymbol{\eta}(\vartheta)|$ defines the standard ULA beam pattern centered at the angle ϑ, expressed as [30]:

$$|\boldsymbol{\eta}(\theta_k)^H \boldsymbol{\eta}(\vartheta)| = \frac{1}{M} \left| \frac{\sin \dfrac{\pi(\theta_k - \vartheta)M}{2}}{\sin \dfrac{\pi(\theta_k - \vartheta)}{2}} \right| \triangleq B_M(\theta_k - \vartheta) \qquad (10.25)$$

which is the square root of the Fejér kernel of order M. Fig. 10.6 shows the beam pattern $B_M(\theta_k - \vartheta)$. As seen in the figure, the beam pattern ripples in a diminishing fashion as θ_k deviates from the beam center angle ϑ, and has nulls at every multiple of $2/M$. The main lobe width of the beam pattern becomes narrow as M increases, but as $M \to \infty$:

$$B_M(\theta_k - \vartheta) \to \left| \frac{2 \sin \dfrac{\pi\delta}{2}}{\pi\delta} \right|, \quad \text{if } |\theta_k - \vartheta| = \frac{\delta}{M} \quad \text{for some } \delta > 0.$$

Furthermore, we have the following observation:

$$B_M(\theta_k - \vartheta) \to 0 \quad \text{for fixed } \vartheta \text{ and } \theta_k,$$

as $M \rightarrow \infty$. From the above observations, we can see that if the number K of users is fixed, the rate $\mathcal{R}_1$ of the scheme diminishes to zero as $M \rightarrow \infty$. On the other hand, if the number K of users as a function of M is sufficiently large so that there exists a user whose propagation path angle is within the main lobe of the transmit beam pattern with a high probability, the scheme yields a nontrivial rate performance due to the MU diversity. The following theorem provides the number of active users required for a nontrivial MU diversity gain for the considered simple scheduling scheme under the UR-SP channel model.

Theorem 10.1. ([30]) *For $K = M^q$ and any given $q \in (0, 1)$, asymptotic upper and lower bounds of $\mathcal{R}_1$ in Eq. (10.24) are given by:*

$$\log\left(1 + M^{2q-1-\epsilon}\right) \lesssim \mathcal{R}_1 \lesssim \log\left(1 + M^{2q-1+\epsilon}\right) \tag{10.26}$$

for any sufficiently small $\epsilon > 0$, where $x \lesssim y$ indicates $\lim_{M \rightarrow \infty} x/y \leq 1$. Furthermore, for $K = M^q$ and any given $q \in \left(\frac{1}{2}, 1\right)$:

$$\lim_{M \rightarrow \infty} \frac{\mathcal{R}_1}{\mathbb{E}\left[\log\left(1 + M \max_k |\alpha_k|^2\right)\right]} = 2q - 1, \tag{10.27}$$

where $\mathbb{E}\left[\log\left(1 + M \max_k |\alpha_k|^2\right)\right]$ is the optimal beamforming rate based on perfect CSI.

Theorem 10.1 states that the simple SU selection method based on single beam transmission and partial CSI feedback has nontrivial asymptotic performance, i.e., $\mathcal{R}_1$ scales as $\log M$ as $M \rightarrow \infty$, when $K = M^q$ with $q \in \left(\frac{1}{2}, 1\right)$. On the contrary, the performance of the scheme is trivial, i.e., $\mathcal{R}_1 \rightarrow 0$ as $M \rightarrow \infty$, when $K = M^q$ with $q \in \left(0, \frac{1}{2}\right)$. Consequently, if the number K of users increases faster than $K = \Theta(\sqrt{M})$, a nontrivial MU diversity gain is obtained by the simple scheduling based on partial CSI. Furthermore, in this case, the simple scheduling method achieves $2q - 1$ fraction of the perfect beamforming rate based on perfect CSI at the BS.

In the first scheme, only one training and data transmission beam is used. However, in most RBF-type scheduling methods multiple training beams are used. Now consider the second training and scheduling scheme based on partial CSI feedback under the UR-SP channel model. In the second scheme, S asymptotically orthogonal training beams with a random angle offset are used as:

$$\mathbf{u}_b = \boldsymbol{\eta}(\vartheta_b) = \boldsymbol{\eta}\left(\vartheta + \frac{2(b-1)}{S}\right) \quad \text{for } b = 1, \ldots, S, \tag{10.28}$$

where the random offset ϑ is given by $\vartheta \sim \text{Unif}[-1, 1]$. Note that the overall normalized AoD range of two is divided by S. When $S = o(M)$, the asymptotic orthogonality among the training beams is guaranteed. During the training period, $\mathbf{u}_1, \ldots, \mathbf{u}_S$ are transmitted sequentially in time, and each user k computes $|\mathbf{h}_k^H \mathbf{u}_i|, i = 1, 2, \ldots, S$ and feed backs the largest value $\max_i |\mathbf{h}_k^H \mathbf{u}_i|$ and the corresponding beam index. After the feedback, the BS selects the pair of user and beam that has the maximum received signal power and transmits a data stream to the selected user with the selected beam. Then, the expected rate in this case is given by:

$$\mathcal{R}_2 = \mathbb{E}\left[\log\left(1 + \max_{1 \le k \le K} \max_{1 \le b \le S} |\alpha_k|^2 M |\mathbf{a}(\theta_k)^H \mathbf{a}(\vartheta_b)|^2\right)\right]. \tag{10.29}$$

The following theorem explains the relationship between the training effect and the MU-diversity in this case.

Theorem 10.2 ([30]) *For $K = M^q$, $S = M^\ell$ and any $\ell, q \in (0, 1)$ such that $\ell + q < 1$, asymptotic upper and lower bounds of $\mathcal{R}_2$ in Eq. (10.29) are given by:*

$$\log\left(1 + M^{2q+2\ell-1-\epsilon}\right) \lesssim \mathcal{R}_2 \lesssim \log\left(1 + M^{2q+2\ell-1+\epsilon}\right) \tag{10.30}$$

for any sufficiently small $\epsilon > 0$. Furthermore, for $K = M^q$, $S = M^\ell$ and any $\ell, q \in (0, 1)$ such that $\dfrac{1}{2} < \ell + q < 1$:

$$\lim_{M \to \infty} \frac{\mathcal{R}_2}{\mathbb{E}[\log(1 + M \max_k |\alpha_k|^2)]} = 2(q + \ell) - 1. \tag{10.31}$$

Theorem 10.2 shows the impact of training on the MU diversity gain. Simply, q in Theorem 10.1 is replaced with $q + \ell$ in Theorem 10.2. Thus, the lack of users in the cell can be compensated for by more training. Although the rate performance in the second scheme is the same if $q + \ell$ is the same, there exists more training time overhead when small q is compensated for by large ℓ. On the other hand, in the case of large q and small ℓ, there exists larger feedback overhead than in the previous case.

As mentioned in Section 10.3, the true gain of the multiple antenna transmission technology lies in providing the multiplexing gain. In the previous two cases of SU selection for data transmission, only the power gain is achieved. Now, consider the third scheduling scheme of RBF [18] applied to the asymptotically orthonormal beam set $\{\mathbf{u}_1, \mathbf{u}_2, \dots, \mathbf{u}_S\}$ in Eq. (10.28). (See Section 10.3.1 for RBF. Note here that S may not be the same as M.) In this scheme, the BS schedules a user that has the maximum SINR value for each beam $\mathbf{u}_b$, $b = 1, \dots, S$, and transmits S data streams to the S selected users. Under the simplifying assumption of $|\alpha_k| = 1$, $\forall k$, the rate of user κ_b selected for beam $\mathbf{u}_b$ is given by [30]:

$$\mathcal{R}_{\kappa_b} = \mathbb{E}\left[\log\left(1 + \max_{1 \le k \le K} \mathrm{SINR}_{k,b}\right)\right]$$

$$= \mathbb{E}\left[\log\left(1 + \frac{\rho M |\boldsymbol{\eta}(\theta_{\kappa_b})^H \boldsymbol{\eta}(\vartheta_b)|^2}{1 + \sum_{b' \ne b} \rho M |\boldsymbol{\eta}(\theta_{\kappa_b})^H \boldsymbol{\eta}(\vartheta_{b'})|^2}\right)\right], \tag{10.32}$$

where $\rho = P_t/S = 1/S$ is the per-user power for each scheduled user, and the expected sum rate of this scheme is given by:

$$\mathcal{R}_3 = \sum_{b=1}^{S} \mathcal{R}_{\kappa_b}. \tag{10.33}$$

The following theorem shows the performance of RBF under the UR-SP channel model with $|\alpha_k| = 1$ for all k:

Theorem 10.3 ([30]) *For $K = M^q$, $S = M^\ell$ with $q \in (0, 1)$ and $\ell \in \left(0, q - \dfrac{\epsilon}{2}\right)$, the per-user rate is asymptotically bounded as:*

$$\log\left(1 + M^{2q-1-\ell-\epsilon}\right) \lesssim \mathcal{R}_{\kappa_b} \lesssim \log\left(1 + M^{2q-1-\ell+\epsilon}\right) \tag{10.34}$$

for any sufficiently small $\epsilon > 0$.

Based on Theorem 10.3, the sum rate of RBF in massive MIMO under the UR-SP channel model can be computed. The sum rate is given by $\mathcal{R}_3 = \sum_{b=1}^{S} \mathcal{R}_{\kappa_b} = M^\ell \log\left(1 + M^{2q-1-\ell}\right)$ when $2q - 1 - \ell > 0$. As ℓ and q converge to 1, i.e., S and K converge to M, the sum rate $\mathcal{R}_3$ converges to $M \log 2$, which is the capacity of the K-user MU-MISO downlink network with $P_t = 1$ and $K \geq M$ under the UR-SP channel model. (Note from Eq. (10.32) that the best SINR value for any user is one when S converges to M with $P_t = 1$.) Thus, RBF asymptotically yields optimal performance under the UR-SP channel model only if K increases linearly with M. This result is surprisingly different from the behavior of RBF under the i. i.d. Rayleigh fading channel model, i.e., RBF achieves linear sum rate scaling with respect to M under the i.i.d. Rayleigh fading channel model when K increases exponentially with respect to M, as seen in Eq. (10.20). Consequently, user scheduling based on partial CSI feedback such as RBF can be useful in mmWave massive MIMO, where the propagation is highly directional and there exist only a few propagation paths, whereas user scheduling based on partial CSI feedback provides few gains in rich scattering environments, as discussed in Section 10.4.

Performance under the UR-SP channel model in the sparse user regime

It has now been shown that RBF achieves optimal performance under the UR-SP channel model for large M when K is in the same order of M. However, this required number K of active users for optimal performance of RBF still may be large for mmWave massive MIMO since M is in the order of hundreds. The situation where the number of active users is not large is called a *sparse user regime*. To compare the relative performance of scheduling algorithms under the UR-SP channel model in the sparse user regime, the *fractional rate order (FRO)* γ is defined as [30]:

$$\gamma \triangleq \lim_{M \to \infty} \frac{\log \mathcal{R}}{\log M} \tag{10.35}$$

with $K = M^q$, $0 \leq q \leq 1$. Basically, a nonzero γ means that the system sum rate behaves as $\mathcal{R} = \Theta(M^\gamma)$ from which the name follows. The FROs of $\mathcal{R}_1$, $\mathcal{R}_2$, and $\mathcal{R}_3$ of the three scheduling schemes in the previous section can be computed as [30]:

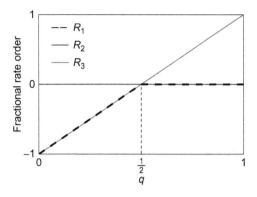

FIG. 10.7

Fractional rate order versus q ($K = M^q$).

Adapted from G. Lee, Y. Sung, J. Seo, Randomly-directional beamforming in millimeter-wave multi-user MISO downlink, IEEE Trans. Wireless Commun. 15 (2) (2016) 1086–1100.

$$\gamma_1 := \lim_{M \to \infty} \frac{\log \mathcal{R}_1}{\log M} = \begin{cases} 0, & \text{for } q \in \left(\frac{1}{2}, 1\right) \\ 2q - 1, & \text{for } q \in \left(0, \frac{1}{2}\right) \end{cases} \tag{10.36}$$

$$\gamma_2 := \lim_{M \to \infty} \frac{\log \mathcal{R}_2}{\log M} = 0, \qquad \text{for} \quad q \in (0, 1) \tag{10.37}$$

$$\gamma_3 := \lim_{M \to \infty} \frac{\log \mathcal{R}_3}{\log M} = 2q - 1, \quad \text{for} \quad q \in (0, 1). \tag{10.38}$$

Fig. 10.7 shows the FROs of the three schemes versus $q \in (0, 1)$. Note that the relative performance of the three scheduling schemes depends on q determining the sparsity of active users. RBF with S beams (γ_3) yields the best performance among the three schemes for $q \in \left(\frac{1}{2}, 1\right)$ implying more users in the cell, whereas the M-training-beam SU selection scheme (γ_2) is the best $q \in \left(0, \frac{1}{2}\right)$ implying less users in the cell. γ_1 is a lower bound on both γ_2 and γ_3. Interestingly, RBF with S beams performs worse than the M-training-beam SU selection scheme for $q \in \left(0, \frac{1}{2}\right)$. This is because RBF requires the computation of SINR for each beam at each user and thus all S training beams should be used for data transmission of S streams. If some of the S training beams are unused for data transmission, then SINR cannot be computed at the user side. Consequently, the number $S = M^\ell$ of training beams should be reduced to be smaller than the number $K = M^q$ of users in the cell. As q decreases to zero, S becomes small and there are not enough training beams in the network. This insufficiency of training beams results in the poor performance of RBF for small q. On the other hand, for large q near one, there exist many users, but the M-training-beam SU selection scheme still selects only one user for the service. Thus, the scheme loses the multiplexing gain, which is exploited by RBF. From this example, one can recognize that both appropriate training and spatial multiplexing should be used for good performance of user scheduling based on partial CSI feedback in mmWave massive MIMO.

10.5.2 EXTENSION TO A GENERAL SPARSE mmWave CHANNEL MODEL

The UR-SP channel model is limited in the sense that it has only one propagation path. Even in the mmWave band, it is observed that there exist a few paths from the BS to a user in general [26, 27]. To reflect more realistic channel environments in the mmWave band, the UR-SP channel model can be generalized to include multiple propagation paths so that the channel vector $\mathbf{h}_k$ from the BS to user k is given by:

$$\mathbf{h}_k = \sqrt{\frac{M}{L}}\sum_{i=1}^{L}\alpha_{k,i}\boldsymbol{\eta}(\theta_{k,i}), \tag{10.39}$$

where L is the number of multiple propagation paths, $\alpha_{k,i} \overset{\text{i.i.d.}}{\sim} \mathcal{CN}(0,1)$ is the path gain of the ith path, and $\theta_{k,i} \overset{\text{i.i.d.}}{\sim} \text{Unif}[-1,1]$ is the normalized AoD of the ith path. In this channel model, the channel vector $\mathbf{h}_k$ is the sum of L uniform random multiple paths with complex Gaussian gains. Thus, this channel model is referred to as the *uniform-random multipath (UR-MP) channel model* [31]. An example of the channel model is shown in Fig. 10.8. Depending on the number L of multiple paths, the UR-MP channel can capture different levels of channel scattering sparsity. When $L = 1$, the UR-MP channel model reduces to the UR-SP channel model in the previous section. On the other hand, it is shown that the UR-MP channel model converges to the i.i.d. Rayleigh fading channel model for large L when L is more than M [31]. When $1 < L < M$, the UR-MP channel model lies in-between the UR-SP channel model and the i.i.d. Rayleigh fading channel model.

The sparse nature of multipaths in mmWave massive MIMO channels can be modeled by the asymptotic scenario that $L/M \to 0$ as $M \to \infty$. The performance

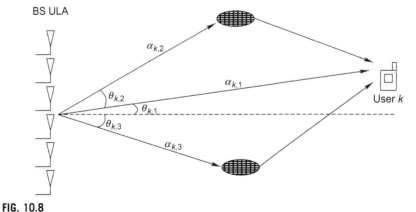

FIG. 10.8

An example with $L = 3$ of the UR-MP channel model.

Table 10.1 Sufficient Number of Users for Linear Sum Rate Scaling With Respect to M (c: Some Positive Constant)

Channel Sparsity	Sufficient Number of Users
L fixed	$K \sim M$ (linear)
$L = \log(M)$	$K \sim M^{1+c}$ (polynomial)
$L = M^{\beta}, 0 < \beta < 1$	$K \sim e^{cM^{\beta}}$ (subexponential)
$L = M$ or faster	$K \sim e^{cM}$ (exponential)

G. Lee, Y. Sung, M. Kountouris, *On the performance of random beamforming in sparse millimeter wave channels, IEEE J. Sel. Topics Signal Process. 10 (3) (2016) 560–575.*

of RBF is investigated under the UR-MP channel model under several channel sparsity levels and a sufficient number of active users in the cell for RBF to achieve linear sum rate scaling with respect to the number M of transmit antennas in MU-MISO downlink is identified for different channel sparsity levels in Lee et al. [31]. The result is shown in Table 10.1. When the number of multipaths is fixed regardless of M, RBF can achieve a linear sum rate scaling with respect to M when the number of active users increases linearly with M. The performance of RBF under the UR-SP channel model in the previous section agrees with this result. When the number of multipaths increases faster than M, the UR-MP channel model converges to the i.i.d. Rayleigh fading channel model and hence the number of active users in the cell should increase exponentially with M for RBF to yield a linear sum rate scaling with respect to M, as already mentioned in Section 10.4.

10.5.3 EFFICIENT SCHEDULING METHODS FOR SPARSE mmWave MIMO CHANNELS

The sufficient number of active users in a cell for RBF to yield a linear sum rate scaling with respect to M shown in Section 10.5.2 is still large for mmWave massive MIMO networks, where the number M of transmit antennas is in the order of hundreds, especially for small cell networks which will be the main application in the mmWave band. It is highly likely that small cell mmWave massive MIMO networks are operated in the sparse user regime. In such cases, the direct use of RBF is not desirable and more efficient user scheduling algorithms requiring reasonable feedback overhead should be considered for better performance in mmWave massive MIMO under the general multipath scenario. In this section, we introduce several scheduling algorithms suitable for sparse mmWave massive MIMO presented in Lee et al. [31].

In Lee et al. [31], to remove the loss associated with insufficient training, full training is assumed, i.e., all M orthogonal training beams in Eq. (10.28) with $S = M$ available with M transmit antennas are used in the training period. Since all M beams of the training period may not be used for later data transmission in the sparse user regime[2] and the indices of data transmission beams are unknown

---

[2]There may be no users in some directions in the sparse user regime.

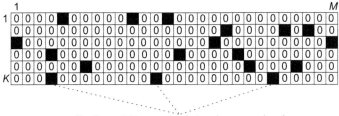

(the largest N_{FB} received signal power values)

FIG. 10.9

A two-dimensional table of user and beam for the received signal power.

during the training period, the SINR, $SINR_{k,b}$ for each beam $\mathbf{u}_b$ cannot be computed at each user k during the training period. A reasonable way of partial CSI feedback in this case is the feedback of the largest N_{FB} received signal power values and the corresponding beam indices from each user to the BS with consideration of the possible existence of multiple propagation paths. Here, N_{FB} can be much smaller than M and incorporate few dominant multipaths in sparse mmWave massive channels. Once the feedback is over, a two-dimensional table consisting of the user and beam indices is constructed and filled with the feedbacked signal power values leaving the unreported box values as zero, as shown in Fig. 10.9. Once the table is constructed, joint user and beam selection can be performed. First, consider the case in which we assign only one beam for a selected user. In this case, we can apply a suboptimal greedy sequential scheduling algorithm as presented in Vicario et al. [48]. The basic idea is as follows. We first select a user-and-beam pair (k_1, b_1) that has the maximum signal power $|\mathbf{h}_{k_1}^H \mathbf{u}_{b_1}|^2$ over all beam-and-user pairs in the table, and eliminate b_1 from the candidate beam set and k_1 from the candidate user set. Then, we select the second user-and-beam pair (k_2, b_2) that has the maximum sum rate among all the remaining user-and-beam pairs and eliminate (k_2, b_2) from the candidate sets. We proceed in this way until the procedure cannot continue [48]. After the user-and-beam pair selection is finished, the BS transmits data to user k_j with beam $b_j, j = 1, ..., J_{STOP}$, where J_{STOP} is the number of total users selected by the procedure.

The user-and-beam selection method as outlined above has low computational complexity due to its sequential nature and is effective in the case of UR-SP channels, since it assigns only one beam for one served user. However, such a beam allocation scheme is not effective in the case that multiple propagation paths exist from the BS to each user as in the UR-MP channel model. In such cases, existing multiple paths should be exploited for better performance based on advanced beamforming, such as maximum ratio transmit (MRT) beamforming. Thus, in Lee et al. [31] another user-and-beam selection method is proposed to harness the MRT gain by transmitting a combined beam to a selected user. The proposed scheduling method is described as follows [31]:

1. The BS initializes the set of candidate user and the set of candidate beams:

$$\mathcal{K} = \{1,2,...,K\}, \qquad \mathcal{B} = \{1,2,...,M\}, \qquad (10.40)$$

and $j = 1$.

2. During the training period, each user k computes N_{FB} dominant training beam powers $\{|\mathbf{h}_k^H \mathbf{u}_b|^2\}$ out of the M training beams

3. and feeds back the magnitudes $\{|\mathbf{h}_k^H \mathbf{u}_b|, b \in \mathcal{B}_k\}$ and phases $\{\angle(\mathbf{h}_k^H \mathbf{u}_b), b \in \mathcal{B}_k\}$ of the N_{FB} dominant training beams and the corresponding beam indices $\mathcal{B}_k$ to the BS.

4. After the feedback, the BS computes:

$$\kappa_j = \arg\max_{k \in \mathcal{K}} \sum_{b \in \mathcal{B}_k} |\mathbf{h}_k^H \mathbf{u}_b|^2 \qquad (10.41)$$

$$\mathbf{x}_{\kappa_j} = \frac{\hat{\mathbf{x}}_{\kappa_j}}{||\hat{\mathbf{x}}_{\kappa_j}||} \quad \text{with} \quad \hat{\mathbf{x}}_{\kappa_j} = \sum_{b \in \mathcal{B}_k} (\mathbf{h}_{\kappa_j}^H \mathbf{u}_b)^* \mathbf{u}_b, \qquad (10.42)$$

and updates $\mathcal{K} \leftarrow \mathcal{K} \backslash \{\kappa_j\}$ and $\mathcal{B} \leftarrow \mathcal{B} \backslash \mathcal{B}_{\kappa_j}$.

5. To implement rough orthogonality among the selected users, the BS updates the set $\mathcal{K}$ of candidate users as:

$$\mathcal{K} \leftarrow \{k \in \mathcal{K} : |\mathcal{B}_{\kappa_j} \cap \mathcal{B}_k| \leq N_{OL}\}, \qquad (10.43)$$

where N_{OL} is the maximum number of allowed overlapped beams between the sets of dominant beams of any two selected users.

6. If $j < M$, $\mathcal{K} \neq \emptyset$, and $\mathcal{B} \neq \emptyset$, then update $j \leftarrow j + 1$ and go to step 3. Otherwise, the algorithm is finished and the set of selected users is $\{\kappa_1, ..., \kappa_j\}$.

In the above algorithm, N_{FB} phase values in addition to N_{FB} power values from each user is additionally fed back to implement MRT. Some modifications to the proposed algorithms are considered in Lee et al. [31], such as an equal gain combining (EGC) beamforming based user-and-beam selection to reduce the amount of feedback per user, the application of ZF or MMSE beamforming to eliminate inter-user interference in the case of $N_{OL} \geq 1$, the application of the two-phase feedback (TPF) idea to improve the performance, the application of the PF principle to ensure user fairness, and the application of the proposed methods to hybrid beamforming. See Lee et al. [31] for detail.

Fig. 10.10 shows the performance of several the user-and-beam scheduling algorithms discussed under the UR-MP channel model. As a reference of optimal performance, the performance of MMSE beamforming is included in the figure. In the case of MMSE beamforming, it is assumed that the BS knows the full channel matrix $\mathbf{H}_{overall}$ of size $K \times M$ from the BS to all K users in Eq. (10.6) and the MMSE transmit beamformer is designed based on $\mathbf{H}_{overall}$. The TPF based algorithm first selected users based on partial CSI feedback of the indices $\mathcal{B}_k$ of N_{FB} dominant beams and

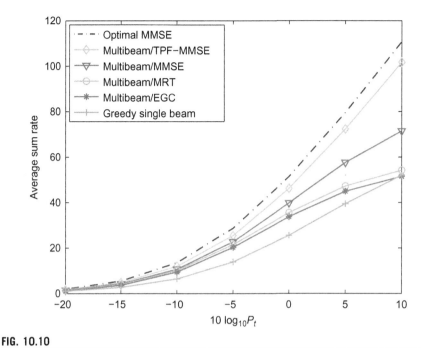

FIG. 10.10

Average sum rate versus P_t: $M = 128$, $K = 20$, $L = 6$, $N_{FB} = 4$, $N_{OL} = 1$.

the power sum $\sum_{b\in\mathcal{B}_k}|\mathbf{h}_k^H\mathbf{u}_b|^2$ of N_{FB} dominant beams from each user in a similar way to that used in the algorithm described in the above and collected CSI, i.e., the magnitudes and phases, of the selected beams from the selected users. Based on the second feedback, an MMSE beamformer was designed and used for data transmission. It is seen that the user-and-beam selection algorithms based on MRT outperform the single-beam based greedy user selection method. Furthermore, it is seen that the TPF-based algorithm almost achieves the performance of the MMSE beamforming based on full CSI.

10.6 CONCLUSIONS

In this chapter, an overview of user scheduling is provided and user scheduling for mmWave massive MIMO is considered. In general MU-MIMO cellular networks, the MU-diversity gain can be obtained by effective user scheduling exploiting wireless channel fading. Such scheduling can also be significant in sparse mmWave massive MIMO channels with highly directional few propagation paths. One aspect not discussed in depth in this chapter is the overhead of training required for channel estimation or partial CSI acquisition. In mmWave massive MIMO systems, this training overhead can be large due to the large size of antenna arrays. More study on efficient training is left for future work to realize mmWave massive MIMO for 5G.

REFERENCES

[1] R. Knopp, P.A. Humblet, Information capacity and power control in single-cell multiuser communication, in: Proceedings of ICC, 1, 1995, pp. 331–335.

[2] D. Tse, P. Viswanath, Fundamentals of Wireless Communication, Cambridge University Press, Cambridge, UK, 2005.

[3] H. Kushner, P. Whiting, Asymptotic properties of proportional-fair sharing algorithms, in: Proceedings of Annual Allerton Conference, 2002.

[4] A. Jalali, R. Padovani, R. Pankaj, Data throughput of CDMA-HDR a high efficiency-high data rate personal communication wireless system, in: Proceedings of IEEE VTC, 2000.

[5] R. Agrawal, A. Bedekar, R. La, V. Subramanian, A class and channel–condition based weighted proportionally fair scheduler, in: Proceedings of ITC, 2001.

[6] P. Viswanath, D. Tse, R. Laroia, Opportunistic beamforming using dumb antennas, IEEE Trans. Inf. Theory 48 (6) (2002) 1277–1294.

[7] F. Capozzi, G. Piro, L. Grieco, G. Boggia, P. Camarda, Downlink packet scheduling in LTE cellular networks: key design issues and a survey, IEEE Commun. Surveys Tuts. 15 (2) (2013) 678–700.

[8] C. Wong, R. Cheng, K. Letaief, R. Murch, Multiuser OFDM with adaptive subcarrier, bit, and power allocation, IEEE J. Sel. Areas Commun. 17 (10) (1999) 1747–1758.

[9] J. Jang, K. Lee, Transmit power adaptation for multiuser OFDM systems, IEEE J. Sel. Areas Commun. 21 (2) (2003) 171–178.

[10] L. Hoo, B. Halder, J. Tellado, J. Cioffi, Multiuser transmit optimization for multicarrier broadcast channels: asymptotic FDMA capacity region and algorithms, IEEE Trans. Commun. 52 (6) (2004) 922–930.

[11] K. Seong, M. Mohseni, J. Cioffi, Optimal resource allocation for OFDMA downlink systems, in: Proceedings of IEEE ISIT, 2006, pp. 1394–1398.

[12] J. Huang, V. Subramanian, R. Agrawal, R. Berry, Downlink scheduling and resource allocation OFDM systems, IEEE Trans. Wirel. Commun. 8 (1) (2009) 288–296.

[13] J. Huang, V. Subramanian, R. Agrawal, R. Berry, Joint scheduling and resource allocation in uplink OFDM systems for broadband wireless access networks, IEEE J. Sel. Areas Commun. 27 (2) (2009) 226–234.

[14] R. Gozali, R. Buehrer, B. Woerner, The impact of multiuser diversity on space-time block coding, IEEE Commun. Lett. 7 (5) (2003) 213–215.

[15] J. Chung, C. Hwang, K. Kim, Y. Kim, A random beamforming techniques in MIMO systems exploiting multiuser diversity, IEEE J. Sel. Areas Commun. 21 (5) (2003) 848–855.

[16] N. Sharma, L. Ozarow, A study of opportunism for multiple-antenna systems, IEEE Trans. Inf. Theory 51 (5) (2005) 1804–1814.

[17] N. Jindal, A. Goldsmith, Dirty-paper coding versus TDMA for MIMO broadcast channels, IEEE Trans. Inf. Theory 51 (2005) 1783–1794.

[18] M. Sharif, B. Hassibi, On the capacity of MIMO broadcast channels with partial side information, IEEE Trans. Inf. Theory 51 (2) (2005) 506–522.

[19] T. Yoo, A. Goldsmith, On the optimality of multiantenna broadcast scheduling using zero-forcing beamforming, IEEE J. Sel. Areas Commun. 24 (3) (2006) 528–541.

[20] E.G. Larsson, F. Tufvesson, O. Edfors, T.L. Marzetta, Massive MIMO for next generation wireless systems, IEEE Commun. Mag. 52 (2) (2014) 186–195.

[21] A. Tomasoni, G. Caire, M. Ferrari, S. Bellini, On the selection of semi-orthogonal users for zero-forcing beamforming, in: Proceedings of IEEE ISIT, 2009.

[22] H. Hur, A.M. Tulino, G. Caire, Network MIMO with linear zero-forcing beamforming: large system analysis, impact of channel estimation, and reduced-complexity scheduling, IEEE Trans. Inf. Theory 58 (5) (2012) 2911–2934.

[23] J. Nam, A. Adhikary, J. Ahn, G. Caire, Joint spatial division and multiplexing: opportunistic beamforming, user grouping and simplified downlink scheduling, IEEE J. Sel. Topics Signal Process. 8 (5) (2014) 876–890.

[24] H.Q. Ngo, E.G. Larsson, T.L. Marzetta, Aspects of favorable propagation in massive MIMO, in: Proceedings IEEE EUSIPCO, 2014, pp. 76–80.

[25] G. Lee, Y. Sung, A new approach to user scheduling in massive multi-user MIMO broadcast channels 2014, available at http://arxiv.org/pdf/1403.6931.pdf.

[26] T.S. Rappaport, E. Ben-Dor, J.N. Murdock, Y. Qiao, 38 GHz and 60 GHz angle-dependent propagation for cellular & peer-to-peer wireless communications, in: Proceedings IEEE ICC, 2012.

[27] S. Sun, T.S. Rappaport, Wideband mmWave channels: implications for design and implementation of adaptive beam antennas, in: Proceedings IEEE IMS, 2014.

[28] G. Lee, Y. Sung, M. Kountouris, On the performance of randomly directional beamforming between line-of-sight and rich scattering channels, in: Proceedings SPAWC, 2015.

[29] G. Lee, Y. Sung, J. Seo, How many users are needed for non-trivial performance of random beamforming in highly-directional mm-wave MIMO downlink, in: Proceedings IEEE ITW, 2015, pp. 244–248.

[30] G. Lee, Y. Sung, J. Seo, Randomly-directional beamforming in millimeter-wave multi-user MISO downlink, IEEE Trans. Wirel. Commun. 15 (2) (2016) 1086–1100.

[31] G. Lee, Y. Sung, M. Kountouris, On the performance of random beamforming in sparse millimeter wave channels, IEEE J. Sel. Topics Signal Process. 10 (3) (2016) 560–575.

[32] H. Weingarten, Y. Steinberg, S. Shamai, The capacity region of the Gaussian MIMO broadcast channel, in: Proceedings IEEE ISIT, 2004.

[33] Z. Tu, R. Blum, Multiuser diversity for a dirty paper approach, IEEE Commun. Lett. 7 (8) (2003) 370–372.

[34] G. Dimic, N. Sidiropoulos, On downlink beamforming with greedy user selection: performance analysis and a simple new algorithm, IEEE Trans. Signal Process. 24 (3) (2005) 506–522.

[35] U. Erez, S. ten Brink, A close-to-capacity dirty paper coding scheme, IEEE Trans. Inf. Theory 51 (10) (2005) 3417–3432.

[36] M. Maddah-Ali, M. Ansari, A. Khandani, Broadcast in MIMO systems based on a generalized QR decomposition: signaling and performance analysis, IEEE Trans. Inf. Theory 54 (3) (2008) 1124–1138.

[37] N. Jindal, MIMO broadcast channels with finite-rate feedback, IEEE Trans. Inf. Theory 11 (2006) 5045–5060.

[38] T. Yoo, N. Jindal, A. Goldsmith, Multi-antenna downlink channels with limited feedback and user selection, IEEE J. Sel. Areas Commun. 25 (7) (2007) 1478–1491.

[39] G. Caire, N. Jindal, M. Kobayashi, N. Ravindran, Multiuser MIMO achievable rates with downlink training and channel state feedback, IEEE Trans. Inf. Theory 56 (6) (2010) 2845–2866.

[40] M. Trivellato, F. Boccardi, H. Huang, On transceiver design and channel quantization for downlink multiuser MIMO systems with limited feedback, IEEE J. Sel. Areas Commun. 6 (8) (2008) 1494–1504.

[41] R. Zakhour, D. Gesbert, A two-stage approach to feedback design in multi-user MIMO channels with limited channel state information, in: Proceedings of IEEE PIMRC, 2007.

[42] W. Xu, C. Zhao, Two-phase multiuser scheduling for multiantenna downlinks exploiting reduced finite-rate feedback, IEEE Trans. Veh. Technol. 59 (3) (2010) 1367–1380.

[43] M. Kountouris, D. Gesbert, T. Salzer, Enhanced multiuser random beamforming: dealing with the not so large number of users case, IEEE J. Sel. Areas Commun. 26 (8) (2008) 1536–1545.

[44] B. Hochwald, T. Marzetta, V. Tarokh, Multiple-antenna channel hardening and its implications for rate feedback and scheduling, IEEE Trans. Inf. Theory 04 (9) (2004) 1893–1909.

[45] N. Ravindran, N. Jindal, Multi-user diversity vs. accurate channel state information in MIMO downlink channels, IEEE Trans. Wirel. Commun. 11 (9) (2012) 3037–3046.

[46] S. Sun, T. Rappaport, R. Heath, A. Nix, S. Rangan, MIMO for millimeter-wave wireless communications: beamforming, spatial multiplexing, or both? IEEE Commun. Mag. 52 (2014) 110–121.

[47] A. Sayeed, J. Brady, Beamspace MIMO for high-dimensional multiuser communication at millimeter-wave frequencies, in: Proceedings of IEEE Globecom, 2013, pp. 3679–3684.

[48] J.L. Vicario, B. Bosisio, C. Anton-Haro, U. Spagnolini, Beam selection strategies for orthogonal random beamforming in sparse networks, IEEE Trans. Wirel. Commun. 7 (2008) 3385–3396.

Enhanced multiple-access for mmWave massive MIMO[a]

11

M. Nasiri Khormuji

Huawei Technologies Sweden AB, Stockholm, Sweden

CHAPTER OUTLINE

---

[a]The material in the book chapter in part is presented at VTC-spring 2015.

mmWave Massive MIMO. http://dx.doi.org/10.1016/B978-0-12-804418-6.00011-X

11.1 INTRODUCTION

11.1.1 BACKGROUND ON mMIMO

Marzetta in [1] introduced a communication solution using massive antenna arrays for multiuser transmission in a cellular system in which base stations are equipped with an abundant numbers of active antenna elements while each user has a single antenna. To cope with the overhead due to channel training, uplink pilot symbols are used in a time-division duplex (TDD) protocol to enable the channel reciprocity to learn the uplink as well as the downlink channels of the users. (See Fig. 11.1 for an illustration of the multiuser uplink transmission.) In Marzetta's work [1] it is shown that simple processing using matched filtering (MF) results in asymptotically interference-free multiuser communications, as long as the users' pilot symbols do not interfere with one another. Additionally, it is shown that imperfections due to the additive white Gaussian noise (AWGN) and small-scale fading vanish as the number of antennas at the base station increases. See Huh et al. [2], Larsson et al. [3], Rusek et al. [4], Hoydis et al. [5], and You et al. [6] for a survey on some recent developments in massive MIMO (mMIMO) field. To avoid intracell pilot contamination, Marzetta [1] employs orthogonal pilot transmission, which sets a limit on the number of users that can be scheduled in a cell within a coherence interval and leads to a notable overhead for uplink transmission (see also Fernandes et al. [7] and Jose et al. [8]).

11.1.2 SCOPE AND CONTRIBUTIONS OF THE CHAPTER

In this chapter, we introduce a new multiple-access transmission protocol (i.e., uplink transmission from the users to the base station) referred to as semiorthogonal multiple-access (SOMA), which enables the scheduling of more users, as compared

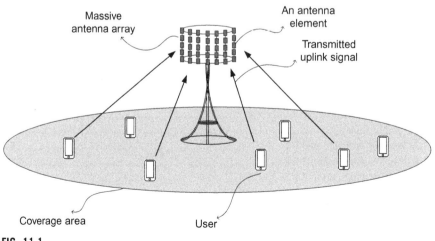

FIG. 11.1

Illustration of multiuser uplink transmission.

to the conventional TDD in Marzetta's work [1], as well as the transmittal of additional uplink data to enhance the spectral efficiency. We investigate the multiuser single-cell uplink scenario where each user terminal is equipped with a single antenna and wishes to communicate to a common base station with a large antenna array. We choose the aggregate throughput as a well-accepted measure to assess the performance. Toward this end, using information theoretic tools we analyze the aggregate throughput for the case of finite and infinite numbers of base station antennas. The results indicate a notable increase in the spectral efficiency as compared to that with conventional TDD transmission. We then extend SOMA to the so-called generalized SOMA (GSOMA) and analyze its throughput.

11.1.3 ORGANIZATION OF THE CHAPTER

The remainder of the chapter is organized as follows. Section 11.2 first reviews the conventional TDD transmission protocol of mMIMO and then addresses the problem of uplink capacity shortage. Section 11.3 presents the new multiple-access solution with a semiorthogonal feature and studies the asymptotic performance for large numbers of antennas and shows 100% gain in spectral efficiency as compared to the conventional TDD. Section 11.4 presents an achievable sum-rate of the proposed SOMA scheme for a finite number of antennas. Section 11.5 discusses numerical results of SOMA. Section 11.6 explains the concept of GSOMA. Section 11.7 tailors GSOMA to two-group GSOMA and discusses potential receivers and the achievable aggregate rates. Section 11.8 presents some representative numerical evaluations of GSOMA and Section 11.9 finally concludes the chapter.

11.2 UPLINK CAPACITY SHORTAGE OF mMIMO

In the sequel, we consider the communication setup illustrated in Fig. 11.1 with K single-antenna users and a single receiver with a massive antenna array. To estimate the channel between two antenna ports, the transmitting node sends pilot symbols, which are known at the receive node (the time-frequency location and the associated value are generally preset). The pilot symbols from each user should have the periodicity of $N = T_c B_c$ symbols in order to track the channel variation over the time and frequency where T_c and B_c denote the coherence time and bandwidth, respectively. In TDD protocol [1, 3–5], the transmission time over each coherence interval of N symbols is divided into four nonoverlapping phases:

-   Channel training: to learn the channel between the users and the base station, each user transmits some known pilot symbols. For example for the case of mMIMO, one pilot symbol per coherence interval per user is enough to acquire the channel. So for K users, K *orthogonal* pilot symbols are required over the time-frequency grid over which the channel is approximately constant.
-   Uplink data: the uplink data of all users are transmitted over the same time-frequency resources in a *nonorthogonal* fashion such that the base station

receives a superposition of all transmitted symbols. The superimposed received signals are then separated using space-division multiple-access (SDMA) at the receiver.
- Processing time: the time that is needed to perform the channel estimation and precode the users' data for the downlink transmission. If this time is less than the symbol duration one may ignore this processing time.
- Downlink data: finally the downlink data of all users are precoded using the estimated channels and transmitted to the users over the same time-frequency resources in a *nonorthogonal* fashion.

Fig. 11.2 illustrates the above four phases of the transmission and reception. The duration of uplink and downlink transmissions may vary and can be adjusted based on the amount of the users' data and the traffic demands. In the event that there is no downlink data, the transmission protocol contains only the two first phases—orthogonal channel training followed by nonorthogonal uplink transmission of all users' data.

The problem is how to coordinate as many users as possible for uplink transmission such that the receiver can perform spatial-division multiple-access (SDMA); that is, the receiver is able to obtain interference-free signals associated to each user via spatial filtering. We next compute the maximum number of active users to maximize the aggregate rate according to the baseline TDD protocol as illustrated in Fig. 11.1 (see also Merzetta [1]). In mMIMO, the receiver can separate the uplink data of different users without interuser interference if the spatial channels are, for example, asymptotically independent [1]. Hence, ignoring the processing time, the optimal number of the coordinated users, under successful decoding, can be obtained by maximizing the total number of transmitted data symbols that are separable at the receiver, which is given by,

$$\max_{1 \le K \le B_c T_c - 1} K(B_c T_c - K) \tag{11.1}$$

where the maximum is taken over the active number of users, K, for a given coherence interval spanning N symbols (cf. the number of dashed symbols in Fig. 11.2). Here it is assumed that all data symbols contain equal information; that is, all users employ the same modulation order. The above optimization leads to the conclusion that the coherence interval should be equally divided between the channel training and data transmission (see also Björnson et al. [9]). So the optimal number of active users operating over the same time-frequency resources should be set to $\left\lfloor \frac{1}{2}N \right\rfloor$ and the total number of data symbols that can be transmitted hence becomes $\left\lfloor \frac{1}{2}N \right\rfloor \cdot \left\lceil \frac{1}{2}N \right\rceil$ symbols.

The problem that we address in this paper is how to improve the maximum number of data symbols (i.e., increasing the aggregate throughput) by (a) transmitting

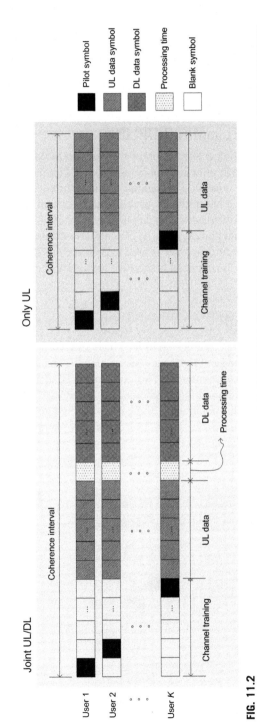

FIG. 11.2

TDD transmission protocol.

additional uplink data and (b) coordinating a simultaneous connection of a *larger* number of active users beyond $\left\lfloor \frac{1}{2}T_cB_c \right\rfloor$ yet enabling SDMA at the receiver with a negligible or no interuser interference.

11.3 SOMA: NOVEL mMIMO UPLINK

We next describe the uplink transmission of the users in SOMA. In SOMA, additional uplink data after each pilot symbol is transmitted. Uplink channel estimation and decoding of users is done in a sequential manner, through sequential channel estimation with successive interference cancellation. Pilot symbols of the first user are transmitted orthogonally to both pilot and data symbols of all other users and any other users data is nonorthogonally, that is, concurrently transmitted with only one user's pilot symbols ("semiorthogonal" is used to describe this feature). We can also add new users because we do not need to sacrifice the uplink data transmission of the earlier users.

11.3.1 TRANSMITTER

Specifically, the scheme coordinates up to $K \leq N - 1$ users, where $N = T_cB_c$ is the number of resources in time and frequency over which the channel is approximately constant, that is, the coherence interval. This number can be identified depending on the communication environment and the mobility of the users or using a channel sounding that estimates this parameter for the users. The coordination among users may be accomplished by a scheduler. Fig. 11.3 depicts an example of how the uplink transmission is configured in which $K = N - 1$ users are scheduled. Over each

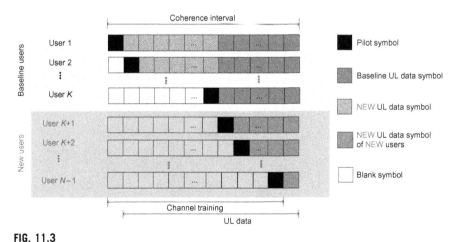

FIG. 11.3

Illustration of semiorthogonal multiple-access transmission.

coherence block, the user $1 \leq j \leq N - 1$, transmits one pilot symbol and $N - j$ data symbols such that the two following properties are satisfied:

- The pilot symbols of users are transmitted over orthogonal time-frequency resources. This prevents pilot contaminations. That is, the pilot symbols do not interfere with one another.
- The data symbols of user j, for all $1 \leq j \leq N - 1$, consume all the time-frequency resources of users $j + 1 \leq k \leq N - 1$, which are used for both pilot and data transmission.

The received symbols at the base station in a coherence window are illustrated in Fig. 11.3. In a given time-slot, some of the users are silent and appear *orthogonal*, while some other users transmit *nonorthogonally*. For example, in time slot one, only user one transmits its pilot and all other users appear orthogonal and in the second time slot the first user transmits its data symbol and second user transmits its pilot symbol and so forth; cf. *semiorthogonal* feature.

So the received signal over the ith time-frequency resource element is given by,

$$y_i = \begin{cases} h_i x_{p_i} + \sum_{j=1}^{i-1} h_j x_{d_{j,i}} + z_i, & 1 \leq i \leq N - 1 \\ \sum_{j=1}^{i-1} h_j x_{d_{j,i}} + z_i, & i = N \end{cases} \tag{11.2}$$

$$=: h_i x_{p_i} \mathbb{I}_{i \neq N} + \sum_{j=1}^{i-1} h_j x_{d_{j,i}} + z_i, \quad 1 \leq i \leq N$$

where y_i denotes the received noisy signal vector of dimension $n_t \times 1$, x_{p_i} denotes the transmitted pilot symbol from user i, $x_{d_{j,i}}$ denotes the transmitted data symbol from user j over the ith time-frequency resource element, and h_i denotes the channel vector between user i and the receiver, which has the dimension $n_t \times 1$. Here n_t denotes number of base station antennas and $\mathbb{I}$ denotes an indicator function.

Compared to the state-of-the-art TDD transmission protocol, in the new proposal, the pilot symbols are *contaminated* by the data symbols of the coordinated users; cf. Figs. 11.2 and 11.3. So the channel estimation should handle the interference from the other users and the data detector should handle the interference from the other users as well as the pilot symbols. We next show the abovementioned SOMA can be combined with a receiver that alleviates the adverse effects of the introduced additional interferences.

11.3.2 **RECEIVER**

For the above described transmitter, the receiver is constructed using sequential spatial filtering with successive interference cancellation. In the following, we employ MF to separate the signals and discuss how many asymptotically error-free data symbols with a fixed rate, can be multiplexed as the number of antennas n_t increases.

Having received superimposed noisy baseband signals, the receiver employs a demultiplexer, minimum mean-square error (MMSE) channel estimator followed by spatial MF to decompose the received noisy baseband signals associated to the transmitted pilot and each modulated data symbol and proceed as follows for each user:

First user

Using the known pilot symbol and the associated received noisy copy, the first channel estimator obtains an estimate of the channel vector for the first user. The received noisy pilot symbol of the first user is given by,

$$y_1 = h_1 x_{p_1} + z_1 \tag{11.3}$$

Over the first time-frequency resource element the other users are silent and the first channel estimator can estimate the channel vector h_1 without any interference. The estimated channel vector can be written as $\hat{h}_1 = h_1 - z_{e_1}$, where z_{e_1} denotes the channel estimation error. The estimated channel vector is then used for spatial filtering of the data of the first user. Now the receiver using the estimated channel vector, $\hat{h}_1$, of the first user obtained via the signal vector y_1, can perform the normalized MF for $2 \le i \le N$, as follows:

$$
\begin{aligned}
\tilde{y}_i &= \frac{1}{n_t} \hat{h}_1^\dagger y_i \\
&= \underbrace{\frac{1}{n_t} h_1^\dagger h_1 x_{d_{1,i}}}_{\text{Information of the first user}} + \underbrace{\frac{1}{n_t} h_1^\dagger h_i x_{p_i} \mathbb{I}_{i \neq N}}_{\text{Interference due the pilot of } i \text{th user}} \\
&\quad + \underbrace{\frac{1}{n_t} \sum_{j=2}^{i-1} h_1^\dagger h_j x_{d_{j,i}}}_{\text{Interference due to other users data}} + \underbrace{\frac{1}{n_t} h_1^\dagger z_i}_{\text{AWGN}} \\
&\quad \underbrace{- \frac{1}{n_t} z_{e_1}^\dagger \left(h_i x_{p_i} \mathbb{I}_{i \neq N} + \sum_{j=1}^{i-1} h_j x_{d_{j,i}} + z_i \right)}_{\text{Propagation of the channel estimation error}}
\end{aligned}
\tag{11.4}
$$

As the number of antennas increases, that is, $n_t \to \infty$, for i.i.d. channels (whose means are Lebesgue integrable) with unit variance and by the orthogonality principle of the MMSE channel estimator (i.e., estimation error is uncorrelated to the actual channel realization) [10, 11], and strong law of large numbers [11] we have the following near-definite results:

$$\frac{1}{n_t} h_i^\dagger h_j \to \delta_{ij}$$

$$\frac{1}{n_t} h_i^\dagger z_i \to 0$$

$$\frac{1}{n_t} z_{e_1}^\dagger h_i \to 0$$

$$\frac{1}{n_t} z_{e_1}^\dagger z_i \to 0, \quad i \ge 2$$

This yields,

$$\tilde{y}_i \to x_{d_{1,i}} \quad \text{as} \quad n_t \to \infty \quad \text{for} \quad 2 \leq i \leq N \tag{11.5}$$

This is the receiver asymptotically retrieving the transmitted modulated data symbols of the first user.

Second user

Using the known pilot symbol transmitted from the second user and the received noisy copy of the pilot symbol, the second channel estimator obtains an estimate of the channel vector between the second user and the base station. The received signal over the second channel use contains the pilot signal of the second user, which is given by,

$$y_2 = h_2 x_{p_2} + h_1 x_{d_{1,2}} + z_2 \tag{11.6}$$

The transmitted pilot signal of the second user is contaminated by the data of the first user. In order to perform the channel estimation the receiver needs to cancel the interfering signal from the first user. Toward this end, the receiver needs to know the channel vector h_1 and the transmitted data symbol of first user $x_{d_{1,2}}$ transmitted over the second time-frequency resource element. However, the receiver using the first channel estimator, knows an estimate of h_1 and using the first decoder (or detector) has recovered the modulated symbol $x_{d_{1,2}}$. Thus, despite a *nonorthogonal* data and pilot symbol transmission, an interference cancellation is possible and the second channel estimator can find an estimate of the second channel vector after the interference cancellation by forming the following signal:

$$\begin{aligned} \check{y}_2 &= y_2 - \hat{h}_1 x_{d_{1,2}} \\ &= h_2 x_{p_2} + z_{e_1} x_{d_{1,2}} + z_2 \end{aligned} \tag{11.7}$$

Now using the signal $\check{y}_2$, the channel estimate can be written as $\hat{h}_2 = h_2 - z_{e_2}$ where the estimation noise z_{e_2} is uncorrelated from $\{h_i\}_{i=1}^{N-1}$, $\{z_i\}_{i=3}^{N-1}$ when the MMSE channel estimator is used. Now using the same approach as that for user one, by applying MF we obtain,

$$\tilde{y}_i = \frac{1}{n_t} \hat{h}_2^\dagger y_i \to x_{d_{2,i}} \quad \text{as} \quad n_t \to \infty \quad \text{for} \quad 3 \leq i \leq N \tag{11.8}$$

Thus modulated data symbols of the second user can be obtained using the semi-orthogonal feature.

User j

We next consider user j, for $3 \leq j \leq N - 1$ and explain the corresponding signal processing. In order to recover the transmitted data of the jth user, the receiver needs to estimate the channel vector associated to the jth user. The known pilot symbol from the jth user is transmitted over the jth channel use and the received noisy copy of the pilot symbols is given by,

$$y_j = h_j x_{p_j} + \sum_{k=1}^{j-1} h_k x_{d_{k,j}} + z_j \tag{11.9}$$

The transmitted pilot signal of user j is contaminated by the data of the first user up to user $j - 1$. In a similar fashion to the second user, in order to perform a good channel estimation the receiver needs to cancel the interfering signals from all users. The receiver can accomplish this task in a similar fashion as that for the second user. The receiver using the channel estimator 1 to $j - 1$ knows estimates of channel vectors $\boldsymbol{h}_1$ to $\boldsymbol{h}_{j-1}$ and, using the decoder/detector 1 up to $j - 1$, has recovered $\{x_{d_{k,j}}\}_{k=1}^{j-1}$. Hence, the jth channel estimator can estimate the channel vector $\boldsymbol{h}_j$ and can perform the interference cancellation to obtain,

$$\breve{\boldsymbol{y}}_j = \boldsymbol{y}_j - \sum_{k=1}^{j-1} \hat{\boldsymbol{h}}_k x_{d_{k,j}}$$

$$= \boldsymbol{h}_j x_{p_j} + \sum_{k=1}^{j-1} \boldsymbol{z}_{e_k} x_{d_{k,j}} + \boldsymbol{z}_j \qquad (11.10)$$

Similarly using $\breve{\boldsymbol{y}}_j$, the channel estimate can be written $\hat{\boldsymbol{h}}_j = \boldsymbol{h}_j - \boldsymbol{z}_{e_j}$ where the estimation noise $\boldsymbol{z}_{e_j}$ is uncorrelated from $\{\boldsymbol{h}_i\}_{i=1}^{N-1}$, $\{\boldsymbol{z}_i\}_{i=j+1}^{N-1}$. Now using the same approach as that for user one, by applying MF we obtain,

$$\tilde{\boldsymbol{y}}_i = \frac{1}{n_t} \hat{\boldsymbol{h}}_j^\dagger \boldsymbol{y}_i \to x_{d_{j,i}} \quad \text{as } n_t \to \infty \text{ for } j < i \leq N \qquad (11.11)$$

Therefore, SOMA with very high number of antennas can schedule K users where user k can transmit $N - k$ asymptotically error-free symbols. Thus the total of error-free symbols of the users is,

$$\sum_{k=1}^{K} (N - k) = K\left(N - \frac{K+1}{2}\right)$$

$$\leq \frac{1}{2} N(N - 1) \qquad (11.12)$$

where the upper bound is achieved when $K = N - 1$. For the baseline scheme with orthogonal training and optimal number of users the total number of asymptotically error-free symbols is $\left\lfloor \frac{1}{2} N \right\rfloor \cdot \left\lceil \frac{1}{2} N \right\rceil$. Thus SOMA *asymptotically* nearly doubles the throughput as compared to the baseline TDD with an optimal number of users. However, if the system wants to serve the maximum number of users, then the gain linearly increases with N.

The construction of SOMA according to Fig. 11.3 is a way in which an unequal number of resources are assigned to each user in a given coherence interval. However, by reordering the users over multiple-coherence intervals, one can achieve a fair assignment of the available resources such that they receive roughly the same throughput on the average.

11.4 SUM-RATE CHARACTERIZATION OF SOMA

The complete characterization of the K-user scheme can be done using the rate region, which is the convex closures of all rate tuples given by $(R_1, R_2, \ldots, R_k)$ where R_k denotes an achievable rate for user k for all $k \in [1 : K]$ [12]. The rate region, that is a subset of $\mathbb{R}^{+K}$ for K users, is generally difficult to envision and to exploit for practical usage when K is large. We, therefore, choose the sum-rate; that is, $\sum_{k=1}^{K} R_k$ to characterize the scheme.

11.4.1 BOUNDS ON SOMA

In this section, we present the uplink throughput when the receiver has n_t antennas and the channels are i.i.d. Rayleigh fading with unit variance and the coherence interval of length N. The following proposition presents our main result, which is the sum-rate of SOMA with $K = N - 1$ users when the sequential channel estimation as explained in Eq. (11.10) is used.

Proposition 11.1 (Sum-Rate of SOMA). *SOMA with sequential channel estimation achieves the sum-rate,*

$$R_{\text{LB}} = \sum_{k=1}^{N-1} R_k \tag{11.13}$$

where,

$$R_k = \frac{1}{N} \sum_{l=k+1}^{N} \log \left(1 + \frac{(1 - N_{e_k}) P_{d_k}}{\dfrac{N_0}{n_t - 1} + \dfrac{P_{p_l} \mathbb{I}_{l \neq N} + N_{e_k} P_{d_k}}{n_t - 1} + \dfrac{\sum_{i=1, i \neq k}^{l-1} P_{d_i}}{n_t - 1}} \right) \tag{11.14}$$

and,

$$N_{e_k} = \frac{N_0 + \sum_{j=1}^{k-1} N_{e_j} P_{d_j}}{N_0 + P_{p_k} + \sum_{j=1}^{k-1} N_{e_j} P_{d_j}} \tag{11.15}$$

Here R_k, P_{p_k}, P_{d_k}, and N_{e_k} respectively denote the achievable rate, the average power consumed for the pilot and data and the variance of channel estimation error for user k, for all $1 \leq k \leq N - 1$, and N_0 denotes the variance of AWGN.

Proof See Appendix 11.A. □

For comparison, we use the following upper bound when SOMA protocol and its associated channel estimation are fixed.

Proposition 11.2 (Upper Bound on SOMA). *The sum-rate of SOMA is upper bounded as,*

$$R_{\text{SOMA}} \leq R_{\text{UB}} := \sum_{k=1}^{N-1} R_k^{\text{ub}} \tag{11.16}$$

where,

$$R_k^{\text{ub}} = \frac{1}{N} \sum_{l=k+1}^{N} \log\left(1 + \frac{n_t(1 - N_{e_k})P_{d_k}}{N_0 + P_{p_l}\mathbb{I}_{l \neq N} + N_{e_k}P_{d_k}}\right) \tag{11.17}$$

Proof See Appendix 11.B. □

The next result discusses the tightness of the bounds.

Proposition 11.3 (Tightness of Bounds). *The upper bound in Proposition 11.2 maintains a constant gap to the lower bound in Proposition 11.1 as a function of the number of antennas; that is, $R_{\text{UB}}(n_t) - R_{\text{LB}}(n_t) \leq m$, where m does not depend on n_t.*

Proof See Appendix 11.C. □

11.4.2 BOUNDS ON CONVENTIONAL TDD

Additionally, we present bounds on the sum-rate of the baseline TDD solution for $1 \leq K \leq N - 1$ users as benchmarks for numerical evaluations. Two of the following results are from Ngo et al. [13], which we recite for completeness.

Proposition 11.4 (Baseline Sum-Rate With MF Without SIC [13]). *The baseline scheme with MF for $K \leq N - 1$ users can achieve the sum-rate,*

$$R_{\text{b,Ngo}}^{\text{MF}} = \frac{N-K}{N} \sum_{k=1}^{K} \log\left(1 + \frac{(1 - N_{e_k})P_{d_k}}{\dfrac{N_0}{n_t - 1} + \dfrac{\sum_i N_{e_i}P_{d_i}}{n_t - 1} + \dfrac{\sum_{i \neq k}P_{d_i}}{n_t - 1}}\right) \tag{11.18}$$

where,

$$N_{e_k} = \frac{N_0}{N_0 + P_{p_k}} \tag{11.19}$$

Proposition 11.5 (Baseline Sum-Rate With MF Without SIC). *The baseline scheme with MF for $K \leq N - 1$ users can achieve the sum-rate,*

$$R_{\mathrm{b}}^{\mathrm{MF}} = \frac{N-K}{N} \sum_{k=1}^{K} \log \left(1 + \frac{(1-N_{e_k})P_{d_k}}{\dfrac{N_0}{n_{\mathrm{t}}-1} + \dfrac{N_{e_k}P_{d_k}}{n_{\mathrm{t}}-1} + \dfrac{\sum_{i \neq k}P_{d_i}}{n_{\mathrm{t}}-1}} \right) \quad (11.20)$$

where N_{e_k} is given in Eq. (11.19).

Proof See Appendix 11.D.   □

The achievable rate of Eq. (11.20) is strictly higher than Eq. (11.18). In deriving the new bound, the channel estimation error in direction of the spatial channel of target user is considered as opposed to that in Ngo et al. [13].

Proposition 11.6 (Baseline Sum-Rate With MF and SIC). *The baseline scheme with MF for $K \leq N - 1$ users and sequential interference cancellation can achieve the sum-rate*

$$R_{\mathrm{b,SIC}}^{\mathrm{MF}} = \frac{N-K}{N} \sum_{k=1}^{K} \log \left(1 + \frac{(1-N_{e_k})P_{d_k}}{\dfrac{N_0}{n_{\mathrm{t}}-1} + \dfrac{\sum_{i=1}^{k}N_{e_i}P_{d_i}}{n_{\mathrm{t}}-1} + \dfrac{\sum_{i=k+1}^{K}P_{d_i}}{n_{\mathrm{t}}-1}} \right) \quad (11.21)$$

where N_{e_k} is given in Eq. (11.19).

Proof See Appendix 11.E.   □

We finally recite the sum-rate of TDD with zero forcing (ZF) from Ngo et al. [13].

Proposition 11.7 (Baseline Sum-Rate With ZF [13]). *The baseline scheme with ZF for $K \leq N - 1$ users can achieve the sum-rate,*

$$R_{\mathrm{b}}^{\mathrm{ZF}} = \frac{N-K}{N} \sum_{k=1}^{K} \log \left(1 + \frac{(1-N_{e_k})P_{d_k}}{\dfrac{N_0}{n_{\mathrm{t}}-K} + \dfrac{\sum_{i=1}^{K}N_{e_i}P_{d_i}}{n_{\mathrm{t}}-K}} \right) \quad (11.22)$$

where N_{e_k} is given in Eq. (11.19).

From the above results, we observe that all the imperfections due to interuser interference, the AWGN noise and the channel estimation error are linearly reduced by the factor of $\dfrac{1}{n_{\mathrm{t}}-1}$ and $\dfrac{1}{n_{\mathrm{t}}-K}$ using MF and ZF, respectively. Comparing the above results, we observe that SOMA schedules more users at the cost of additional interferences from nonorthogonal pilots and higher channel estimation error.

However, these additional interferences and the estimation errors are similarly reduced by the factor $\dfrac{1}{n_t - 1}$ using MF.

11.5 NUMERICAL EVALUATIONS OF SOMA

In this section, we discuss numerical examples. For this purpose, we consider the achievable sum-rate of SOMA scheme in Eq. (11.13) and the upper bound on the sum-rate of SOMA given in Eq. (11.16). For the conventional TDD, we consider four bounds: the achievable sum-rate with MF given in Eqs. (11.18), (11.20), and (11.21) and the achievable sum-rate achievable with ZF given in Eq. (11.22). Figs. 11.4–11.9 plot the bounds for SOMA and the conventional TDD for a channel with the coherence interval of N symbols. For SOMA the number of users is set to $K = N - 1$ (which is the maximum number of users) and for the conventional TDD we choose $K = \dfrac{N}{2}$ (which is optimal for TDD when $n_t \gg 1$) and $K = N - 1$ which is the maximum number of users. For all figures, $P_{d_k} = P$ and $P_{p_k} = P + \Delta$, where Δ is the power

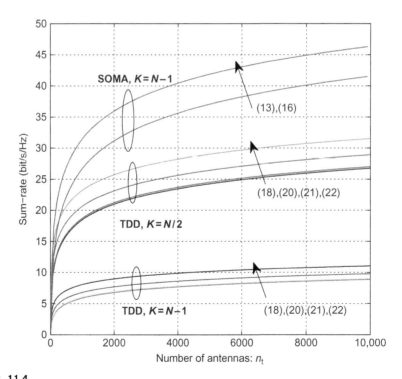

FIG. 11.4

The sum-rate of the schemes for $N = 10$, $\text{SNR}_d = -10$ dB, $\Delta = 0$ dB.

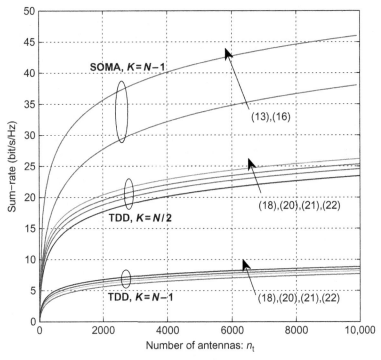

FIG. 11.5

The sum-rate of the schemes for $N = 10$, $SNR_d = -10$ dB, $\Delta = 5$ dB.

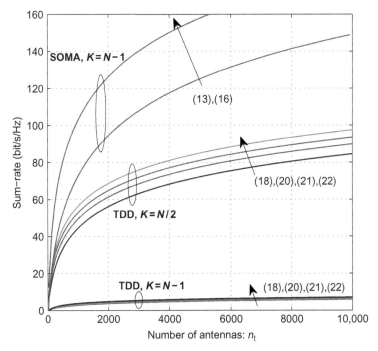

FIG. 11.6

The sum-rate of the schemes for $N = 10$, $SNR_d = -10$ dB, $\Delta = 10$ dB.

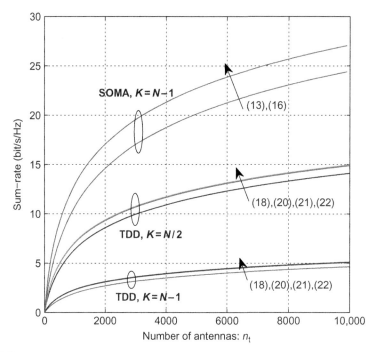

FIG. 11.7

The sum-rate of the schemes for $N = 10$, $\text{SNR}_d = 0$ dB, $\Delta = 0$ dB.

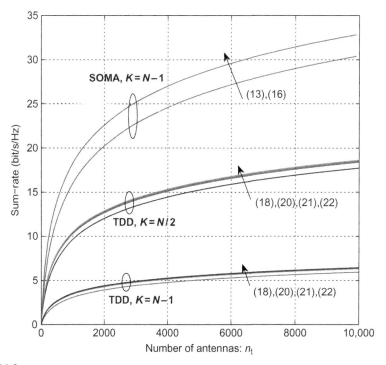

FIG. 11.8

The sum-rate of the schemes for $N = 10$, $\text{SNR}_d = 0$ dB, $\Delta = 10$ dB.

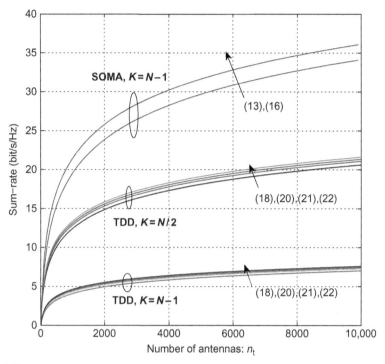

FIG. 11.9

The sum-rate of the schemes for $N = 50$, $SNR_d = -10$ dB, $\Delta = 10$ dB.

offset parameter to study the effect of the quality of channel estimation on the sum-rate, and we set the variance of AWGN at the receiver to $N_0 = 0$ dB. For these figures, we define $SNR_d := P/N_0$.

Figs. 11.4–11.6 show the sum-rate for $N = 10$, $SNR_d = -10$ dB and $\Delta = 0, 5, 10$ dB, respectively. Figs. 11.7 and 11.8 illustrate the sum-rate for $N = 10$, $SNR_d = 0$ dB and $\Delta = 0, 10$ dB, respectively. Fig. 11.9 shows the sum-rate for $N = 50$, $SNR_d = -10$ dB and $\Delta = 10$ dB, respectively. The smaller N corresponds to a fast-varying channel and the larger N corresponds to a slow-varying channel, where for the latter, a higher number of users are coordinated.

From Figs. 11.4 to 11.9 we observe that for shown range of antenna numbers, n_t, SOMA can provide up to nearly 70% gain in throughput with respect to the conventional TDD with optimal number of users. The asymptotic gain of SOMA (i.e., gain for $n_t \to \infty$) in sum-rate with respect to the conventional TDD for different cases increases with the number of antennas and can reach up 100% (cf. the analysis in Section 11.3). We additionally see that the baseline TDD solution performs very poorly when it is configured with the maximum number of users, because it does not efficiently utilize the available degrees of freedom. However, the proposed SOMA scheme performs much better by cleverly coordinating the transmission in a semiorthogonal fashion. The higher complexity at the receiver, in particular for the channel estimation, is the cost for obtaining this performance improvement.

We also observe that increasing the power of pilots enhances the sum-rate, because it improves the quality of channel estimation, which in turn improves the quality of spatial filtering. This also makes the lower bound approach the upper bound. In other words, a high-quality channel estimation allows for the removal of the interuser interference and the coherent combination of the signals in the direction of the desired channel. From the figures, we see that the lower bound maintains a nearly constant gap to the upper bound as a function of the number of antennas, which is in agreement with the result in Proposition 11.3.

For low-power transmission, we additionally see that ZF and SIC do not notably improve performance and MF without SIC provides a performance comparable to that of ZF. However, ZF is more useful for higher power cases or when the number of users is very high (i.e., slow fading). The sum-rate for slow-fading is much higher compared to that in fast-fading, since slower variations in the radio channel allows for the scheduling of more users in the coherence block, which dramatically enhances the sum-rate.

11.6 GENERALIZED SOMA

We next present the concept of GSOMA. Fig. 11.10 depicts the multiuser GSOMA transmission in which the users are grouped into J groups where each group contains k_j users for $j \in \{1, 2, ..., J\}$ and $K = \sum_{j=1}^{J} k_j$ is the total number of the users. The user i in group j employs the pilot sequence s_i for all $j \in \{1, 2, ..., J\}$. That is the pilot sequences are reused by different user groups. The pilot sequences within each group are mutually orthogonal such that it allows interference-free channel estimation for the users in each group. Code-division multiplexing (CDM) can be used for pilot transmission of each group. The maximum number of pilot sequences, therefore, is $\max_j k_j$. The conventional reuse of pilot sequences where the pilot symbols interfere with one another results in " pilot contamination," which severely degrades the performance of the users. However, with this new solution, it is possible to reuse the pilot sequences in a *controlled* fashion. The pilot reuse is performed *semiorthogonally* to boost the spectral efficiency of the network. However, the interference is controlled by a transmission of a *resource-offset* (e.g., timing-offset or frequency-offset) such that the received packets at the mMIMO access node have the following structure:

- The pilot signals of different groups are received nonorthogonally (e.g., nonoverlapping time slots with timing-offset).
- The pilot sequences of the group $j \in \{1, 2, ..., J\}$, only experience interference from data symbols of the users in groups 1 to $j - 1$. That is, the pilot sequences of the first group are received interference-free. This implies that the other users appear silent at the receiver side when the users in the first group transmit their pilot signals.

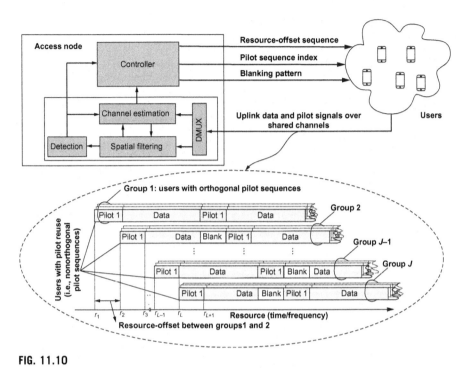

FIG. 11.10

Generalized semiorthogonal multiple-access (GSOMA) transmission.

With GSOMA, the user may use partial blanking which has the same granularity as the length of pilot region to control the intergroup interference. GSOMA includes both the conventional TDD solution proposed by Marzetta [1] and SOMA scheme as special cases. When each group contains one user and no blanking is used, GSOMA reduces to SOMA. When there is only one group with maximum number of users, then GSOMA reduces to the conventional TDD, wherein mutually orthogonal pilot sequences are used. Therefore, a properly designed GSOMA can combine the advantages of both SOMA and the conventional TDD. The advantage of GSOMA with respect to the conventional TDD is that it schedules more groups, which enhances the aggregate rate. Since within each group the pilot sequences are mutually orthogonal, as in the conventional TDD, it allows for the performance of joint channel estimation without interference for all users in each group and for the employment of joint spatial filtering, such as ZF, per group to suppress the interuser interference among the users in each group. This also increases the aggregate rate.

11.7 TWO-GROUP GSOMA

In this section, we tailor the design of GSOMA to the cases of the two groups and discuss two modes of transmission.

11.7.1 MODE 1: WITHOUT BLANKING

Fig. 11.11 shows the transmission protocol and the corresponding receiver for two groups where there is no blanking. The users are grouped into two groups such that each group is designed according the conventional TDD wherein half of the coherence time, that is, $\frac{1}{2}T_c$, is used for pilot transmission and the other half is consumed for the data transmission. To enable the channel estimation without interference, a time-offset equal to $\frac{1}{2}T_c$ is used. The receiver first estimates the channel of the users in the first group using the knowledge of the orthogonal pilot sequences and then performs joint spatial filtering to decode the data of the first group. The decoded data are fed back to the channel estimator of the second group to perform interference cancellation prior to the channel estimation. After the interference cancellation, the channels of the second group are estimated using the known orthogonal pilot sequences used for the users in the second group. The decoded data of the second group are used to cancel the interference over the pilot symbols of first group. This sequential channel estimation and decoding is iteratively continued until all data are successfully decoded.

We next present the single-cell uplink throughput for the protocol in Fig. 11.12 with L consecutive subframes when the receiver has n_t antennas and the channels are i.i.d. Rayleigh fading with unit variance. We first consider MF and then ZF. The proof can be obtained using similar approaches as those used with SOMA.

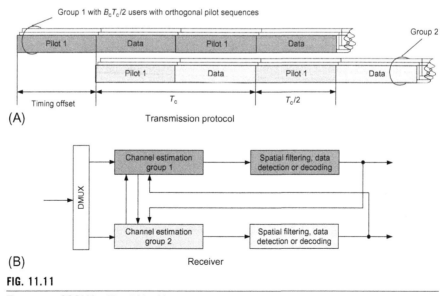

FIG. 11.11

Two-group GSOMA without blanking.

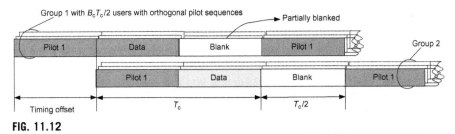

FIG. 11.12

Two-group GSOMA with blanking.

Proposition 11.8. *The transmission protocol in Fig. 11.11 where each group contains K users and the receiver employs MF achieves the sum-rate,*

$$R^{\mathrm{MF}}_{\Sigma,1} = \frac{K}{2L+1} \sum_{l=1}^{L} (R_{1,l} + R_{2,l}) \tag{11.23}$$

where,

$$R_{1,l} = \log\left(1 + \frac{(n_t - 1)(1 - N_{e_{1,l}})P_{d_{1,l}}}{N_0 + N_{e_{1,l}}P_{d_{1,l}} + P_{p_{2,l}} + (K-1)P_{d_{1,l}}}\right)$$

$$R_{2,l} = \log\left(1 + \frac{(n_t - 1)(1 - N_{e_{2,l}})P_{d_{2,l}}}{N_0 + N_{e_{2,l}}P_{d_{2,l}} + P_{p_{1,l+1}} + (K-1)P_{d_{2,l}}}\right)$$

$$N_{e_{1,l}} = \frac{N_0 + KN_{e_{2,l-1}}P_{d_{2,l-1}}}{N_0 + P_{p_{1,l}} + KN_{e_{2,l-1}}P_{d_{2,l-1}}}$$

$$N_{e_{2,l}} = \frac{N_0 + KN_{e_{1,l}}P_{d_{1,l}}}{N_0 + P_{p_{2,l}} + KN_{e_{1,l}}P_{d_{2,l}}}$$

and $P_{p_{j,l}}$, $P_{d_{j,l}}$ *denote the average power consumed for the pilot and data symbols of the users in group* $j \in \{1, 2\}$ *and subframe* $l \in \{1, 2, ..., L\}$, *and* $N_{e_{j,l}}$ *and* N_0 *respectively denote variance of AWGN at the receiver and channel estimation error, and* $N_{e_{1,0}} = P_{d_{1,0}} = P_{p_{1,L+1}} = 0$.

Proposition 11.9. *The transmission protocol in Fig. 11.11 where each group contains K users and the receiver employs ZF achieves the sum-rate,*

$$R^{\mathrm{ZF}}_{\Sigma,1} = \frac{K}{2L+1} \sum_{l=1}^{L} \log\left(1 + \frac{(n_t - K)(1 - N_{e_{1,l}})P_{d_{1,l}}}{N_0 + KN_{e_{1,l}}P_{d_{1,l}} + P_{p_{2,l}}}\right)$$
$$+ \log\left(1 + \frac{(n_t - K)(1 - N_{e_{2,l}})P_{d_{2,l}}}{N_0 + KN_{e_{2,l}}P_{d_{2,l}} + P_{p_{1,l+1}}}\right) \tag{11.24}$$

where $N_{e_{j,l}}$ *is defined in Proposition 11.8.*

11.7.2 MODE 2: WITH BLANKING

For the case the array contains not so many antenna elements with respect to the scheduled number of users in each group, it is beneficial to partially blank some part of subframes to enhance the channel estimation and consequently to improve the performance of spatial filtering, which in turn improves the spectral efficiency of the system. The blanking pattern, should be chosen based on the interuser interference, when for higher interference a blanking pattern with a higher density is selected and, similarly, a sparser blanking pattern should be selected for lower interference to improve the spectral efficiency of the system. Fig. 11.12 shows the transmission protocol for the two groups where the blanking is used. The users are organized into two groups and transmission is arranged such that the pilot symbols of first group do not experience any interference. This is useful to enhance the channel estimation for the users in the first group, this also enhances the interference cancellation for channel estimation of the second group. Additionally, this arrangement is also useful where first and second groups compromise far and near users, respectively.

Proposition 11.10. *The transmission protocol in Fig. 11.12 where each group contains K users and the receiver employs MF achieves the sum-rate,*

$$
\begin{aligned}
R^{\text{MF}}_{\sum,2} &= \frac{1}{3}K\log\left(1 + \frac{(n_t-1)(1-N_{e_1})P_{d_1}}{N_0 + N_{e_1}P_{d_1} + P_{p_2} + (K-1)P_{d_1}}\right) \\
&+ \frac{1}{3}K\log\left(1 + \frac{(n_t-1)(1-N_{e_2})P_{d_2}}{N_0 + N_{e_2}P_{d_2} + (K-1)P_{d_2}}\right)
\end{aligned}
\tag{11.25}
$$

where P_{p_j} and P_{d_j} denote the average power consumed for the pilot and data symbols of the users in group $j \in \{1, 2\}$, and N_0 denotes the variance of AWGN at the receiver. The quantity $N_{e_j} = N_{e_{j,1}}$ denotes the variance of channel estimation error for users in group $j \in \{1, 2\}$.

Proposition 11.11. *The transmission protocol in Fig. 11.12 where each group contains K users and the receiver employs ZF achieves the sum-rate,*

$$
\begin{aligned}
R^{\text{ZF}}_{\sum,2} &= \frac{1}{3}K\log\left(1 + \frac{(n_t-K)(1-N_{e_1})P_{d_1}}{N_0 + KN_{e_1}P_{d_1} + P_{p_2}}\right) \\
&+ \frac{1}{3}K\log\left(1 + \frac{(n_t-K)(1-N_{e_2})P_{d_2}}{N_0 + KN_{e_2}P_{d_2}}\right)
\end{aligned}
\tag{11.26}
$$

where $N_{e_j} = N_{e_{j,1}}$ for $j \in \{1, 2\}$ is given in Proposition 11.10.

From the sum-rates in Propositions 11.8–11.11, one can see that the imperfections due to interuser interference, the channel estimation error and AWGN are linearly reduced by the factor of $\dfrac{1}{n_t - 1}$ and $\dfrac{1}{n_t - K}$ for MF and ZF, respectively.

We additionally see that ZF removes the interuser interference in each group due the data transmission, which is one of the advantages of GSOMA with respect to SOMA.

11.8 NUMERICAL EVALUATIONS OF GSOMA

For comparison, we consider time-shared TDD as a baseline when the resources are shared between two groups of users when each group is designed according to the conventional TDD. This, using ZF, gives the sum-rate,

$$
R_{\Sigma,\mathrm{TS}}^{\mathrm{ZF}} = \frac{1}{4} K \log \left(1 + \frac{(n_t - K)(1 - N_{e_1}) P_{d_1}}{N_0 + K N_{e_1} P_{d_1}} \right)
$$
$$
+ \frac{1}{4} K \log \left(1 + \frac{(n_t - K)(1 - N_{e_2}) P_{d_2}}{N_0 + K N_{e_2} P_{d_2}} \right)
$$

(11.27)

where $N_{e_j} = \dfrac{N_0}{N_0 + P_{p_j}}$.

We next discuss two numerical examples of the sum-rate. Figs. 11.13 and 11.14 show the sum-rate of the schemes as a function of the number of antennas for

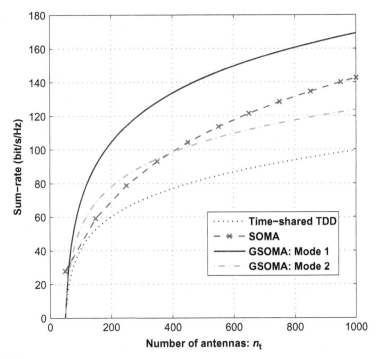

FIG. 11.13

The sum-rate of the schemes for $P_{d_1} = 0$, $P_{d_2} = -20$ dB.

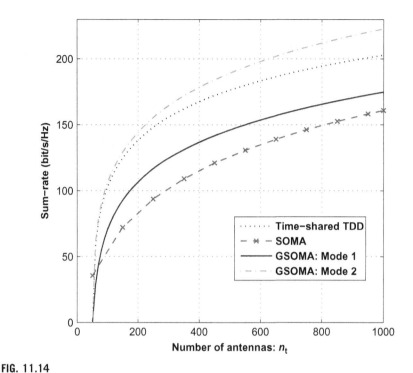

FIG. 11.14

The sum-rate of the schemes for $P_{d_1} = 10$, $P_{d_2} = -5$ dB.

$N = B_c T_c = 100$, where the number of users in each group is $K = \frac{1}{2}N = 50$. We set $P_{d_{1,l}} = P_{d_1}, P_{d_{2,l}} = P_{d_2}$ (i.e., uniform power allocation across the subframes) and $N_0 = 0$ dB, and the power of the associated pilots are set 10 dB higher than the data for all users to ensure a good channel estimation. In each figure, four schemes are considered: time-shared TDD with ZF, SOMA with MF, GSOMA with transmission protocol in Fig. 11.11 (i.e., Mode 1) with ZF and $L = 100$, GSOMA with transmission protocol in Fig. 11.12 with ZF (i.e., Mode 2). In Fig. 11.13 we set $P_{d_1} = 0, P_{d_2} = -20$ dB and in Fig. 11.14 we set $P_{d_1} = 10, P_{d_2} = -5$ dB. In both cases, GSOMA provides an enhanced aggregate rate as compared to the time-shared TDD. The gain is more pronounced for the groups with a higher difference in the received signal strength for which Mode 1 performs better. Fig. 11.14, Mode 2 performs better than Mode 1 due to the fact that the channel estimation for the high-power users in group 1 is less degraded as compared to that in Mode 1.

11.9 CONCLUSIONS

We presented an enhanced multiple-access solution for uplink mMIMO communications and characterized its sum-rate for arrays with infinite numbers of antennas as well as for finite numbers of antennas. The results showed that the scheme can nearly double the aggregate throughput with respect to the conventional TDD for large scale arrays. The extensions of the current work include investigation and optimization of the scheme for multicell scenarios.

APPENDIX
11.A PROOF OF PROPOSITION 11.1

We next prove the result in Proposition 11.1. The decoding is performed sequentially; that is, the data of user j are decoded after user 1 to $j - 1$. Consider the following series of inequalities for user j:

$$R_j \overset{(a)}{=} \frac{1}{N} \mathbb{E}I\left(x_{p_j}, \{x_{d_{j,k}}\}_{k>j}; \{y_k\}_{k\geq j} | \{x_{d_{k,i}}, \hat{h}_k\}_{k<j}\right)$$

$$\overset{(b)}{=} \frac{1}{N} \mathbb{E}I\left(\{x_{d_{j,k}}\}_{k>j}; \{y_k\}_{k\geq j} | x_{p_j}, \{x_{d_{k,i}}, \hat{h}_k\}_{k<j}\right)$$

$$\overset{(c)}{=} \frac{1}{N} \mathbb{E}I\left(\{x_{d_{j,k}}\}_{k>j}; \{y_k\}_{k>j} | x_{p_j}, y_j, \{x_{d_{k,i}}, \hat{h}_k\}_{k<j}\right)$$

$$\overset{(d)}{\geq} \frac{1}{N} \sum_{i=j+1}^{N} \mathbb{E}I\left(x_{d_{j,i}}; \{y_k\}_{k>j} | x_{p_j}, \{x_{d_{j,l}}\}_{l<i}, y_j, \{x_{d_{k,i}}, \hat{h}_k\}_{k<j}\right)$$

$$\overset{(e)}{\geq} \frac{1}{N} \sum_{i=j+1}^{N} \mathbb{E}I\left(x_{d_{j,i}}; y_i | x_{p_j}, \{x_{d_{j,l}}\}_{l<i}, y_j, \{x_{d_{k,i}}, \hat{h}_k\}_{k<j}\right)$$

$$\overset{(f)}{=} \frac{1}{N} \sum_{i=j+1}^{N} \mathbb{E}I\left(x_{d_{j,i}}; y_i | x_{p_j}, \breve{y}_j, \{x_{d_{j,l}}\}_{l<i}, y_j, \{x_{d_{k,i}}, \hat{h}_k\}_{k<j}\right)$$

$$\overset{(g)}{\geq} \frac{1}{N} \sum_{i=j+1}^{N} \mathbb{E}I\left(x_{d_{j,i}}; y_i | x_{p_j}, \breve{y}_j\right)$$

$$\overset{(h)}{=} \frac{1}{N} \sum_{i=j+1}^{N} \mathbb{E}I\left(x_{d_{j,i}}; y_i | \mathbb{E}[h_j | x_{p_j}, \breve{y}_j], x_{p_j}, \breve{y}_j\right)$$

$$\overset{(i)}{\geq} \frac{1}{N} \sum_{i=j+1}^{N} \mathbb{E}I\left(x_{d_{j,i}}; y_i | \hat{h}_j = \mathbb{E}[h_j | x_{p_j}, \breve{y}_j]\right)$$

$$\overset{(j)}{\geq} \frac{1}{N} \sum_{i=j+1}^{N} \mathbb{E}I\left(x_{d_{j,i}}; \hat{h}_j^\dagger y_i | \hat{h}_j\right)$$

$$\overset{(k)}{=} \frac{1}{N} \sum_{i=j+1}^{N} \mathbb{E}I\left(x_{d_{j,i}}; \|\hat{\boldsymbol{h}}_j\|^2 x_{d_{j,i}} + z_{\mathrm{equ}_i}|\hat{\boldsymbol{h}}_j\right)$$

$$= \frac{1}{N} \sum_{i=j+1}^{N} \mathbb{E}I\left(x_{d_{j,i}}; \|\hat{\boldsymbol{h}}_j\| x_{d_{j,i}} + \frac{z_{\mathrm{equ}_i}}{\|\hat{\boldsymbol{h}}_j\|}\Big|\hat{\boldsymbol{h}}_j\right)$$

$$\overset{(l)}{\geq} \frac{1}{N} \sum_{i=j+1}^{N} \mathbb{E}I\left(x_{d_{j,i}}; \|\hat{\boldsymbol{h}}_j\| x_{d_{j,i}} + z_{\mathrm{G}_i}|\hat{\boldsymbol{h}}_j\right)$$

$$\overset{(m)}{\geq} \frac{1}{N} \sum_{i=j+1}^{N} \mathbb{E}\log\left(1 + \frac{\|\hat{\boldsymbol{h}}_j\|^2 P_{d_j}}{\mathbb{V}\mathrm{ar}(z_{\mathrm{G}_i})}\right)$$

$$= \frac{1}{N} \sum_{i=j+1}^{N} \mathbb{E}\log\left(1 + \frac{P_{d_j}}{\frac{1}{\|\hat{\boldsymbol{h}}_j\|^2}\mathbb{V}\mathrm{ar}(z_{\mathrm{G}_i})}\right)$$

$$\overset{(n)}{\geq} \frac{1}{N} \sum_{i=j+1}^{N} \log\left(1 + \frac{P_{d_j}}{\mathbb{E}\left[\frac{1}{\|\hat{\boldsymbol{h}}_j\|^2}\right]\mathbb{V}\mathrm{ar}(z_{\mathrm{G}_i})}\right)$$

$$\overset{(o)}{=} \frac{1}{N} \sum_{i=j+1}^{N} \log\left(1 + \frac{(1-N_{e_j})P_{d_j}}{\left[\frac{1}{n_{\mathrm{t}}-1}\right]\mathbb{V}\mathrm{ar}(z_{\mathrm{G}_i})}\right)$$

$$\overset{(p)}{=} \frac{1}{N} \sum_{i=j+1}^{N} \log\left(1 + \frac{(1-N_{e_j})P_{d_j}}{\frac{N_0}{n_{\mathrm{t}}-1} + \frac{P_{p_i}\mathbb{I}_{i\neq N} + N_{e_j}P_{d_j}}{n_{\mathrm{t}}-1} + \frac{\sum_{l=1,l\neq j}^{i-1}P_{d_l}}{n_{\mathrm{t}}-1}}\right) \tag{11.28}$$

where,

(a) follows by a sequential decoding where the data symbols as well as the estimated channels for users $1, 2, \ldots, j-1$ are assumed as side-information prior to decoding of user j. Here the mutual information is computed over one coherence block for which the channel remains constant. The expectation outside the mutual information expression is because the channel is ergodic, and the rate is normalized by the number of channel uses, which is N;

(b) follows by the chain rule and the fact that the pilot symbol is independent from the data symbols;

(c) follows by the chain rule and the fact that the signal $\boldsymbol{y}_j$ is independent of the transmitted data symbol of user j;

(d) follows by the chain rule for mutual information and using the nonnegativity of the mutual information;

(e) same as (d);

(f) follows by data processing inequality and defining the following signal constructed by the side information,

$$\check{y}_j = y_j - \sum_{k=1}^{j-1} \hat{h}_k x_{d_{k,j}}$$

$$= h_j x_{p_j} + \sum_{k=1}^{j-1} z_{e_k} x_{d_{k,j}} + z_j \tag{11.29}$$

(g) holds since conditioning reduces the entropy;

(h) follows by data processing inequality and forming the MMSE channel estimate $\hat{h}_j$ from the side information;

(i) same as (g);

(j) follows by data processing inequality;

(k) follows by defining the residual signals after MF as an additive noise denoted by z_{equ};

(l) follows by using the fact that the worst uncorrelated noise is Gaussian under the variance constraint [14];

(m) follows by using random Gaussian codebooks whose elements are generated i.i.d. according to $\mathcal{CN}(0, P_{d_j})$;

(n) follows by applying Jensen inequality for a $\cup$-concave function; that is, $\mathbb{E}f(x) \geq f(\mathbb{E}x)$ where,

$$f(x) := \log\left(1 + \frac{1}{x}\right) \tag{11.30}$$

(o) follows by the fact that $\dfrac{1}{\|\hat{h}_j\|^2}$ is distributed according the inverse-Wishart distribution [15] because each element of the vector $\hat{h}_j$ is i.i.d. distributed according to $\mathcal{CN}(0, 1 - N_{e_j})$ where,

$$N_{e_j} = \frac{N_0 + \sum_{l=1}^{j-1} N_{e_l} P_{d_l}}{N_0 + P_{p_j} + \sum_{l=1}^{j-1} N_{e_l} P_{d_l}} \tag{11.31}$$

(p) follows by substituting $\mathbb{V}\text{ar}(z_G)$, which completes the proof of the proposition.

11.B PROOF OF PROPOSITION 11.2

We, in this appendix, prove the result in Proposition 11.2. In the following, we fix SOMA protocol as illustrated in Fig. 11.3 and its channel estimation procedure with SIC. Having obtained the spatial vector channels of the coordinated users, the best one can do is to completely remove the interuser interference and coherently combine the signals in the direction of the desired channel of each user.

This yields,

$$R_{\text{SOMA}} \leq \sum_{k=1}^{N-1} R_k^{\text{ub}} \tag{11.32}$$

such that,

$$R_k^{ub} \leq \frac{1}{N} \sum_{l=k+1}^{N} \mathbb{E} \log \left(1 + \frac{\|\hat{\boldsymbol{h}}_k\|^2 P_{d_k}}{N_0 + P_{p_l} \mathbb{I}_{l \neq N} + N_{e_k} P_{d_k}} \right)$$

$$\overset{(a)}{\leq} \frac{1}{N} \sum_{l=k+1}^{N} \log \left(1 + \frac{\mathbb{E} \|\hat{\boldsymbol{h}}_k\|^2 P_{d_k}}{N_0 + P_{p_l} \mathbb{I}_{l \neq N} + N_{e_k} P_{d_k}} \right)$$

$$\overset{(b)}{=} \frac{1}{N} \sum_{l=k+1}^{N} \log \left(1 + \frac{n_t (1 - N_{e_k}) P_{d_k}}{N_0 + P_{p_l} \mathbb{I}_{l \neq N} + N_{e_k} P_{d_k}} \right) \tag{11.33}$$

where,

(a) follows by applying Jensen inequality for a ∩-convex function; that is, $\mathbb{E} f(x) \leq f(\mathbb{E} x)$, which is defined as,

$$f(x) := \log (1 + x) \tag{11.34}$$

(b) holds because each element of the random vector $\hat{\boldsymbol{h}}_k$ is i.i.d. distributed according to $\mathcal{CN}(0, 1 - N_{e_k})$, which completes the proof.

11.C PROOF OF PROPOSITION 11.3

We next prove the result in Proposition 11.3. Consider R_k^{ub} in Proposition 11.2 and R_k in Proposition 11.1. We can write the following series of inequalities:

$$\delta(n_t) = R_k^{ub}(n_t) - R_k(n_t)$$

$$= \frac{1}{N} \sum_{l=k+1}^{N} \log \left(1 + \frac{n_t (1 - N_{e_k}) P_{d_k}}{N_0 + P_{p_l} \mathbb{I}_{l \neq N} + N_{e_k} P_{d_k}} \right)$$

$$- \frac{1}{N} \sum_{l=k+1}^{N} \log \left(1 + \frac{(n_t - 1)(1 - N_{e_k}) P_{d_k}}{N_0 + P_{p_l} \mathbb{I}_{l \neq N} + N_{e_k} P_{d_k} + \sum_{i=1, i \neq k}^{l-1} P_{d_i}} \right)$$

$$\overset{(a)}{\leq} \frac{1}{N} \sum_{l=k+1}^{N} \log \left(1 + \frac{n_t (1 - N_{e_k}) P_{d_k}}{N_0 + P_{p_l} \mathbb{I}_{l \neq N} + N_{e_k} P_{d_k}} \right)$$

$$- \frac{1}{N} \sum_{l=k+1}^{N} \log \left(1 + \frac{(n_t - 1)(1 - N_{e_k}) P_{d_k}}{N_0 + P_{p_l} \mathbb{I}_{l \neq N} + N_{e_k} P_{d_k} + \sum_{i=1}^{l-1} P_{d_i}} \right)$$

$$\overset{(b)}{=} \frac{1}{N} \sum_{l=k+1}^{N} \log \left(1 + \frac{n_t (1 - N_{e_k}) P_{d_k}}{N_{eq,l}} \right)$$

$$-\frac{1}{N}\sum_{l=k+1}^{N}\log\left(1+\frac{(n_{\mathrm{t}}-1)(1-N_{e_k})P_{d_k}}{N_{eq,l}+I_l}\right)$$

$$=\frac{1}{N}\sum_{l=k+1}^{N}\log\left(\frac{N_{eq,l}+n_{\mathrm{t}}(1-N_{e_k})P_{d_k}}{N_{eq,l}+I_l+(n_{\mathrm{t}}-1)(1-N_{e_k})P_{d_k}}\cdot\frac{N_{eq,l}+I_l}{N_{eq,l}}\right)$$

$$=\frac{1}{N}\sum_{l=k+1}^{N}\log\left(\frac{N_{eq,l}+n_{\mathrm{t}}(1-N_{e_k})P_{d_k}}{N_{eq,l}+I_l-(1-N_{e_k})P_{d_k}+n_{\mathrm{t}}(1-N_{e_k})P_{d_k}}\cdot\frac{N_{eq,l}+I_l}{N_{eq,l}}\right)$$

$$\overset{(c)}{=}\frac{1}{N}\sum_{l=k+1}^{N}\log\left(\frac{N_{eq,l}+n_{\mathrm{t}}(1-N_{e_k})P_{d_k}}{N_{eq,l}+\tilde{I}_l+n_{\mathrm{t}}(1-N_{e_k})P_{d_k}}\cdot\frac{N_{eq,l}+I_l}{N_{eq,l}}\right)$$

$$\overset{(d)}{=}\frac{1}{N}\sum_{l=k+1}^{N}\log\left(g(n_{\mathrm{t}})\cdot\frac{N_{eq,l}+I_l}{N_{eq,l}}\right)$$

$$\overset{(e)}{\leq}\frac{1}{N}\sum_{l=k+1}^{N}\log\left(\max_{n_{\mathrm{t}}\geq 1}g(n_{\mathrm{t}})\cdot\frac{N_{eq,l}+I_l}{N_{eq,l}}\right)$$

$$\overset{(f)}{\leq}\frac{1}{N}\sum_{l=k+1}^{N}\log\left(\frac{N_{eq,l}+I_l}{N_{eq,l}}\right) \tag{11.35}$$

where,

(a) follows because $\sum_{i=1,i\neq k}^{l-1}P_{d_i}\leq\sum_{i=1}^{l-1}P_{d_i}$;
(b) follows by defining $N_{eq,l}:=N_0+P_{p_l}\mathbb{I}_{l\neq N}+N_{e_k}P_{d_k}$ and $I_l:=\sum_{i=1}^{l-1}P_{d_i}$;
(c) follows by defining $\tilde{I}_l:=\sum_{i=1}^{l-1}P_{d_i}-N_{e_k}P_k$;
(d) holds by defining,

$$g(n_{\mathrm{t}}):=\frac{N_{eq,l}+n_{\mathrm{t}}(1-N_{e_k})P_{d_k}}{N_{eq,l}+\tilde{I}_l+n_{\mathrm{t}}(1-N_{e_k})P_{d_k}}$$

(e) since $\log(x)$ is a nondecreasing function;
(f) follows since $\max_{n_{\mathrm{t}}\geq 1}g(n_{\mathrm{t}})\leq 1$, which follows from $\tilde{I}_l\geq 0$, $N_{eq,l}\geq 0$ and $0<N_{e_k}\leq 1$. This completes the proof of the proposition because the upper bound does not depend on number of antennas n_{t}.

11.D PROOF OF PROPOSITION 11.5

We next prove the result in Proposition 11.5. There are K users where in each coherence interval each user consumes one channel use (i.e., a time-frequency resource element) for pilot transmission and $N-K$ channel uses for data transmission. The base station estimates the channel vectors using the received noisy pilot signals and performs a normalized MF filtering. Consider user k in the following. The processed received signal for the channel use $K + 1 \leq i \leq N$ can be written as:

$$
\tilde{y}_i = \frac{\hat{h}_k^\dagger}{\|\hat{h}_k\|} y_i
$$

$$
= \underbrace{\frac{\hat{h}_k^\dagger \hat{h}_k}{\|\hat{h}_k\|} x_{d_k,i}}_{\text{Information of user } k} + \underbrace{\sum_{j=1, j \neq k}^{K} \frac{\hat{h}_1^\dagger h_j x_{d_j,i}}{\|\hat{h}_k\|}}_{\text{Interference due to other users data}} + \underbrace{\frac{\hat{h}_k^\dagger z_i}{\|\hat{h}_k\|}}_{\text{AWGN}} + \underbrace{\frac{\hat{h}_k^\dagger z_{e_1}}{\|\hat{h}_k\|} x_{d_k,i}}_{\text{Propagation of the channel estimation error}}
$$

(11.36)

Using the fact that worst-case uncorrelated noise is Gaussian, we replace the above uncorrelated components with a Gaussian noise with the same variance. This yields,

$$
R_k = \frac{N - K}{N} \mathbb{E} \log \left(1 + \frac{\|\hat{h}_k\|^2 P_{d_k}}{N_0 + N_{e_k} P_{d_k} + \sum_{i \neq k} P_{d_i}} \right)
$$

$$
= \frac{N - K}{N} \mathbb{E} \log \left(1 + \frac{P_{d_k}}{\frac{1}{\|\hat{h}_k\|^2} \left(N_0 + N_{e_k} P_{d_k} + \sum_{i \neq k} P_{d_i} \right)} \right)
$$

$$
\geq \frac{N - K}{N} \log \left(1 + \frac{P_{d_k}}{\mathbb{E} \left[\frac{1}{\|\hat{h}_k\|^2} \right] \left(N_0 + N_{e_k} P_{d_k} + \sum_{i \neq k} P_{d_i} \right)} \right)
$$

$$
= \frac{N - K}{N} \log \left(1 + \frac{(1 - N_{e_k}) P_{d_k}}{\frac{N_0}{n_t - 1} + \frac{N_{e_k} P_{d_k}}{n_t - 1} + \frac{\sum_{i \neq k} P_{d_i}}{n_t - 1}} \right)
$$

(11.37)

where the inequality follows by Jensen inequality similar as that in Proposition 11.1 and,

$$
N_{e_k} = \frac{N_0}{N_0 + P_{p_k}}
$$

(11.38)

denotes the variance of the channel estimation error for user k. Therefore, we obtain the sum-rate in the proposition, which completes the proof.

11.E PROOF OF PROPOSITION 11.6

We next prove the sum-rate in Proposition 11.6. The proof is similar to that of Proposition 11.5, but here the earlier decoded signals are used to cancel the interference prior to the decoding of the next users; that is, successive interference cancellation (SIC). The spatially processed received signal at the ith channel for user k, after the interference cancellation of the interfered data from users 1 to $k - 1$ is given by,

$$\tilde{y}_i = \frac{\hat{h}_k^\dagger}{\|\hat{h}_k\|}\left(y_i - \sum_{j=1}^{k-1}\hat{h}_j x_{d_{j,i}}\right)$$

$$= \underbrace{\frac{\hat{h}_k^\dagger \hat{h}_k}{\|\hat{h}_k\|}x_{d_{k,i}}}_{\text{Information of user }k} + \underbrace{\sum_{j=k+1}^{K}\frac{\hat{h}_1^\dagger h_j x_{d_{j,i}}}{\|\hat{h}_k\|}}_{\text{Interference due to undecoded users }k+1,\dots,N} + \underbrace{\frac{\hat{h}_k^\dagger z_i}{\|\hat{h}_k\|}}_{\text{AWGN}}$$

$$+ \underbrace{\frac{\hat{h}_k^\dagger z_{e_1}}{\|\hat{h}_k\|}x_{d_{k,i}}}_{\text{Propagation of the channel estimation error}} + \underbrace{\sum_{j=1}^{k-1}\frac{\hat{h}_k^\dagger z_{e_j}}{\|\hat{h}_k\|}x_{d_{j,i}}}_{\text{Propagation of the channel estimation error due to SIC}} \qquad (11.39)$$

Thus, the following rate is achievable,

$$R_k = \frac{N-K}{N}\mathbb{E}\log\left(1 + \frac{\|\hat{h}_k\|^2 P_{d_k}}{N_0 + \sum_{i=1}^k N_{e_i}P_{d_i} + \sum_{i=k+1}^K P_{d_i}}\right)$$

$$\geq \frac{N-K}{N}\log\left(1 + \frac{P_{d_k}}{\mathbb{E}\left[\frac{1}{\|\hat{h}_k\|^2}\right]\left(N_0 + \sum_{i=1}^k N_{e_i}P_{d_i} + \sum_{i=k+1}^K P_{d_i}\right)}\right) \qquad (11.40)$$

$$= \frac{N-K}{N}\log\left(1 + \frac{(1-N_{e_k})P_{d_k}}{\frac{N_0}{n_t - 1} + \frac{\sum_{i=1}^k N_{e_i}P_{d_i}}{n_t - 1} + \frac{\sum_{i=k+1}^K P_{d_i}}{n_t - 1}}\right)$$

where the inequality similarly follows by using Jensen inequality as that in the Proposition 11.1 and,

$$N_{e_k} = \frac{N_0}{N_0 + P_{p_k}} \qquad (11.41)$$

is the variance of the channel estimation error. Therefore, we obtain the sum-rate in Proposition 11.6 which completes the proof.

REFERENCES

[1] T.L. Marzetta, Noncooperative cellular wireless with unlimited numbers of base station antennas, IEEE Trans. Wirel. Commun. 9 (11) (2010) 3590–3600.
[2] H. Huh, G. Caire, H.C. Papadopoulos, S.A. Ramprashad, Achieving "massive MIMO" spectral efficiency with a not-so-large number of antennas, IEEE Trans. Wirel. Commun. 11 (9) (2012) 3226–3239.
[3] E.G. Larsson, F. Tufvesson, O. Edfors, T.L. Marzetta, Massive MIMO for next generation wireless systems, IEEE Commun. Mag. 52 (2) (2014) 186–195.

[4] F. Rusek, D. Persson, B.K. Lau, E.G. Larsson, T.L. Marzetta, O. Edfors, F. Tufvesson, Scaling up MIMO: opportunities and challenges with very large arrays, IEEE Signal Process. Mag. 30 (1) (2013) 40–60.

[5] J. Hoydis, S.T. Brink, M. Debbah, Massive MIMO in the UL/DL of cellular networks: how many antennas do we need? IEEE J. Sel. Areas Commun. 31 (2) (2013) 160–171.

[6] L. You, X. Gao, X.-G. Xia, N. Ma, Y. Peng, Massive MIMO transmission with pilot reuse in single cell, in: Proceedings of the IEEE International Conference on Communications (ICC), 2014, pp. 4783–4788.

[7] F. Fernandes, A. Ashikhmin, T.L. Marzetta, Inter-cell interference in noncooperative TDD large scale antenna systems, IEEE J. Sel. Areas Commun. 31 (2) (2013) 192–201.

[8] J. Jose, A. Ashikhmin, T.L. Marzetta, S. Vishwanath, Pilot contamination and precoding in multi-cell TDD systems, IEEE Trans. Wirel. Commun. 10 (8) (2011) 2640–2651.

[9] E. Björnson, E.G. Larsson, M. Debbah, Optimizing multi-cell massive MIMO for spectral efficiency: how many users should be scheduled? arXiv preprint, arXiv:1410.3522 (2014).

[10] D. Tse, P. Viswanath, Fundamentals of Wireless Communication, Cambridge University Press, Cambridge, UK, 2005.

[11] A. Papoulis, S.U. Pillai, Probability, Random Variables, and Stochastic Processes, Tata McGraw-Hill Education, Noida, India, 2002.

[12] A.E. Gamal, Y.-H. Kim, Network Information Theory, Cambridge University Press, Cambridge, UK, 2011.

[13] H.Q. Ngo, E.G. Larsson, T.L. Marzetta, Energy and spectral efficiency of very large multiuser MIMO systems, IEEE Trans. Commun. 61 (4) (2013) 1436–1449.

[14] B. Hassibi, B.M. Hochwald, How much training is needed in multiple-antenna wireless links? IEEE Trans. Inform. Theory 49 (4) (2003) 951–963.

[15] T.W. Anderson, An Introduction to Multivariate Statistical Analysis, third edition, Wiley, Hoboken, NJ, 2003.

Fronthaul design for mmWave massive MIMO

12

Z. Gao*, L. Dai†, X. Gao†, M.Z. Shakir‡, Z. Wang†

Beijing Institute of Technology, Beijing, China* Tsinghua University, Beijing, China†
University of the West of Scotland, Paisley, United Kingdom‡

CHAPTER OUTLINE

mmWave Massive MIMO. http://dx.doi.org/10.1016/B978-0-12-804418-6.00012-1

12.1 **INTRODUCTION**

Explosive traffic demand is challenging current cellular networks, including the most advanced fourth generation (4G) network. It has been the consensus that future 5G networks should realize the goals of 1000-fold system capacity, 100-fold energy efficiency, and 10-fold lower latency over the next 10 years [1]. To realize such an aggressive 5G vision, the millimeter-wave (mmWave) massive multiple-input multiple-output (MIMO)-based heterogeneous network (HetNet) has been considered a promising solution to enable gigabit-per-second user experience, seamless coverage, and green communications [2].

For such an HetNet as shown in Fig. 12.1, the macrocell base station (BS) provides the control signaling service for the large coverage area using the conventional low-frequency band, while the small BSs densely deployed in hot spots (e.g., office buildings, shopping malls, residential apartments) are specialized for high-speed data transmission with limited coverage area. Such an HetNet can provide traffic offloading from macrocells because the large majority of traffic demand comes from these hot spots [2]. Moreover, the centralized radio access network architecture can be integrated into the HetNet for the improved physical layer transmission, handover, scheduling, and so on, where the ultradense small cells are regarded as the remote ratio

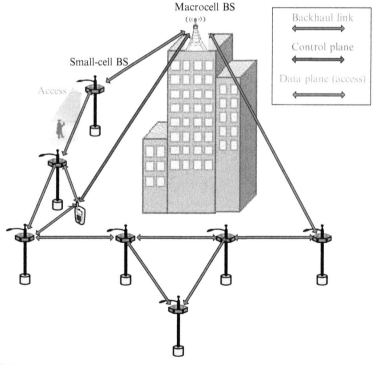

FIG. 12.1

In HetNet, mmWave massive MIMO can be used for both fronthaul and access.

head (RRH) and macrocell BS is considered the baseband unit (BBU) [2]. Here, the transmission link between the RRH and BBU is referred to as fronthaul. However, the fronthaul in HetNet is a challenging issue for the implementation of HetNet. It is clear that with the increasing number of ultradense small cells, an high throughput fronthaul network with good scalability is required to accommodate the exponentially growing mobile data demand. Moreover, the 1000-fold capacity increase of the 5G cellular network in the future decade indicates the cellular capacity growth projection of roughly 115.44% compound average growth rate (CAGR). Hence, the installation cost of the fronthaul network with very high throughput can be nonnegligible. So far, fiber has been widely adopted in existing cellular networks for the macrocell fronthaul. However, the installation and leasing for a fiber fronthaul network can lead to high cost, and it may not be suitable for 5G HetNet [3].

On the other hand, wireless fronthaul has been considered the backbone for the fronthaul in HetNet. For the wireless fronthaul, there is no unified standard so far, and the existing solutions range from sub-6 GHz nonline-of-sight (NLOS) to 70 GHz E-band line-of-sight (LOS). For the wireless fronthaul working at low-frequency bands, its deployment is easy due to the good channel propagation characteristics of low-frequency bands for NLOS signals and thus the longer fronthaul link. Unfortunately, the scarce microwave frequency band at sub-6 GHz indicates the trade-off between the spectrum resources for the access link and those for the fronthaul link. For the high-frequency bands, especially the mmWave frequency bands at E-band and V-band, a large amount of underutilized spectrum resources can be used for the fronthaul link, however, without mature solutions for the fronthaul in HetNet. Additionally, due to the small wavelength of mmWave, a large number of antennas can easily be employed for mmWave fronthaul, which can improve the signal directivity and link reliability [4–7].

Against this background, the mmWave massive MIMO has been emerging as a promising candidate for both the fronthaul and access in future 5G cellular networks [1,2]. Inheriting the advantages of conventional massive MIMO, mmWave massive MIMO has the flexible beamforming, spatial multiplexing, and diversity, still with low hardware cost and power consumption. Hence, mmWave massive MIMO brings not only the improved reliability of the fronthaul link, but also new architecture of a fronthaul network, including the flexible network topology and scheduling scheme. The advantages of mmWave massive MIMO-based fronthaul for the ultradensely deployed small-cell BSs are listed as follows:

High-capacity and inexpensive: The large amount of underutilized mmWave, including unlicensed V-band (57–67 GHz) and lightly licensed E-band (71–76 GHz and 81–86 GHz) (the specific regulation may vary from country to country), can provide potential gigahertz transmission bandwidth [4]. For example, more than 1 Gbps fronthaul capacity can be supported over a 250 MHz channel at E-band [8].

Immunity to interference: The transmission distance comfort zone for E-band is up to several kilometers due to rain attenuation, while that for V-band is about 500–700 m due to both the rain and oxygen attenuation [1]. Due to high path loss, mmWave is suitable for the ultradense deployment of small cells in future HetNet, where improved frequency reuse and reduced intercell interference are expected. It should be pointed out that rain attenuation is not a big issue for mmWave used in fronthaul. If we consider very heavy rainfall of 25 mm/h, the rain attenuation is only around 2 dB at E-band over the distance of 200 m fronthaul link in typical urban HetNet [4].

Small form factor: The small wavelength of mmWave implies that massive antennas can easily be equipped at both macrocell BSs and small-cell BSs, which can improve the signal directivity and compensate severe path loss of mmWave to achieve larger coverage in turn [8]. Hence, compact mmWave fronthaul equipment can easily be deployed with low-cost sites (e.g., light poles, building walls, and bus stations) and short installation time [2].

Spatial multiplexing with low cost: Compared with the conventional mmWave multiantenna systems limited to point-to-point transmission, the emerging technique of mmWave massive MIMO can support multiuser transmission and multistream transmission for each node, which can effectively realize the spatial multiplexing [1]. Moreover, the transceiver structure with the much smaller number of baseband chains than that of the antennas can substantially reduce the hardware cost and power consumption of mmWave massive MIMO systems [6].

In this chapter, we will first present a survey of the existing fronthaul solutions in Section 12.2. In Section 12.3, we will investigate the market requirements of the fronthaul network for future 5G HetNet. In Section 12.4, we will present the concept of mmWave mesh fronthaul network, where some issues including antenna techniques, beamforming design, duplexing protocol, in-band fronthaul, and so on, will be discussed in detail. Finally, conclusions will be drawn in Section 12.5.

12.2 A SURVEY OF EXISTING FRONTHAUL SOLUTIONS

In this section, we will provide the survey of existing fronthaul solutions. Specifically, we will first investigate the category of existing fronthaul solutions. Then, we will discuss the fronthaul network topology. Moreover, we will introduce the pros and cons of different spectrum resources used for fronthaul. Finally, we will elaborate why and how we will exploit the mmWave band for fronthaul.

12.2.1 CATEGORY OF FRONTHAUL SOLUTIONS

For the mmWave massive MIMO HetNets, the ultradensed small-cell BSs exploit wired and/or wireless fronthaul to transport traffic data to the core network [9]. For the wired fronthaul network, as we discussed in Section 12.1, the fiber will only make up the minority of the fronthaul link in HetNet due to the expensive installation and lease. It should be pointed out that the macrocell and/or small cell with fiber availability, which are usually developed from existing 2/3/4G BSs, are essential to the deployment of the ultradense small-cell BSs as well as the design of the fronthaul network because these macrocell and/or small-cell BSs with fiber availability provide the locations for traffic aggregation points to the core network for the fronthaul network. On the other hand, the wireless fronthaul will dominate the fronthaul link in HetNet, which is the focus of this chapter. The category of wireless fronthaul solutions is listed in Table 12.1. Especially for the wireless fronthaul network, the spectrum of the fronthaul options can range from the microwave (sub-6 GHz) to mmWave. According to the fronthaul network topology, the fronthaul options for the "last mile" can be divided into point-to-point (PtP), point-to-multipoint (PtMP), and mesh; according to the propagation quality of the wireless channels, the fronthaul options can be divided into LOS and NLOS, which will be further illustrated in Section 12.2.2. Among those solutions, the sub-6 GHz NLOS fronthaul solution can greatly improve the flexibility of small-cell site deployment. Moreover, the fronthaul can share the same transmission resource (e.g., LTE carrier) with the access used by users, which can substantially reduce the cost of fronthaul deployment with the aid of the emerging licensed model, called authorized shared access (ASA). ASA allows the fronthaul network to access the underutilized spectrum on a shared basis without interfering with incumbent spectrum holders. The fronthaul network with ASA enjoys exclusive spectrum rights of use where and when the spectrum is not used by the incumbent. This framework provides a predictable quality of service to the incumbent as well as the fronthaul with ASA licensee. ASA can potentially release hundreds of megahertz of high-quality spectrum for the fronthaul network. However, with the scarcer and scarcer spectrum in sub-6 GHz, the mmWave-based fronthaul network may provide the more sustainable strategy for future cellular networks.

Table 12.1 Category of Existing Fronthaul Solutions [9]

	Spectrum	**Network Topology**	**Spectrum Licensing**
NLOS	Television white space, TVWS (sub-800 MHz)	PtP, PtMP	Dynamic and light
	Sub-6 GHz	PtP, PtMP, mesh	Area licensed, unlicensed, or ASA
LOS	Microwave band (6–57 GHz)	PtP, PtMP	Link licensed (PtP) or area licensed (PtMP)
	V-band (57–67 GHz)	PtP	Unlicensed
	E-band (71–76 GHz, 81–86 GHz)	PtP	Light licensed

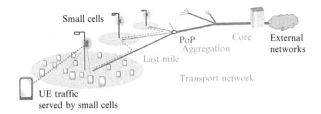

FIG. 12.2

The entire fronthaul network can be divided into tree architecture and branch architecture [10].

12.2.2 FRONTHAUL NETWORK TOPOLOGY

The entire fronthaul network can be divided into tree architecture and branch architecture as shown in Fig. 12.2 [10], which is not the emphasis of this chapter. What we focus on is the "last mile" connectivity in the fronthaul network, which can be mainly divided into PtP, PtMP, and mesh, as shown in Fig. 12.3 [9].

For the PtP topology, there is a dedicated radio frequency (RF) channel for each link between the aggregation point and the endpoint. By contrast, for the PtMP topology, the RF channel at the aggregation point is shared by the link connecting multiple endpoints. Furthermore, the topology structures can be extended to the chain, tree, and mesh topologies, which are determined by the requirements in the fronthaul network. For the cases in which there is not a direct fronthaul link connecting the small-cell BS and the aggregate point due to physical blockage, the multihop fronthaul link can be considered, where topologies like chain and mesh can be used. These kinds of topologies typically require the small-cell BSs to simultaneously support multiple fronthaul links and at least one of the small-cell BSs to be connected to the core

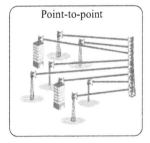 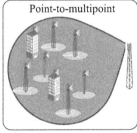 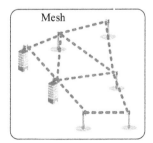

FIG. 12.3

The "last mile" connectivity in the fronthaul network can be divided into PtP, PtMP, and mesh [9].

network, via a macrocell BS for instance, for the further connectivity of the fronthaul network among these small-cell BSs.

12.2.3 PROS AND CONS OF DIFFERENT SPECTRUM RESOURCES FOR FRONTHAUL

In this section, we will provide the pros and cons of different spectrum resources for the fronthaul link.

Television white space (TVWS) (sub-800 MHz) can provide the long link range due to its good propagation properties. The available channel in TVWS has the bandwidth of 6/7/8 MHz varying from country to country [9]. However, the availability of TVWS is a big issue, and the bandwidth is limited, which may not meet the requirements of fronthaul for the ultradense deployment of small cells.

Sub-6 GHz licensed spectrum can provide the NLOS fronthaul link with both PtP and PtMP topologies [9]. However, the available licensed spectrum in sub-6 GHz is expensive and limited. Hence, the sub-6 GHz licensed spectrum can only support limited fronthaul throughput due to the narrow bandwidth. In PtMP topology, due to the good propagation properties, the limited frequency reuse can further reduce the available throughput of the fronthaul link.

Sub-6 GHz unlicensed spectrum is more popular for current industrial practices. Actually, there have been many fronthaul solutions operating at 2.4 and 5 GHz over PtMP protocol. Especially the solutions with NLOS sub-6 GHz unlicensed spectrum have high business case scalability from the short-term perspective. However, the precious sub-6 GHz unlicensed spectrum may not be sufficient to support the 1000-fold capacity increase of radio access networks required by future cellular networks. On one level, the interference in these extensively used bands has been very high, which implies the disappointing performance of fronthaul with sub-6 GHz unlicensed spectrum, especially for the ultradense deployment of small cells in urban areas. On another level, the potential unlicensed spectrum at sub-6 GHz may be authorized to some novel 5G techniques, including massive MIMO. Consequently, the sub-6 GHz unlicensed spectrum can be exploited as the stopgap or backup when the sub-6 GHz licensed spectrum or LOS fronthaul is unavailable [9].

Microwave PtP/PtMP (6–57 GHz) has been widely used in existing 3G and 4G cellular fronthaul, but the requirements of fronthaul in future 5G cellular networks will quickly deplete its available capacity. Compared with the sub-6 GHz spectrum, microwave PtP/PtMP requires NLOS as the solutions working at V-band and E-band. However, it may suffer from larger antenna array, less fronthaul link throughput and more expensive spectrum when compared with those working at V-band and E-band. On the other hand, microwave PtP/PtMP can provide longer link distance compared with that at V-band and E-band, which can support the longer fronthaul link (e.g., in rural small-cell deployments) at the cost of the reduced reuse spectrum, especially in unban scenarios.

V-band (57–67 GHz) has attracted the most attention because it is unlicensed. The atmospheric attenuation and oxygen attenuation limit the fronthaul link supported at V-band [9]. However, this high path loss makes it compatible for the ultradense deployment of small cells in future 5G cellular networks due to less interference and better spectrum reuse. Moreover, there is no explicit statement about the channelization and antenna design at V-band, which indicates the more flexible fronthaul solution and much smaller equipment form factor. However, the required LOS may challenge the deployment of small-cell BSs. Moreover, because it is unlicensed, the fronthaul equipment requires dynamic interference avoidance algorithms to mitigate and tolerate interference without supervision of the regulator.

E-band (71–76 GHz and 81–86 GHz) is a lightly licensed spectrum with specific channelization and antenna designs [9]. For instance, FCC regulation requires that the minimum antenna gain at E-band is 43 dBi, which practically defines the size of the antenna to be around 1 ft and leads to bigger antennas compared with those at V-band. On the other hand, because it is a licensed spectrum, the fronthaul link at E-band is licensed and registered in the database, which can effectively avoid and manage the interference.

12.2.4 WHY AND HOW WILL WE EXPLOIT mmWave FOR FRONTHAUL?

The exponential increase of 5G throughput requirement drives the spectrum used for fronthaul from conventional microwave band to mmWave spectrum. It has been the consensus that the unlicensed V-band and lightly licensed E-band are promising candidates to support the fronthaul for future 5G cellular networks due to the large underutilized unlicensed or lightly licensed spectrum, LOS fronthaul link with high frequency reuse as well as low interference, and small RF components. Even for the NLOS scenarios, multihop fronthaul network or other appreciated fronthaul topologies (chain, mesh) can be exploited to handle this issue with adequate quality of service (QoS).

Table 12.2 shows the approximate industry coverage of fronthaul solutions for small cells. It can be observed that there are no available PtMP and mesh solutions in mmWave to support the flexible fronthaul network. In Section 12.3, this gap will be addressed from the specific physical layer techniques to the network architecture.

By far, microwave PtP and mmWave PtP have been used to support the high throughput for macrocell fronthaul for more than 20 years. In recent years, IEEE 802.11n and the emerging IEEE 802.11ac have been exploited to realize the mesh

Table 12.2 Industry Coverage of Fronthaul Solutions [9]

	PtP	PtMP	Mesh
Sub-6 GHz licensed	☆☆	☆	
Sub-6 GHz unlicensed	☆	☆☆	☆
Microwave	☆☆☆		☆
V-band	☆☆☆	★	★
E-band	☆☆	★	★

☆, *Few companies;* ☆☆, *several companies;* ☆☆☆, *many companies;* ★, *industry need.*

solution in the unlicensed sub-6 GHz. Compared with IEEE 802.11n using 40 MHz channel, 802.11ac offers significant data improvement with the channel of 80 and 160 MHz. Moreover, by introducing the 256 QAM modulation, the peak rate achieved by 802.11ac can be higher than that in 802.11n. However, both 802.11n and 802.11ac operate at 5 GHz, which is challenged by the more and more crowded spectrum and interference. Hence, the fronthaul performance (e.g., the use of 256 QAM) and throughput may not be guaranteed in the unlicensed sub-6 GHz band [9]. For the mmWave fronthaul solution, both V-band and E-band are promising fronthaul candidates. We will elaborate why V-band and E-band can be appreciated for wireless fronthaul [8].

V-band is driven by Internet Service Providers (ISPs) and the Wi-Fi Alliance, who want to upgrade the existing devices using 60 GHz to seize business opportunities of fronthaul in cellular networks. By leveraging the four channels in 802.15.3c with the range from 57 to 67 GHz, the aggregate throughput for fronthaul can be up to 17 Gbps with 16 QAM and 25 Gbps with 64 QAM [8]. However, the regulation of V-band in some regions confines the emitted power at power amplifier output to 10 dBm [equivalently, the 40 dBm effective isotropic radiated power (EIRP)], which may limit the longest distant of the fronthaul link supported. It should be pointed out that, because V-band is unlicensed, the fronthaul devices have to rely on dynamic interference avoidance algorithms to mitigate and tolerate interference.

E-band is suggested by small-to-medium enterprises providing discrete high-end components and their integration in mmWave modules, large equipment providers, and operators. Their motivation is to reduce the cost of fronthaul devices and provide a competitive small-cell fronthaul business case. The advantage of E-band is its large, lightly licensed spectrum resource with the range of 71–76 GHz and 81–86 GHz for most regions. For example, the European Telecommunications Standards Institute (ETSI) subdivides the large bandwidth at E-band into multiple channels of 250 MHz with a yearly fee for each link and channel. It has been reported that the commercialized fronthaul device with 43.5 dBi antennas and 64-QAM can provide more than 1 Gbps throughput over the link distance of 3.5 km [11]. Note that E-band enjoys interference protection; that is, if there are two links in the same location and one interferes with another, the link that was installed first is protected and the second link will be reconfigured to prevent the interference or removed.

Two big issues for the mmWave fronthaul are rain attenuation and atmospheric attenuation, which do not make an obviously negative impact on the fronthaul for small-cell scenarios due to the short distance of the fronthaul link. For the heavy rain level (25 mm/h), the path loss can be 10 dB/km, as shown in Fig. 12.4. If we consider very heavy rainfall of 25 mm/h, the rain attenuation is only around 2 dB at E-band if we consider the distance of a fronthaul link is 200 m in the typical ultradense small cells [4]. Typically, the transmission distance comfort zone for E-band is up to several kilometers due to rain attenuation, while that for V-band is about 500–700 m due to both the rain and oxygen attenuation as shown in Fig. 12.5. For the practical fronthaul network designed for the ultradense deployment of small cells in future 5G cellular networks, our conclusion is *both V-band and E-band should be combined*

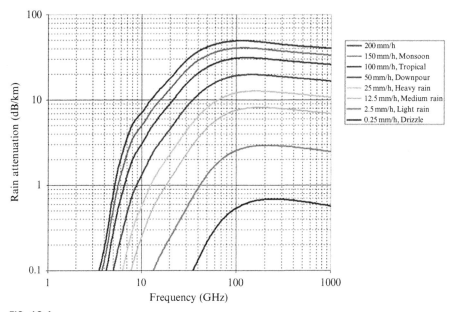

FIG. 12.4

The rain attenuation with different rainfall [4].

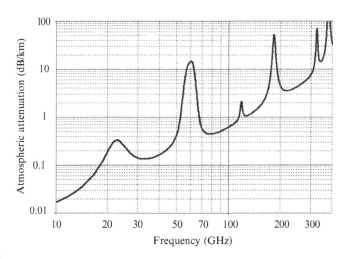

FIG. 12.5

The atmospheric attenuation against different frequency [4].

Table 12.3 Characteristics of V-Band and E-Band Used for Wireless Fronthaul

Characteristics	E-Band	V-Band
Regulation implications	Lightly licensed Simple registration with light license fee; avoid interference protection scheme	Unlicensed No registration without license fee; dynamic interference avoidance algorithms are required to mitigate the mutual interference
Distance implications	Several kilometers due to rain attenuation	Up to 500–700 m due to both rain and oxygen attenuation
Form factor implications	Can be compact and dictated by typical 1 ft antenna due to the requirement of minimum antenna gain. Suitable for rooftops, towers, and masts	Can be tiny and blended on the street level, building walls, light poles, bus stations, traffic lights
Applications	Fiber extension for businesses; macro fronthaul Small-cell fronthaul in some particular cases; aggregation	Security (CCTV, traffic radars) WiFi fronthaul Small-cell fronthaul GTTH—Fiber extension to customer premises

for the fronthaul design by sufficiently considering their respective characteristics, provided in Table 12.3.

12.3 MARKET REQUIREMENTS OF mmWave FRONTHAUL

In this section, we will investigate the market requirements of mmWave fronthaul for the ultradense deployment of small cells, including the total cost of ownership (TCO), latency, data rate, and fronthaul link range (intersite distance). These requirements will enlighten the technical direction of the fronthaul solution for the mmWave massive MIMO-based fronthaul solutions in Section 12.4.

12.3.1 TOTAL COST OF OWNERSHIP (TCO)

In this section, we will investigate the main factors that may have an impact on the final TCO of the fronthaul network, and the advantages of mmWave fronthaul on TCO are also presented.

With the increasingly dense deployment of small cells, PtMP is superior to PtP due to the financial advantage [9]. Specifically,

- For licensed spectrum, the spectrum is paid by region in PtMP solutions while paid by link in PtP solutions. Hence, the spectrum fee in PtMP solutions does not increase with the density of the small cells.
- PtP solutions require the LOS link, which may increase the deployment cost of planning and installation.
- The device cost in PtP solutions is more expensive because a dedicated RF channel is required for each endpoint.

The different TCO for NLOS and LOS options in PtMP solutions is due to the different costs of devices and installation. Specifically, PtMP-LOS solutions suffer from the more costly installation to align the antennas to the desired direction at both ends of each link when LOS transmission is considered. By contrast, PtMP-NLOS solutions can facilitate the installation thanks to the good propagation characteristics, which can facilitate the deployment of small cells and reduce the installation cost. Moreover, it has been reported that the cost of PtMP-LOS solutions is around 35% higher than that of PtMP-NLOS solutions [12]. On the other hand, spectrum cost may be not a big issue when the spectrum above 3 GHz is considered and per-link pricing is avoided [12,13]. Consequently, to reduce the TCO, the fronthaul network should meet the following several principles [9]:

1. Easy deployment
2. Exploit PtMP for reduced fee in licensed spectrum and avoid the per-link pricing
3. Avoid fiber due to its high cost of installation and leasing

It should be pointed out that the spectrum at sub-6 GHz has been more and more crowded, and it may be even assigned for radio access networks dedicated to the emerging 5G techniques such as massive MIMO. However, mmWave fronthaul, especially with the aid of the emerging techniques of mmWave massive MIMO, will facilitate the installation and deployment. For example, the mmWave massive MIMO with adaptively beamforming algorithms may build the fronthaul link without human intervention, which can reduce the installation cost and avoid sudden link blockage. Moreover, the multistream and multipoint transmission supported by the mmWave massive MIMO can naturally match the requirement of PtMP and mesh solutions, which can facilitate the deployment for NLOS scenarios [1]. To sum up, *mmWave massive MIMO-based fronthaul solutions with high fronthaul throughput can provide the PtMP and mesh fronthaul topologies, and it can reduce the cost of deployment and cost due to flexible beamforming.*

12.3.2 THROUGHPUT REQUIREMENT OF FRONTHAUL NETWORK

In this section, we will investigate the throughput requirement of the fronthaul network for the ultradense deployment of small cells.

As we mentioned in Section 12.1, the wireless link will dominate the future fronthaul network, and it can be costly to deploy the new fiber for fronthaul due to its high cost of installation and leasing. Hence, the wireless fronthaul network will aggregate the fronthaul data of multiple small cells in the same cluster from/to the existing fiber point-of-presence (PoP). For example, Fig. 12.6 provides the star and

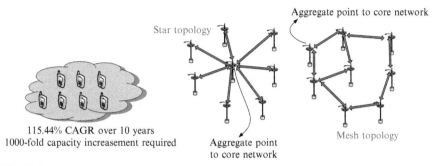

Star topology

Aggregate point to core network

Mesh topology

115.44% CAGR over 10 years
1000-fold capacity increasement required

Aggregate point
to core network

FIG. 12.6

Throughput of fronthaul network has to meet the rapid increase of mobile data.

mesh topologies, where the data of all small cells in the same fronthaul network will finally flow to an existing fiber PoP. Clearly, for such a fronthaul network, the nodes closer to the fiber PoP should support more fronthaul throughput due to the cumulative fronthaul data from other small cells. On the other hand, the number of existing fiber PoP will not increase, while the deployment of small cells will become denser and denser; that is, the density of PoP per area will not increase, while the density of fronthaul throughput per area will increase. As a result, the throughput of fronthaul network will also accommodate the cellular capacity increase prediction by around 105.98% CAGR or 1000 × increase over the next 10 years. Currently, an average fronthaul throughput is about 35 Mbps. for current small-cell business cases [9]. According to the 1000-fold capacity increase over the next 10 years, the fronthaul throughout should be up to multi-Gbps [8]. It should be pointed out that, although there have been several wireless solutions that can meet the fronthaul throughput of this level, those solutions are typically limited to the tower mounted PtP fronthaul with long planning and deployment cycles. For the future ultradense deployment of small cells, what we expect is the cost-effective PtMP solution with the multi-Gpbs fronthaul throughput and high business case scalability.

12.3.3 TRAFFIC CLASSES AND LATENCY

In this section, we will investigate the latency required by the fronthaul network for the ultradense deployment of small cells. Latency is an essential parameter that determines the quality of experience (QoE) for the end-to-end services, and the typical latency for various services is shown in Table 12.4 [9]. It is clear that the fronthaul network will add the additional latency, which must be considered in the fronthaul network design. In this section, we will evaluate the latency required by the fronthaul network according to the latency requirements of the most common scenarios and most stringent scenarios. Specifically, we provide the table of the packet delay budgets for different services, where the delay considers the single-trip link from user termination to the packet gateway according to 3GPP TS 23.203 [15], which includes the single-trip latency of 20 ms assumed for the latency between the radio BS segment and the packet gateway. Here,

Table 12.4 Latency for Various Services [9]

Service	CAGR 2011–2016 [14]	Packet Delay That Applies to Radio Interference (Single Trip [15])
Real-time gaming	63%	50 ms
Voice (VoIP)	19%	100 ms
Video (live streaming) Interactive gaming	90% (live and buffered combined)	150 ms
Video (buffered streaming)		300 ms
TCP-based (e.g., www, email, chat, ftp, p2p file sharing, progressive video, etc.)	19%	300 ms

we consider the most stringent scenarios of gaming as an example; a typical latency for the delay segments from the user equipment to the core network may be 20 ms when air interface conditions are good. This implies that only 10 ms is available for the additional latency for the fronthaul network, which is within the reasonable range with the unbound of 20 ms suggested by the recent next generation mobile networks (NGMN) Alliance that "The overall fronthaul delay budget in one direction from small cell connection point to the core network equipment should not exceed 20 ms, for 98% packets for high priority Classes of Service or in uncongested conditions. We note that the fronthaul latency must fit into the operator's overall end-to-end (E2E) latency budget for the service(s) being offered" [10]. The single-trip fronthaul from the RAN to the core network can typically be 5 ms, resulting in only an additional 5–10 ms remaining for the single-trip fronthaul in the "last mile" for the most stringent scenarios. *Hence, our conclusion is the latency of the "last mile" fronthaul should be as low as possible and under 5 ms for a single trip.*

12.3.4 INTERCELL/INTERSITE DISTANCE (ISD)

The trend of small-cell deployment is smaller and smaller ISD due to the limited spectrum for better spatial reuse in fronthaul and access. From the perspective of the mmWave access link, as shown in Fig. 12.7, the ultradense deployment of small cells should address the issue of the blockage effect in the access link. Because each small cell is equipped with a fronthaul device to connect the core network, the blockage effect in the mmWave access link also has a significant impact on the design of the fronthaul network. It has been demonstrated that the signal strength of the LOS paths can be dozens of decibels stronger than that of the NLOS paths [4]. To solve the blockage effect, the small cells should be ultradensely deployed to guarantee that at least one LOS communication link between the user and the small cell or the collection of multiple NLOS access links can provide the satisfactory channel quality. On the other hand, from the perspective of the reliable fronthaul link, the high path loss

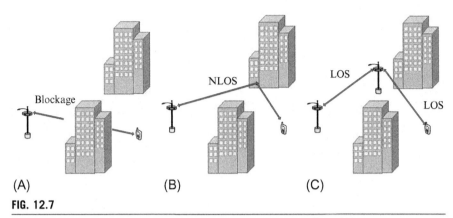

FIG. 12.7

Ultradense deployment of small cells can avoid the blockage effect: (A) blockage due to buildings; (B) NLOS link can be used to serve users at the cost of high path loss; (C) LOS link can be used to serve users thanks to the ultradense deployment of small cells.

and LOS link required in mmWave fronthaul also limit the distance of the fronthaul link. As we have mentioned, the fronthaul distance comfort zone for E-band is up to several kilometers, while that for V-band is up to 500–700 m.

In sum, by comprehensively considering link reliability, spatial reuse of the spectrum resource, and the latency of mmWave access in ultradense deployment of small cells, the ISD of small cells in future cellular networks can be 50–200 m [9,11].

12.4 mmWave MASSIVE MIMO-BASED FRONTHAUL SOLUTION

In this section, we describe the mmWave massive MIMO-based fronthaul solution. Moreover, its feasibility based on the requirements discussed in Section 12.3, and some key issues including antenna techniques, channel estimation, precoding, and so on, will be also addressed.

12.4.1 CONCEPT OF mmWave MASSIVE MIMO-BASED FRONTHAUL

The mmWave massive MIMO based fronthaul can be illustrated in Fig. 12.8, which has several distinctive features and provisions:

- Compared with conventional mmWave multiantenna systems limited to PtP communications, the emerging mmWave massive MIMO technique can support multiuser transmission, which enables the mesh fronthaul network topology. Moreover, the mmWave massive MIMO such as the lens antennas can provide highly directional transmission to mitigate path loss with low interference.
- The mesh fronthaul topology can provide the flexible fronthaul network architecture, which can effectively accommodate various deployments of ultradense small cells. Moreover, we suggest less than half an hour installation

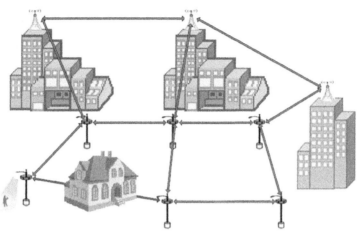

FIG. 12.8

mmWave massive MIMO based fronthaul network [9].

time on street furniture with a minimally trained installer, and the fronthaul device is self-configured to build the fronthaul network supporting plug-and-pull. Hence, the mmWave mesh fronthaul has high business scalability because new small cells can be added at the low cost of fronthaul planning.

- The mesh fronthaul topology can offer multihop fronthaul links. As a result, the long fronthaul links are not necessary with the reduced TCO, the reliable LOS fronthaul link can be guaranteed, and the robust fronthaul links can be guaranteed due to the various route options.

- By exploiting the adaptive beamforming algorithms and self-configuration techniques, for the cases in which existing fronthaul links are congested, or some links are blocked, each node can automatically build the optimal links to its neighbors and further establish the optimal route to the PoP based on QoS requirements, including fronthaul throughput, latency and packet-error rate. The mesh fronthaul is intelligent and can self-organize in real time, without human intervention.

- The exploitation of V-band and E-band can provide immunity against interference. Moreover, automatic interference management is necessary to overcome any overlap in directional beams and external interference.

- Full-duplex is dedicated for mesh fronthaul to simultaneously support transmitting and receiving for each node.

12.4.2 **ANTENNA TECHNIQUES**

The cost of the mmWave massive MIMO-based mesh fronthaul mainly depends on the mmWave fronthaul devices because the cost on licensed spectrum can be negligible. Table 12.5 summarizes existing typical antenna techniques for mmWave massive MIMO.

Table 12.5 Features of Several Typical mmWave Communication Transceivers

mmWave Massive MIMO	Features
mmWave multiantenna systems [1]	Support single-stream transmission, can be used for PtMP with TDM scheme. Variable beamwidth with flexible beam direction. Analog phase-shift network may lead to high power consumption and large volume
mmWave massive MIMO with analog phase-shift network [1]	Multistream transmission, can be used for PtMP and even mesh network with both TDM and BDM schemes. Variable beamwidth with flexible beam direction. Analog phase-shift network may lead to high power consumption and large volume
mmWave massive MIMO with discrete lens array [16]	Multistream transmission, can be used for PtMP and even mesh network with both TDM and BDM schemes. Fixed beamwidth with fixed beam direction. Highly directional beamspace may challenge the problem of beam alignment

So far, the 30 dBi antennas have been commercialized at both V-band and E-band by using the horn and/or dielectric lenses solutions [8]. However, these solutions may suffer from high cost and bulky equipment. To solve this problem, there have been extensive studies on cost-competitive alternatives such as discrete lenses, Fabry-Perot antennas and planar antenna arrays based on substrate-integrated waveguides [8]. On the other hand, tight integration of the transceiver and the antenna is required to simplify the interfaces for further reduced losses and costs. Electronic beamsteering/switching antennas can provide the narrow angular sector and allow the beam alignment at the installation stage. Moreover, it can compensate the small misalignment during operations due to wind or other causes. In the long term, beamsteering or beam-switching antennas over a large angular sector would enable mesh fronthaul networks.

Currently, most mmWave multiantenna systems are limited to the single RF chain, where only single-stream transmission can be supported at the same time. To enable the mesh fronthaul topology, the time division multiplexing (TDM) scheme has been proposed to support the PtMP and mesh solutions. However, the mesh fronthaul network in which each node employs the single RF chain device by only using the TDM scheme may suffer from large system latency. To solve this issue, the fronthaul device with multi-RF chains is suggested, so that each node can simultaneously connect multiple nodes using different beams. In [1], the proposed mmWave massive MIMO-based fronthaul network can realize beam division multiplexing (BDM)-based scheduling due to the flexible spatial multiplexing and hybrid beamforming. In a practical fronthaul network, according to the fronthaul load, each node can flexibly combine TDM and BDM to support the mesh fronthaul network. For instance, links with heavy load or without the LOS paths can be assigned more beam resources, while multiple links with light load can multiplex in the time domain.

12.4.3 RELIABLE CHANNEL ESTIMATION SCHEMES

In this section, we will briefly introduce the channel estimation for mmWave massive MIMO. Conventionally, mmWave is combined with the single-RF chain and multiantenna systems used for fronthauling, where the beam alignment is finished during its installation and channel estimation is not required as the conventional cellular network. Moreover, the conventional fronthaul link is long, and the signal-to-noise-ratio (SNR) before beamforming is conventionally considered to be low due to the high path loss, which leads to the challenging channel estimation in mmWave communications.

To solve the issue of low SNR before beamforming, beamforming scanning-based channel estimation schemes have been widely adopted in many standards, including IEEE 802.15.3c (TG3c) for WPAN and IEEE 802.11.ad for WLAN [1]. For example, in IEEE 802.15.3c, a codebook that predefines the set of potential beamforming spaces with different beam directions and beam widths is designed together with an interaction searching scheme. The basic concept of this scheme is to divide the whole beamspace into several pairs for both the transmitter and receiver with low resolution at first, and then the iterative ping-pong searching scheme is employed to find the best pair of the beamspace. After that, the searched best pair of the beamspace will be divided again with higher resolution for the next round of searching. Such procedure will be continued until some criteria are satisfied. In IEEE 802.11.ad, a single-sided searching scheme is adopted to find the optimal beam by two steps: the combiner is fixed to exhaustively search the best precoder at first, and then this best precoder is fixed to exhaustively search the best combiner.

For the beamforming scanning-based channel estimation, the estimated channel is implicit. Moreover, once the codebook-based channel estimation is finished, both the transmitter and the receiver also finish the beamforming and combining. However, the beamforming-based channel estimation is typically considered in the single-RF multiantenna systems, which may not be suitable for the emerging mmWave massive MIMO systems with multiple RF chains. Moreover, for mmWave massive MIMO channels to support multiple streams transmission, the explicit channel estimation is required for the following beamforming with better performance. Additionally, for the ultradense deployment of small-cell scenarios, we will also clarify the misunderstanding of low SNR before beamforming in an mmWave fronthaul network by comparing the path loss in an ultradense network operating at 30 GHz and that in conventional cellular networks working at 3 GHz. Specifically, considering the multipath fading, signal dispersion, and other loss factors, the path loss component of Friis equation in decibels (dBs) can be provided as [17]

$$\eta = 32.5 + 20\log_{10}(f_c) + 10\alpha\log_{10}(d) + (\alpha_o + \alpha_r)d \tag{12.1}$$

where f_c (MHz) is the carrier frequency, α (dB/km) is the path loss exponent, d (km) is the link distance, α_o (dB/km) is the atmospheric attenuation coefficient, and α_r (dB/km) is the rain attenuation coefficient. For the conventional cellular systems with $f_c = 3$ GHz and $d = 1$ km, for example, we have $\eta = 192.62$ dB, where $\alpha = 2.2$ dB/km is considered in urban scenarios [17], and the atmospheric attenuation and rain attenuation are ignored. By contrast, for the ultradense deployment of small cells with $f_c = 30$ GHz, we have $\eta = 188.27$ dB with $d = 100$ m for the fronthaul link and $\eta = 161.78$ dB with $d = 30$ m for the access link, where $\alpha = 2.2$ dB/km, $\alpha_o = 0.1$ dB/km, and $\alpha_r = 5$ dB/km when the heavy rain with 25 mm/h is considered. Due to the short fronthaul link distance, the path loss in mmWave is even smaller than that in conventional cellular networks, which indicates the appropriate SNR for channel estimation in the mmWave fronthaul even before beamforming.

By far, the existing channel estimation schemes for mmWave massive MIMO-based fronthauling are provided as follows, and their respective features and research directions are also summarized:

Beamforming scanning-based channel estimation has been widely used in existing single-RF mmWave systems, including IEEE 802.15.3c (TG3c) for WPAN and IEEE 802.11.ad for WLAN. However, these schemes usually consider the single-RF transceiver [17,18]. The multiresolution codebook design and its extension to the multi-RF transceiver should be further investigated.

Compressive sensing (CS)-based channel estimation exploits the angle-domain sparsity of mmWave massive MIMO channels [6], where only a small number of measurements can be used to estimate the mmWave channels of high dimension. However, the CS-based channel estimation schemes typically assume the quantized angle of arrival (AoA) or angle of departure (AoD), which may lead to a certain performance loss. Moreover, it requires the array manifold as a priori information. Hence, how to design the codebook (codebook is related to the measurement matrix in CS theory) and the CS-based channel estimation algorithm to estimate the continuous AoA and AoD can be an important research direction for the mmWave massive MIMO-based fronthaul.

Finite rate of innovation (FRI)-based channel estimation can acquire the super-resolution estimation of AoA/AoD of mmWave massive MIMO channels, whereby the sparsity of AoA/AoD in the angular domain can be exploited [1]. Compared with classical CS theory, FRI theory can be considered the analog CS theory, where the support of sparse signals can be continuous other than discrete required by CS theory. However, this scheme still requires the array manifold as a priori information.

Subspace estimation and decomposition (SED)-based channel estimation can directly acquire the dominated eigenmodes used for beamforming rather than the entire channel [19], whereby the low-rank property of the mmWave massive MIMO channel matrix due to the sparsity of mmWave channels is exploited. However, SED-based channel estimation requires multiple ping-pong operations between the BS and users to acquire the dominated eigenmodes, which may introduce too much noise and degrade the final estimation performance, especially in mmWave with low SNR before beamforming. It should be pointed out that the SED-based channel estimation scheme does not require the array manifold as a priori information, which indicates that this scheme can be widely used for mmWave massive MIMO systems with arbitrary array manifold.

12.4.4 FLEXIBLE BEAMFORMING DESIGN

Reliable and accurate beamforming design can maximize the link throughput. The beamforming algorithms of the mmWave massive MIMO for mesh fronthaul should meet the following several aspects:

- The beamforming can meet self-configuration and plug-the-pull for the initial building and future extension of the fronthaul network without human intervention.
- The beamforming can fast track the beams when the beam is misaligned due to wind or other causes.
- The beamforming can support multistream transmission for the flexible BDM scheme.

By far, there are several existing works on beamforming for mmWave massive MIMO-based fronthauling. El Ayach et al. [20] have proposed the spatially sparse precoding for mmWave massive MIMO systems, where the sparsity of mmWave channels is exploited. However, this scheme only considers the PtP multistream transmission. Ghauch et al. [19] have proposed an iterative algorithm to acquire the analog and digital beamforming approximating the dominated eigenmodes, whereby the low-rank property of the mmWave massive MIMO channel matrix is exploited. However, this scheme also only considers the single-user transmission and cannot be directly used in a PtMP fronthaul network. Han et al. [5] and Chiu et al. [21] have proposed the multiuser precoding schemes for mmWave massive MIMO. However, those schemes only consider the single-stream transmission for each user, which may not be suitable for the PtMP fronthaul network. In our paper [1], based on the phased antenna array with multi-RF chains, we proposed an effective hybrid precoding/combining scheme and the associated compressive sensing-based channel estimation scheme. The proposed scheme can guarantee that each node simultaneously connects multiple nodes with multiple streams for

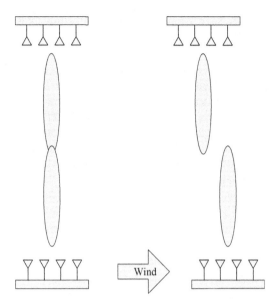

FIG. 12.9

An example that the pole movement can bring the beam delection due to wind [17].

each node. This is essentially different from conventional hybrid precoding/combining used for radio access networks. The proposed scheme can provide a viable approach to realize the desired mesh fronthaul topology and novel BDM-based scheduling, and it may also facilitate in-band backhaul in mmWave.

Besides, the mmWave massive MIMO systems used for fronthaul usually require narrow beams, which increase the sensitivity to movement caused by pole sway and other environmental concerns. To be specific as shown in Fig. 12.9, because the fronthaul devices will be equipped at outdoor structures such as poles, vibration and movement induced by wind flow and gusts have the potential to cause unacceptable outage probability if beam alignment is not frequently performed. To overcome this problem, Hur et al. [18] have proposed an efficient beam alignment technique with adaptive subspace sampling as well as the hierarchical codebooks. Hence, an efficient beam alignment algorithm with reduced overhead to overcome the movement caused by pole sway and other environmental concerns is a practical research direction to be further investigated.

12.4.5 TIME DIVISION DUPLEX (TDD), FREQUENCY DIVISION DUPLEX (FDD), OR FULL DUPLEX?

Full-duplex operation is the optimal duplexing option to enable mesh fronthaul links, where transmitting and receiving are simultaneously achieved by providing enough separation between transmitter and receiver antennas. However, full duplex is still under the phase of research and testing, especially for the full duplex in mmWave. As alternative options, FDD and TDD are considered. For FDD-based mmWave fronthaul, the uplink and downlink have to use different bands. However, the regulation in mmWave may be different in different countries. This indicates that one single device may not be suitable in various countries. In contrast, the uplink and downlink in TDD share the same band. Hence, a single device can be employed in various countries. Moreover, because different operators will employ ultradense small cells in the same areas, the mutual interference of fronthaul networks must be considered. Compared to FDD with different uplink/downlink channels, TDD makes it easier to find a clean spectrum and avoid interference. Moreover, because asymmetric traffic is dominant in a fronthaul network, TDD can flexibly adjust the ratio of time slots in the uplink and downlink according to the traffic requirement [1]. For a practical TDD mmWave massive MIMO-based fronthaul, adaptive interference management is desired to avoid mutual interference of different operators' small cells, and automated configuration solutions are expected for the plug-and-play fronthaul network, especially for unlicensed V-band.

12.4.6 IN-BAND, OUT-BAND, OR HYBRID-BAND

As we have demonstrated, mmWave wireless fronthaul can provide the most cost-effective solution to support the future high-capacity fronthaul in the ultradense deployment of small cells. Meanwhile, with the exponentially increasing throughput requirement on radio access networks, mmWave also draws more attention for the access link. Multiplexing fronthaul and access on the same frequency band, also named in-band wireless fronthaul, has obvious cost benefits from the hardware and frequency reuse perspective, but poses significant technology challenges. Fig. 12.9 shows a future mmWave massive MIMO-based cellular network, where the small-cell BSs use mmWave to support both fronthaul and access sharing the same frequency resource and hardware. For in-band fronthaul, there are two challenges. The first issue is the interference between access link and fronthaul link, where the interference from the access/fronthaul link may significantly degrade the signal-to-interference-ratio (SIR) of the fronthaul/access link. The second issue is the self-interference, where transmit/receive signals will interfere with receive/transmit signals from the same node. To solve this issue, the article [22] proposes the TDM-based in-band scheme to avoid both kinds of interference. However, the TDM between different fronthaul links may suffer from large system latency, and the TDM between the access link and fronthaul link may require large storage in each node for buffering the access data. In [1], we proposed the flexible hybrid beamforming design with the concept of BDM. Obviously, with the proposed mmWave massive MIMO scheme, BDM-based scheduling may be another competitive solution for the in-band mmWave fronthaul with lower latency.

12.5 SUMMARY

Dense small-cell deployment requires cost-effective fronthaul with high capacity to accommodate the 1000-fold increase of cellular capacity over the next 10 years. The fiber fronthaul solution suffers from high cost due to expensive installation and leasing. Existing wireless fronthaul solutions are optimized for current 3G/4G macrocells, which do not meet the requirements of the 5G fronthaul. Against this background, the mmWave massive MIMO provides a promising solution to wireless fronthaul, which is very compatible with the ultradense deployment of small cells because the fronthaul link can be short (typically 50–200 m) to mitigate the high path loss and guarantee the LOS link. Moreover, by using the cutting-edge technique of mmWave massive MIMO supporting multistream transmission, we can easily realize the mesh fronthaul topology, which can facilitate the installation and shorten the deployment circle. Additionally, beamforming techniques of mmWave massive MIMO can make the mesh fronthaul more flexible and intelligent. To be specific, the self-configuring with advanced Media Access Control (MAC) scheduling can provide a flexible fronthaul route meeting the fronthaul requirement without human intervention, and the in-band fronthaul may be possible for further reduced operation cost. Consequently, with the promising technique of mmWave massive MIMO, the mmWave mesh fronthaul has high scalability and is attractive for the design of future 5G fronthaul networks.

REFERENCES

[1] Z. Gao, L. Dai, D. Mi, Z. Wang, M.A. Imran, M.Z. Shakir, MmWave massive MIMO based wireless fronthaul for 5G ultra-dense network, IEEE Wirel. Commun. 21 (6) (2014) 136–143.

[2] R.J. Weiler, et al., Enabling 5G fronthaul and access with millimeter-waves, in: European Conference on Networks and Communications 2014 (EuCNC'14), 23–26 June, 2014, pp. 1–5.

[3] T. Parker, Fierce Broadband Wireless [Online], http://www.fiercebroadbandwireless.com/story/abi-ofdm-non-line-sight-dominate-small-cell-fronthaul-2017/2012-09-12?utm_medium=nl&utm_source=internal, 2012, September.

[4] L. Wei, R.Q. Hu, Y. Qian, G. Wu, Key elements to enable millimeter wave communications for 5G wireless systems, IEEE Wirel. Commun. 21 (6) (2014) 136–143.

[5] S. Han, I. Chih-lin, Z. Xu, C. Rowell, Large-scale antenna systems with hybrid precoding analog and digital beamforming for millimeter wave 5G, IEEE Commun. Mag. 53 (1) (2015) 186–194.

[6] A. Alkhateeb, J. Mo, N. Gonzalez-Prelcic, R.W. Heath, MIMO precoding and combining solutions for millimeter-wave systems, IEEE Commun. Mag. 52 (12) (2014) 122–131.

[7] A. Alkhateeb, O.E. Ayach, G. Leus, R.W. Heath, Channel estimation and hybrid precoding for millimeter wave cellular systems, IEEE J. Sel. Top. Sign. Proces. 8 (5) (2014) 831–846.

[8] C. Dehos, J.L. González, A.D. Domenico, D. Kténas, L. Dussopt, Millimeter-wave access and fronthauling: the solution to the exponential data traffic increase in 5G mobile communications systems? IEEE Commun. Mag. 52 (9) (2014) 88–95.

[9] Small Cell Millimeter Wave Mesh Fronthaul [Online]. http://pan.baidu.com/s/1o8LBFya.

[10] NGNM Alliance, Small Cell Fronthaul Requirements [Online]. http://pan.baidu.com/s/1gfHgn9T.

[11] V. Benedetto, et al., Huawei E-Band RTN380—Field Trial Report [Online], Huawei white paper, 2013. http://pan.baidu.com/s/1bMzSDK.

[12] M. Paolini, S. Fili, A Backhaul Solution for Organic Small-Cell Growth: A TCO Analysis [Online]. http://pan.baidu.com/s/1pKUMRn9.

[13] F. Rayal, Small Cell Wireless Fronthaul Business Case [Online], August 21, 2012 http://pan.baidu.com/s/1c2cALK8.

[14] Cisco, Cisco Visual Networking Index: Forecast and Methodology, 2011-2016. 2012, May. http://pan.baidu.com/s/1hrIjTKo.

[15] 3GPP, LTE; Policy and Charging Control Architecture [Online], 2012, September. ftp://ftp.3gpp.org/Specs/latest/Rel-11/23_series/23203-b70.zip.

[16] P.V. Amadori, C. Masouros, Low RF-complexity millimeter-wave beamspace-MIMO systems by beam selection, IEEE Trans. Commun. 63 (6) (2015) 2212–2223.

[17] S. Hur, T. Kim, D.J. Love, J.V. Krogmeier, T.A. Thomas, A. Ghosh, Multilevel millimeter wave beamforming for wireless fronthaul, in: Proceedings of IEEE GLOBE-COM Workshops 2011, December 5–9, 2011, pp. 253–257.

[18] S. Hur, T. Kim, D.J. Love, J.V. Krogmeier, T.A. Thomas, A. Ghosh, Millimeter wave beamforming for wireless fronthaul and access in small cell networks, IEEE Trans. Commun. 61 (10) (2013) 4391–4403.

[19] H. Ghauch, T. Kim, M. Bengtsson, M. Skoglund, Subspace estimation and decomposition for large millimeter-wave MIMO systems. IEEE J. Sel. Topics Signal Process. 10 (3) (2016) 528–542.

[20] O. El Ayach, S. Rajagopal, S. Abu-Surra, Z. Pi, R.W. Heath, Spatially sparse precoding in millimeter wave MIMO systems, IEEE Trans. Wirel. Commun. 13 (3) (2014) 1499–1513.

[21] L.-K. Chiu, S.-H. Wu, Hybrid radio frequency beamforming and baseband recoding for downlink MU-MIMO mmWave channels, in: Proceedings of IEEE ICC, June 2015.

[22] D.R. Taori, A. Sridharan, Point-to-multipoint in-band mmWave fronthaul for 5G networks, IEEE Commun. Mag. 53 (1) (2015) 195–201.

mmWave cellular networks: Stochastic geometry modeling, analysis, and experimental validation

13

W. Lu, M. Di Renzo

CNRS/Université Paris-Saclay, Gif-sur-Yvette, France

CHAPTER OUTLINE

mmWave Massive MIMO. http://dx.doi.org/10.1016/B978-0-12-804418-6.00013-3

13.1 INTRODUCTION

In spite of common belief, recently conducted channel measurements have shown that millimeter wave (mmWave) frequencies may be suitable for cellular networks, provided that the cell radius is of the order of 100–200 m [1]. Based on these measurements, Akdeniz et al. [2] have recently investigated system-level performance of mmWave cellular networks and have compared them against conventional microwave (μWave) cellular networks. The obtained results have highlighted that mmWave cellular networks may outperform μWave cellular networks, by assuming similar cellular network densities, provided that a sufficient beamforming gain is guaranteed between base stations (BSs) and mobile terminals (MTs). These preliminary but encouraging results have motivated several researchers to investigate potential and challenges of mmWave cellular networks for wireless access, in the light of the large and unused spectrum available at these frequencies [3, 4].

System-level performance evaluation of cellular networks is usually conducted by relying on numerical simulations, which are often time-consuming and do not provide direct insight into the effect of various parameters without analyzing a sufficiently large set of data. The mathematical analysis of cellular networks, on the other hand, is widely recognized to be a difficult task [5]. This is due to the lack of tractable approaches for modeling the locations of the BSs and the other-cell interference. Only recently, a new mathematical methodology has gained prominence due to its analytical tractability, its capability of capturing the inherent performance trends of currently deployed cellular networks, and the possibility of studying next-generation heterogeneous network deployments. This emerging approach exploits results from stochastic geometry and relies on modeling the locations of the BSs as points of a point process [5]. Usually, the Poisson point process (PPP) is used due to its mathematical tractability [6]. Recent results on cellular networks modeling based on stochastic geometry are available in Rangan et al., Ghosh et al., Andrews et al., ElSawy et al., Di Renzo et al. (May 2013, Jul. 2013), Di Renzo and Lu (May 2014, Jan. 2015, Mar. 2015, Jun. 2015), Di Renzo and Guan (Aug. 2014, Sep. 2014), Di Renzo and Lu and, Di Renzo [7–16], to which the reader is referred for a comprehensive literature review.

Motivated by the mathematical flexibility of the PPP-based abstraction model, a few researchers have recently turned their attention to the system-level performance of mmWave cellular networks with the aid of stochastic geometry. The aim is to develop mathematical frameworks specifically tailored to account for the peculiarities of mmWave propagation channels and transmission schemes [17–20]. In fact, currently available mathematical frameworks for modeling μWave cellular networks are not directly applicable to mmWave cellular networks. The main reasons are related to the need for the incorporation of realistic path-loss and blockage models, which are significantly different from μWave transmission. For example, Rappaport et al. [1] and Akdeniz et al. [2] have pointed out that line-of-sight (LOS) and nonline-of-sight (NLOS) links need to be appropriately modeled and may have different distributions, due to the more prominent impact of spatial blockages at mmWave frequencies compared to μWave frequencies. Also, in mmWave cellular networks a new outage state may be present in addition to LOS and NLOS states, which better reflects blockage effects at high frequencies and accounts for the fact that a link may be too weak to be established. In addition, large-scale antenna arrays are expected to be used for directional beamforming in mmWave systems, in order to overcome the increased path-loss at mmWave frequencies and to provide other-cell interference isolation. Therefore, directional beamforming needs to be included in the mathematical modeling and performance evaluation. Recently reported results on stochastic geometry modeling of mmWave cellular networks, however, take these aspects into account only in part [19–22].

The proposed approach is shown to be mathematically tractable thanks to the introduction of a multiball approximation for the link state, which is based on the concept of path-loss intensity matching. As a by-product of this approximation, an important result is unveiled: if the average cell size is of the order of 50 m, no outage state is present and the probability that a generic MT is served via a LOS link is greater than 80%. In this case, in addition, mmWave cellular networks outperform their μWave counterpart not only in terms of average rate, but also in terms of coverage probability. It is shown, furthermore, that a peculiarity of the three-state link statistical model lies in the existence of a nonzero communication blockage probability, which corresponds to a network scenario where all the available BSs are in an outage state and, thus, a generic MT cannot be served. The numerical examples show, however, that the presence of an outage state may be beneficial in dense network deployments, since the BSs in an outage state do not contribute to the other-cell interference, which may result in a better coverage probability.

The proposed modeling approach is validated with the aid of base station locations and building footprints from two publicly available databases (OFCOM and Ordnance Survey) from the United Kingdom. Our study confirms that the abstraction model based on stochastic geometry is capable of accurately modeling the communication performance of mmWave cellular networks in dense urban environments.

The remainder of this chapter is organized as follows. In Section 13.2, system model and modeling assumptions are introduced. In Section 13.3, the statistical distribution of deterministic and random transformations of the path-loss is provided. In Section 13.4, the frameworks for computing coverage and rate of cellular networks are described, by assuming that the network is noise-limited. In Section 13.5, the frameworks in Section 13.4 are generalized, by incorporating the impact of the other-cell interference. In Section 13.6, the analysis is validated via experimental data and numerical simulations. Also, the performance of mmWave and μWave cellular networks are compared against each other. Finally, Section 13.7 concludes this chapter.

13.2 SYSTEM MODEL
13.2.1 PPP-BASED ABSTRACTION MODELING

A bi-dimensional downlink cellular network is considered, where a probe MT is located, without loss of generality thanks to the Slivnyak theorem [23, vol. 1, Theorem 1.4.5], at the origin and the BSs are modeled as points of a homogeneous PPP, denoted by Ψ, of density λ. The MT is assumed to be served by the BS providing the smallest path-loss to it. The serving BS is denoted by $BS^{(0)}$. Similar to [5, Section VI], full-frequency reuse is considered. For notational simplicity, the set of interfering BSs is denoted by $\Psi^{(\backslash 0)} = \Psi \backslash BS^{(0)}$. The distance from a generic BS to the MT is denoted by r.

13.2.2 DIRECTIONAL BEAMFORMING MODELING

Thanks to the small wavelength, mmWave cellular networks are capable of exploiting directional beamforming for compensating for the increased path-loss at mmWave frequencies and for overcoming the additional noise due to the large transmission bandwidth. As a desirable bonus, directional beamforming provides interference isolation, which reduces the impact of the other-cell interference. Thus, antenna arrays are assumed at both the BSs and the MT for performing directional beamforming. For mathematical tractability and similar to Bai and Heath, and Singh et al. [19, 20], the actual antenna array patterns are approximated by a sectored antenna model. In particular, the antenna gain of a generic BS, $G_{BS}(\cdot)$, and of the MT, $G_{MT}(\cdot)$, can be formulated as follows:

$$G_q(\theta) = \begin{cases} G_q^{(\max)} & \text{if } |\theta| \leq \omega_q/2 \\ G_q^{(\min)} & \text{if } |\theta| > \omega_q/2 \end{cases} \tag{13.1}$$

where $q \in \{BS, MT\}$, $\theta \in [-\pi, \pi)$ is the angle off the boresight direction, ω_q is the beamwidth of the main lobe, $G_q^{(\max)}$ and $G_q^{(\min)}$ are the array gains of main and side lobes.

The MT and its serving BS, $BS^{(0)}$, are assumed to estimate the angles of arrival and to adjust their antenna steering orientations accordingly. In the absence of alignment errors, therefore, the maximum directivity gain can be exploited on the intended link. Thus, the directivity gain of the intended link is $G^{(0)} = G_{BS}^{(\max)} G_{MT}^{(\max)}$. The beams of all nonintended links are assumed to be randomly oriented with respect to each other and to be uniformly distributed in $[-\pi, \pi)$. Accordingly, the directivity gains of the interfering links, $G^{(i)}$ for $i \in \Psi^{(\backslash 0)}$, are randomly distributed. Based on Eq. (13.1), their probability density function (PDF) can be formulated as follows:

$$\begin{aligned} f_{G^{(i)}}(g) = & \frac{\omega_{BS}}{2\pi} \frac{\omega_{MT}}{2\pi} \delta\left(g - G_{BS}^{(\max)} G_{MT}^{(\max)}\right) \\ & + \frac{\omega_{BS}}{2\pi}\left(1 - \frac{\omega_{MT}}{2\pi}\right)\delta\left(g - G_{BS}^{(\max)} G_{MT}^{(\min)}\right) \\ & + \left(1 - \frac{\omega_{BS}}{2\pi}\right)\frac{\omega_{MT}}{2\pi}\delta\left(g - G_{BS}^{(\min)} G_{MT}^{(\max)}\right) \\ & + \left(1 - \frac{\omega_{BS}}{2\pi}\right)\left(1 - \frac{\omega_{MT}}{2\pi}\right)\delta\left(g - G_{BS}^{(\min)} G_{MT}^{(\min)}\right) \end{aligned} \tag{13.2}$$

where $\delta(\cdot)$ is the Kronecker's delta function.

13.2.3 LINK STATE MODELING

Let an arbitrary link of length r, that is, the distance from a generic BS to the MT is equal to r. Motivated by recent experimental findings on mmWave channel modeling [2, Section III-D], a three-state statistical model for each link is assumed, according to which a link can be in an LOS, NLOS, or in an outage (OUT) state. An LOS state occurs if there is no blockage between BS and MT. An NLOS state, on the other hand, occurs if the BS-to-MT link is blocked. An outage state occurs if the path-loss between BS and MT is so high that no link between them can be established. In the latter case, the path-loss of the link is assumed to be infinite. In practice, outages implicitly occur if the path-loss in either an LOS or an NLOS state is sufficiently large. In Akdeniz et al. [2, Fig. 7], with the aid of experimental results, they prove that adding an outage state, which is usually not observed for transmission at μWave frequencies, provides a more accurate statistical description of the inherent coverage possibilities at mmWave frequencies.

In the study by Akdeniz et al. [2, Eq. 8], the probabilities of occurrence $p_{LOS}(\cdot)$, $p_{NLOS}(\cdot)$, $p_{OUT}(\cdot)$ of LOS, NLOS, and outage states, respectively, as a function of the distance r can be formulated as follows:

$$p_{OUT}(r) = \max\left\{0, 1 - \gamma_{OUT}e^{-\delta_{OUT}r}\right\}$$
$$p_{LOS}(r) = (1 - p_{OUT}(r))\gamma_{LOS}e^{-\delta_{LOS}r} \qquad (13.3)$$
$$p_{NLOS}(r) = (1 - p_{OUT}(r))\left(1 - \gamma_{LOS}e^{-\delta_{LOS}r}\right)$$

where $(\delta_{LOS}, \gamma_{LOS})$ and $(\delta_{OUT}, \gamma_{OUT})$ are parameters that depend on the propagation scenario and on the carrier frequency being considered. Examples are available in Akdeniz et al. [2, Table I].

Under the assumption that the BSs are modeled as points of a homogeneous PPP and that the events that the BS-to-MT links are in LOS, NLOS, or outage state are independent, Ψ can be partitioned, from a typical MT perspective, into three (one for each link state) independent and nonhomogeneous PPPs, that is, Ψ_{LOS}, Ψ_{NLOS}, Ψ_{OUT}, such that $\Psi = \Psi_{LOS} \cup \Psi_{NLOS} \cup \Psi_{OUT}$. This originates from the thinning property of PPPs [23]. From Eq. (13.3), the densities of Ψ_{LOS}, Ψ_{NLOS}, Ψ_{OUT} are $\lambda_{LOS}(r) = p_{LOS}(r)\lambda$, $\lambda_{NLOS}(r) = p_{NLOS}(r)\lambda$, $\lambda_{OUT}(r) = p_{OUT}(r)\lambda$, respectively.

13.2.4 PATH-LOSS MODELING

Based on the channel measurements in Akdeniz et al. [2], the path-loss of LOS and NLOS links is as follows:

$$l_{LOS}(r) = (\kappa_{LOS}r)^{\beta_{LOS}}$$
$$l_{NLOS}(r) = (\kappa_{NLOS}r)^{\beta_{NLOS}} \qquad (13.4)$$

where r denotes a generic BS-to-MT distance, κ_{LOS} and κ_{NLOS} can be interpreted as the path-loss of LOS and NLOS links at a distance of 1 m, respectively, β_{LOS} and β_{NLOS} denote the power path-loss exponents of LOS and NLOS links, respectively.

As mentioned in Section 13.2.3, the path-loss of the links that are in an outage state is assumed to be infinite, that is, $l_{OUT}(r) = \infty$. This model is usually known as the "close-in" path-loss model [24, 25].

The path-loss model in Eq. (13.4) is general enough for modeling several practical propagation conditions. For example, it can be linked to the widespread used (α, β) or "floating-intercept" path-loss model [1, 2], by setting $\kappa_{LOS} = 10^{\alpha_{LOS}/(10\beta_{LOS})}$ and $\kappa_{NLOS} = 10^{\alpha_{NLOS}/(10\beta_{NLOS})}$, where α_{LOS} and α_{NLOS} are defined in Akdeniz et al. [2, Table I]. It is worth mentioning that in the floating-intercept model, unlike the close-in model, the parameters (α, β) have no physical interpretation and they denote only the floating intercept and the slope of the best linear fit of empirical data [24, 25].

13.2.5 SHADOWING MODELING

In addition to the distance-dependent path-loss of Section 13.2.4, each link is subject to a random power variation, which, for a generic BS-to-MT link, is denoted by $|h|^2$. According to Akdeniz et al. [2], this power variation $|h|^2$ is assumed to follow a log-normal distribution with mean (in dB) equal to $\mu^{(dB)}$ and with standard deviation (in dB) equal to $\sigma^{(dB)}$. Thus, $|h|^2$ takes into account mid-scale fading, that is, shadowing. In general, $\mu^{(dB)}$ and $\sigma^{(dB)}$ for LOS and NLOS links are different [2]. In the sequel, they are denoted by $\mu_s^{(dB)}$, $\sigma_s^{(dB)}$, where $s = \{LOS, NLOS\}$ indicates the link state.

As mentioned in Section 13.2.3, for mathematical tractability (shadowing), correlations between links are ignored. Thus, the random power gains of LOS and NLOS links are assumed to be independent, but nonidentically distributed. As recently remarked and verified with the aid of simulations in Bai and Heath [19], this assumption usually causes a minor loss of accuracy in the evaluation of the statistics of the signal-to-interference-plus-noise-ratio (SINR). For ease of description, fast-fading is neglected, but it may be readily incorporated.

13.2.6 CELL ASSOCIATION CRITERION

In μWave cellular networks, cell association is usually performed based on downlink reference pilot signals, which undergo both path-loss and shadowing. Thus, a typical MT is usually served by the BS providing the highest received power to it, where both path-loss and shadowing are taken into account. This is because shadowing is, in general, a slowly varying effect. In mmWave cellular networks, on the other hand, shadowing is expected to be less slowly varying than in μWave cellular networks. This is mostly due to the more pronounced impact of blockages on the received signal power. In mmWave cellular networks, as a result, the MT may not be able to completely take into account the random fluctuations introduced by shadowing during cell association. Motivated by these considerations, the MT is assumed to be served by the BS providing the smallest path-loss to it. Thus, shadowing is not taken into account for cell association.

Let $L_{\text{LOS}}^{(0)}$, $L_{\text{NLOS}}^{(0)}$, and $L_{\text{OUT}}^{(0)}$ be the smallest path-loss of LOS, NLOS, and OUT links. They can be formulated as follows:

$$L_s^{(0)} = \begin{cases} \min_{n\in\Psi_s}\left\{l_s\left(r^{(n)}\right)\right\} & \text{if } \Psi_s \neq \emptyset \\ +\infty & \text{if } \Psi_s = \emptyset \end{cases}$$

$$L_{\text{OUT}}^{(0)} = \min_{n\in\Psi_{\text{OUT}}}\left\{l_{\text{OUT}}\left(r^{(n)}\right)\right\} = +\infty \tag{13.5}$$

where $s = \{\text{LOS}, \text{NLOS}\}$, $r^{(n)}$ denotes the distance from a generic BS to the MT, and $\emptyset$ denotes an empty set. Hence, the path-loss of the serving BS, $\text{BS}^{(0)}$, can be formulated as $L^{(0)} = \min\left\{L_{\text{LOS}}^{(0)}, L_{\text{NLOS}}^{(0)}, L_{\text{OUT}}^{(0)}\right\}$.

Remark 13.1. Based on the link state model of Section 13.2.3, a link may be in an outage state. Accordingly, the event that all the available BSs are in an outage state may occur with a nonzero probability. By using the notation in Eq. (13.5), this occurs if $\Psi_{\text{LOS}} = \Psi_{\text{NLOS}} = \emptyset$. In this case, no BSs are available to serve the MT and it is said to be in a *communication blockage state*.

13.2.7 PROBLEM FORMULATION

Let $U^{(0)}$ be the intended received power, that is, the power received at the MT and transmitted by the serving BS, $\text{BS}^{(0)}$. If the MT is in a communication blockage state, then $U^{(0)} = 0$. Otherwise, $U^{(0)} > 0$ and it depends on the cell association being used. Thus, it is further detailed in Section 13.4. The SINR of the downlink cellular network under analysis can be formulated as $\text{SINR} = U^{(0)}\left(\sigma_N^2 + I_{\text{agg}}\right)^{-1}$, where σ_N^2 is the noise power and I_{agg} is the aggregate other-cell interference, that is, the total interference generated by the BSs in $\Psi^{(\backslash 0)}$. In particular, σ_N^2 is defined as $\sigma_N^2 = 10^{\sigma_N^2(\text{dBm})/10}$, where $\sigma_N^2(\text{dBm}) = -174 + 10\log_{10}(B_{\text{W}}) + \mathcal{F}_{\text{dB}}$, B_{W} is the transmission bandwidth and $\mathcal{F}_{\text{dB}}$ is the noise figure in dB. The aggregate other-cell interference is defined as $I_{\text{agg}} = \sum_{i\in\Psi^{(\backslash 0)}}\left(PG^{(i)}\left|h^{(i)}\right|^2/L^{(i)}l\left(r^{(i)}\right)\right)$, where P is the transmit power of the BSs and $l(\cdot)$ is the path-loss of Section 13.2.4, which depends on a BS being in LOS, NLOS, or outage.

From the SINR, coverage probability ($P^{(\text{cov})}$) and average rate (R) can be formulated as follows [11]:

$$P^{(\text{cov})}(T) = \Pr\{\text{SINR} \geq T\} \tag{13.6}$$

$$\begin{aligned} R &= \mathbb{E}_{\text{SINR}}\{B_{\text{W}}\log_2(1+\text{SINR})\} \\ &= \frac{B_{\text{W}}}{\ln(2)}\int_0^{+\infty} P^{(\text{cov})}(e^t - 1)dt \\ &= \frac{B_{\text{W}}}{\ln(2)}\int_0^{+\infty} \frac{P^{(\text{cov})}(t)}{t+1}dt \\ &\overset{(a)}{\approx} \frac{B_{\text{W}}}{\ln(2)}\sum_{u=1}^{N_{\text{GCQ}}} w^{(u)}\frac{P^{(\text{cov})}\left(x^{(u)}\right)}{x^{(u)}+1} \end{aligned} \tag{13.7}$$

where $T > 0$ is a reliability threshold and $\mathbb{E}\{\,\cdot\,\}$ denotes the expectation operator. The approximation in (a) follows from the Gauss-Chebyshev Quadrature (GCQ) rule [26, Eq. (25.4.39)], where $w^{(u)}$ and $x^{(u)}$ for $u = 1, 2, \ldots, N_{\mathrm{GCQ}}$ are weights and abscissas of the quadrature, respectively, which are available in closed-form in Di Renzo et al. [8, Eq. (13)]. The approximation in (a) is especially useful if the coverage probability cannot be formulated in a closed-form expression.

It is worth mentioning that if a communication blockage occurs, that is, $U^{(0)} = 0$, then SINR $= 0$, and coverage and rate are equal to zero.

13.3 PRELIMINARIES: ANALYSIS AND APPROXIMATIONS OF TRANSFORMATIONS OF THE PATH-LOSS

In this section, we provide general results for the distribution of transformations of the path-loss of mmWave systems, which account for LOS, NLOS, and outage states. These results are useful for computing coverage and rate in Section 13.4 and Section 13.5.

Lemma 13.1. *Let* $\Phi = \{\Phi_{\mathrm{LOS}}, \Phi_{\mathrm{NLOS}}, \Phi_{\mathrm{OUT}}\}$, *where* $\Phi_{\bar{s}} = \left\{ l_{\bar{s}}\left(r^{(n)} \right), n \in \Psi_{\bar{s}} \right\}$, *for* $\bar{s} \in \{\mathrm{LOS}, \mathrm{NLOS}, \mathrm{OUT}\}$, *are transformations of the path-loss of LOS, NLOS, and OUT BSs, respectively, which is defined in Sections 13.2.3 and 13.2.4.*

Let the link state model in Eq. (13.3). Then, Φ *is a PPP with intensity measure,*

$$\Lambda_\Phi([0, x)) = \Lambda_{\mathrm{LOS}}([0, x)) + \Lambda_{\mathrm{NLOS}}([0, x)) \tag{13.8}$$

where

$$\begin{aligned}
\Lambda_{\mathrm{LOS}}([0, x)) &= Y_0(x; s = \mathrm{LOS}) \\
\Lambda_{\mathrm{NLOS}}([0, x)) &= Y_1(x; s = \mathrm{NLOS}) - Y_0(x; s = \mathrm{NLOS})
\end{aligned} \tag{13.9}$$

and $Y_0(\,\cdot\,;\,\cdot\,)$ *and* $Y_1(\,\cdot\,;\,\cdot\,)$ *are defined as follows:*

$$\begin{aligned}
Y_0(x; s) &= \mathcal{K}_2\left(e^{-W} + We^{-W} - e^{-V_s x^{1/\beta_s}} - V_s x^{1/\beta_s} e^{-V_s x^{1/\beta_s}} \right) \mathcal{H}(x - Z_s) \\
&\quad + \mathcal{K}_1\left(1 - e^{-Q_s x^{1/\beta_s}} - Q_s x^{1/\beta_s} e^{-Q_s x^{1/\beta_s}} \right) \bar{\mathcal{H}}(x - Z_s) \\
&\quad + \mathcal{K}_1\left(1 - e^{-R} - Re^{-R} \right) \mathcal{H}(x - Z_s) \\
Y_1(x; s) &= \pi\lambda\kappa_s^{-2} x^{2/\beta_s} \bar{\mathcal{H}}(x - Z_s) + \pi\lambda\left(\delta_{\mathrm{OUT}}^{-1} \ln(\gamma_{\mathrm{OUT}}) \right)^2 \mathcal{H}(x - Z_s) \\
&\quad + \left(\gamma_{\mathrm{OUT}}^{-1} + \gamma_{\mathrm{OUT}}^{-1} \ln(\gamma_{\mathrm{OUT}}) - e^{-T_s x^{1/\beta_s}} - T_s x^{1/\beta_s} e^{-T_s x^{1/\beta_s}} \right) \\
&\quad \times 2\pi\lambda\delta_{\mathrm{OUT}}^{-2} \gamma_{\mathrm{OUT}} \mathcal{H}(x - Z_s)
\end{aligned} \tag{13.10}$$

where $s = \{\mathrm{LOS}, \mathrm{NLOS}\}$, $\mathcal{H}(\cdot)$ *is the Heaviside function,* $\bar{\mathcal{H}}(x) = 1 - \mathcal{H}(x)$, $\mathcal{K}_1 = 2\pi\lambda\gamma_{\mathrm{LOS}}\delta_{\mathrm{LOS}}^{-2}$, $\mathcal{K}_2 = 2\pi\lambda\gamma_{\mathrm{LOS}}\gamma_{\mathrm{OUT}}(\delta_{\mathrm{LOS}} + \delta_{\mathrm{OUT}})^{-2}$, $R = \delta_{\mathrm{LOS}}\delta_{\mathrm{OUT}}^{-1} \ln(\gamma_{\mathrm{OUT}})$, $W = (\delta_{\mathrm{LOS}} + \delta_{\mathrm{OUT}})\delta_{\mathrm{OUT}}^{-1} \ln(\gamma_{\mathrm{OUT}})$, $Q_s = \delta_{\mathrm{LOS}}\kappa_s^{-1}$, $T_s = \delta_{\mathrm{OUT}}\kappa_s^{-1}$, $V_s = (\delta_{\mathrm{LOS}} + \delta_{\mathrm{OUT}})\kappa_s^{-1}$, $Z_s = \left(\kappa_s\delta_{\mathrm{OUT}}^{-1} \ln(\gamma_{\mathrm{OUT}}) \right)^{\beta_s}$.

Proof. See Appendix 13.A.   □

Corollary 13.1 *Let* $\delta_{OUT} = 0, \gamma_{OUT} = 1, that is, p\ OUT(r) = 0\ in\ Eq.\ (13.3).\ \Lambda_\Phi(\cdot)\ in$
Eq. (13.8) *holds with* $\Upsilon_0(x;s) = \mathcal{K}_1\left(1 - e^{-Q_s x^{1/\beta_s}} - Q_s x^{1/\beta_s} e^{-Q_s x^{1/\beta_s}}\right),\ \Upsilon_1(x;s) =$
$\pi\lambda\kappa_s^{-2}x^2/\beta_s$ *for* $s = \{LOS, NLOS\}$.

Proof. It follows directly from Eq. (13.10), since $Z_s \to +\infty$ for $s = \{LOS, NLOS\}$.□

Lemma 13.2. *Let* $\Phi^{(0)} = \min\{\Phi\}$ *be the smallest element of the PPP* Φ *introduced*
in Lemma 13.1. Its cumulative density function (CDF), that is, $F_{\Phi^{(0)}}(x) =$
$\Pr\left\{\Phi^{(0)} \le x\right\}$, *can be formulated as follows:*

$$F_{\Phi^{(0)}}(x) = 1 - \exp\left(-\Lambda_\Phi([0,x))\right) \tag{13.11}$$

where $\Lambda_\Phi(\cdot)$ *is defined in* **Lemma 13.1.**

Proof. It follows by applying the void probability theorem of PPPs [27, Corollary 6].   □

13.3.1 TWO-BALL APPROXIMATION

It is apparent, from Eqs. (13.9) and (13.10), that the intensity measure of Φ is available in closed-form, but at the cost of a nonnegligible complexity that makes intractable the analysis of coverage and rate in the presence of other-cell interference. In order to overcome this issue, we propose an approximation for modeling LOS, NLOS and outage link states.

The proposed approach consists of computing the link state probabilities based on a "two-ball" approximation of Eq. (13.3). More specifically, the probabilities in Eq. (13.3) are approximated as follows:

$$
\begin{cases}
p_{\bar{s}}(r) \approx p_{\bar{s}}^{(approx)}(r) = q_{\bar{s}}^{[0,D_1]}\mathbf{1}_{[0,D_1)}(r) + q_{\bar{s}}^{[D_1,D_2]}\mathbf{1}_{(D_1,D_2)}(r) + q_{\bar{s}}^{[D_2,\infty]}\mathbf{1}_{[D_2,+\infty)}(r) \\
\displaystyle\sum_{\bar{s}\in\{LOS,NLOS,OUT\}} q_{\bar{s}}^{[0,D_1]} = \sum_{\bar{s}\in\{LOS,NLOS,OUT\}} q_{\bar{s}}^{[D_1,D_2]} = \sum_{\bar{s}\in\{LOS,NLOS,OUT\}} q_{\bar{s}}^{[D_2,\infty]} = 1
\end{cases}
$$

$$\tag{13.12}$$

where $\bar{s}\in\{LOS,NLOS,OUT\}$, $D_2 \ge D_1 \ge 0$ are the radii of the approximating balls, $\mathbf{1}_{[a,b)}(\cdot)$ is the indicator function, which is defined as $\mathbf{1}_{[a,b)}(r) = 1$ if $r \in [a,b)$ and $\mathbf{1}_{[a,b)}(r) = 0$ if $r \notin [a,b)$, and $q_{\bar{s}}^{[a,b]}$ denote the probability that a link of length $r\in[a,b)$ is in state $\bar{s}$. The second equality in Eq. (13.12) guarantees that each link of length r is only in one of the three possible states $\bar{s}\in\{LOS,NLOS,OUT\}$. In what follows, it is referred to as *approximation constraint*.

The rationale behind Eq. (13.12) originates from the visual inspection of Akdeniz et al. [2, Fig. 7]. It is apparent from Akdeniz et al. [2, Fig. 7], in fact, that two breaking distances (D_1 and D_2) emerge for arbitrary values of the link length r, which results in three connectivity regions: the first, for $r \in [0,D_1)$, where the links are most likely to be in LOS or NLOS; the second, for $r \in (D_1,D_2)$, where the links can be in any state; and the third, for $[D_2, +\infty)$, where the links are most likely to be in outage; Eq. (13.12) accounts for this empirical observation for any $q_{\bar{s}}^{[a,b]}$.

Before describing the path-loss intensity matching approach for computing the parameters of the approximation in Eq. (13.12), that is, $\left(D_1, D_2, q_{\bar{s}}^{[0,D_1]}, q_{\bar{s}}^{[D_1,D_2]}, q_{\bar{s}}^{[D_2,\infty]}\right)$ for $\bar{s} \in \{\text{LOS, NLOS, OUT}\}$, Lemma 13.1 needs to be generalized based on the link state model in Eq. (13.12).

Lemma 13.3. *Let $\Phi^{(\text{approx})}$ denote $\left\{\Phi_{\text{LOS}}^{(\text{approx})}, \Phi_{\text{NLOS}}^{(\text{approx})}, \Phi_{\text{OUT}}^{(\text{approx})}\right\}$, where $\Phi_{\bar{s}}^{(\text{approx})} = \left\{l_{\bar{s}}\left(r^{(n)}\right), n \in \Psi_{\bar{s}}^{(\text{approx})}\right\}$, for $\bar{s} \in \{\text{LOS, NLOS, OUT}\}$, are transformations of the path-loss of LOS, NLOS, and OUT BSs, respectively, where the path-loss model is defined in Section 13.2.4 and the link state model is given in Eq. (13.12), that is, $\Psi_{\bar{s}}^{(\text{approx})}$ for $\bar{s} \in \{\text{LOS, NLOS, OUT}\}$ has the same definition as Ψ_s except that Eq. (13.3) is replaced by Eq. (13.12). Then, $\Phi^{(\text{approx})}$ is a PPP with intensity given in Eqs. (13.8) and (13.9), which are obtained by replacing $\Lambda_s(\cdot)$ for $s \in \{\text{LOS, NLOS}\}$ with $\Lambda_s^{(\text{approx})}(\cdot)$ defined as follows:*

$$\Lambda_s^{(\text{approx})}([0,x))$$
$$= \mathcal{G}_s^{(2)} x^{2/\cdot} \beta_s \bar{\mathcal{H}}\left(x - (\kappa_s D_2)^{\beta_s}\right) + \left(\mathcal{G}_s^{(1)} x^{2/\cdot} \beta_s + \mathcal{G}_s^{(3)}\right) \bar{\mathcal{H}}\left(x - (\kappa_s D_1)^{\beta_s}\right)$$
$$- \mathcal{G}_s^{(3)} + \mathcal{G}_s^{(4)} \mathcal{H}\left(x - (\kappa_s D_1)^{\beta_s}\right) + \left(\mathcal{G}_s^{(6)} x^{2/\cdot} \beta_s + \mathcal{G}_s^{(5)}\right) \mathcal{H}\left(x - (\kappa_s D_2)^{\beta_s}\right) \quad (13.13)$$

where $\mathcal{G}_s^{(1)} = \pi \lambda \kappa_s^{-2}\left(q_s^{[0,D_1]} - q_s^{[D_1,D_2]}\right)$, $\mathcal{G}_s^{(2)} = \pi \lambda \kappa_s^{-2} q_s^{[D_1,D_2]}$, $\mathcal{G}_s^{(3)} = \pi \lambda D_1^2 q_s^{[D_1,D_2]}$, $\mathcal{G}_s^{(4)} = \pi \lambda D_1^2 q_s^{[0,D_1]}$, $\mathcal{G}_s^{(5)} = \pi \lambda D_2^2 \left(q_s^{[D_1,D_2]} - q_s^{[D_2,\infty]}\right)$, $\mathcal{G}_s^{(6)} = \pi \lambda \kappa_s^{-2} q_s^{[D_2,\infty]}$.

Proof. The proof follows the same steps as the proof of Lemma 13.1. The only difference lies in replacing $p_s(\cdot)$ with $p_s^{(\text{approx})}(\cdot)$ in Eq. (13.12) and in computing the integrals. □

Remark 13.2. Based on Lemma 13.3, $\Lambda_\Phi(\cdot)$ in Eq. (13.8) can be formulated in closed-form as $\Lambda_\Phi([0,x)) \approx \Lambda_\Phi^{(\text{approx})}([0,x)) = \Lambda_{\text{LOS}}^{(\text{approx})}([0,x)) + \Lambda_{\text{NLOS}}^{(\text{approx})}([0,x))$. This confirms the usefulness of the two-ball approximation for realistic channel models. The CDF of $\min\left\{\Phi^{(\text{approx})}\right\}$ follows from Lemma 13.2 and the approximation $F_{\Phi^{(0)}}(x) \approx 1 - \exp\left(-\Lambda_\Phi^{(\text{approx})}([0,x))\right)$ holds.

We are now in the position of describing the procedure for computing the 11 parameters of the approximation in Eq. (13.12), that is, D_1, D_2 and $\left(q_{\bar{s}}^{[0,D_1]}, q_{\bar{s}}^{[D_1,D_2]}, q_{\bar{s}}^{[D_2,\infty]}\right)$

for $\bar{s} \in \{\text{LOS}, \text{NLOS}, \text{OUT}\}$. Let the PPP of the path-loss $L = \{L_{\text{LOS}}, L_{\text{NLOS}}, L_{\text{OUT}}\}$ based on Eq. (13.3). From Lemma 13.1, its intensity can be formulated by $\Lambda_\Phi([0,x)) = \Lambda_{\text{LOS}}([0,x)) + \Lambda_{\text{NLOS}}([0,x))$, where $\Lambda_{\text{LOS}}(\cdot)$ and $\Lambda_{\text{NLOS}}(\cdot)$ are defined in Eq. (13.9). Let the PPP of the path-loss $L^{(\text{approx})} = \left\{ L_{\text{LOS}}^{(\text{approx})}, L_{\text{NLOS}}^{(\text{approx})}, L_{\text{OUT}}^{(\text{approx})} \right\}$ based on Eq. (13.12). From Lemma 13.3 and Remark 13.2, its intensity is $\Lambda_\Phi^{(\text{approx})}([0,x)) = \Lambda_{\text{LOS}}^{(\text{approx})}([0,x)) + \Lambda_{\text{NLOS}}^{(\text{approx})}([0,x))$, where $\Lambda_{\text{LOS}}^{(\text{approx})}(\cdot)$ and $\Lambda_{\text{NLOS}}^{(\text{approx})}(\cdot)$ are defined in Eq. (13.13). The proposed matching procedure consists of two steps:

1. The first step lies in computing the 11 parameters in Eq. (13.12) as the best fit of the following *unconstrained* optimization problem:

$$\min_{\left(D_1, D_2, q_{\bar{s}}^{[0,D_1]}, q_{\bar{s}}^{[D_1,D_2]}, q_{\bar{s}}^{[D_2,\infty]}\right)} \left\{ \frac{1}{2} \left\| \ln(\Lambda_\Phi([x_m,x_M))) - \ln\left(\Lambda_\Phi^{(\text{approx})}([x_m,x_M))\right) \right\|_F^2 \right\} \quad (13.14)$$

where $x_m \in (0, +\infty), x_M \in (0, +\infty), x_m \ll x_M, \bar{s} \in \{\text{LOS}, \text{NLOS}, \text{OUT}\}$, $\|\cdot\|_F$ denotes the Frobenius norm and the pair (x_m, x_M) is judiciously chosen in order to take the main body of $\Lambda_\Phi(\cdot)$ into account for an accurate matching. The initial point for solving Eq. (13.14) is randomly chosen. The optimization problem is unconstrained, since the approximation constraint in Eq. (13.12) is neglected. The solution of Eq. (13.14) is denoted by

$$\left(\hat{D}_1, \hat{D}_2, \hat{q}_{\bar{s}}^{[0,D_1]}, \hat{q}_{\bar{s}}^{[D_1,D_2]}, \hat{q}_{\bar{s}}^{[D_2,\infty]} \right).$$

2. The second step lies in computing the 11 parameters in Eq. (13.12) as the best fit of the *constrained* optimization problem still formulated as in Eq. (13.14), but by taking into account the approximation constraint in Eq. (13.12) and by assuming as the initial point of the search the solution of the first step, that is,

$$\left(\hat{D}_1, \hat{D}_2, \hat{q}_{\bar{s}}^{[0,D_1]}, \hat{q}_{\bar{s}}^{[D_1,D_2]}, \hat{q}_{\bar{s}}^{[D_2,\infty]} \right).$$

By applying the proposed two-step approximation technique to the empirical three-state link model proposed in Akdeniz et al. [2, Table I], the approximation in Table 13.1 is obtained. The accuracy of this approximation is studied in Section 13.6. Besides being more mathematically tractable without loosing accuracy, the two-ball approximation allows us to draw some interesting conclusions about the connectivity potential of mmWave cellular networks. In particular:

1. If the BS-to-MT distance r is less than (about) 50 m, that is, $r < D_1$, no link outage occurs. In other words, a link can be either in LOS or in NLOS. Also, the probability of being in LOS is greater than 80%.
2. If the BS-to-MT distance r is greater than (about) 50 m but less than (about) 200 m, that is, $r \in [D_1, D_2]$, a link can be in any of the three possible states. Also, most likely, the MT is served by a NLOS BS.
3. If the BS-to-MT distance r is greater than (about) 200 m, that is, $r > D_2$, a link is most likely to be in outage: no communication between BS and MT is possible.
4. The distance D_2 identifies a critical operating regime, which is specific to mmWave systems and that is not observed at µWave frequencies that are characterized by a two-state link model. It is worth noting that D_2 is

Table 13.1 Three-State Link and Path-Loss Models From Akdeniz et al. [2, Table I] and Corresponding Two-Ball Approximation Obtained by Using the Algorithm Described in Section 13.3.1, by Setting $x_m = 70$ dB and $x_M = 300$ dB

Carrier Frequency (F_c)	Three-State Link and Path-Loss Models ([2, Table I], Eqs. (13.3), (13.4))	Two-Ball Approximation
28 GHz	$\alpha_{LOS} = 61.4$ dB, $\beta_{LOS} = 2$	$D_1 = 56.9945$, $D_2 = 201.4371$
	$\alpha_{NLOS} = 72$ dB, $\beta_{NLOS} = 2.92$	$q_{LOS}^{[0,D_1]} = 0.8282$, $q_{NLOS}^{[0,D_1]} = 0.1718$
	$\delta_{LOS} = 1/67.1$, $\gamma_{LOS} = 1$	$q_{LOS}^{[D_1,D_2]} = 0.1216$, $q_{NLOS}^{[D_1,D_2]} = 0.7424$
	$\delta_{OUT} = 1/30$, $\gamma_{OUT} = \exp(5.2)$	$q_{LOS}^{[D_2,\infty]} = 0$, $q_{NLOS}^{[D_2,\infty]} = 0$
73 GHz	$\alpha_{LOS} = 69.8$ dB, $\beta_{LOS} = 2$	$D_1 = 53.6287$, $D_2 = 195.3275$
	$\alpha_{NLOS} = 82.7$ dB, $\beta_{NLOS} = 2.69$	$q_{LOS}^{[0,D_1]} = 0.8670$, $q_{NLOS}^{[0,D_1]} = 0.1330$
	$\delta_{LOS} = 1/67.1$, $\gamma_{LOS} = 1$	$q_{LOS}^{[D_1,D_2]} = 0.1339$, $q_{NLOS}^{[D_1,D_2]} = 0.7889$
	$\delta_{OUT} = 1/30$, $\gamma_{OUT} = \exp(5.2)$	$q_{LOS}^{[D_2,\infty]} = 0$, $q_{NLOS}^{[D_2,\infty]} = 0$

The probabilities of being in an outage state are, by definition, $q_{OUT}^{[0,D_1]} = 1 - q_{LOS}^{[0,D_1]} - q_{NLOS}^{[0,D_1]}$, $q_{OUT}^{[D_1,D_2]} = 1 - q_{LOS}^{[D_1,D_2]} - q_{NLOS}^{[D_1,D_2]}$, and $q_{OUT}^{[D_2,\infty]} = 1 - q_{LOS}^{[D_2,\infty]} - q_{NLOS}^{[D_2,\infty]}$.

approximately equal to 200 m, which is in agreement with the conclusions drawn by Rappaport et al. [1] and Akdeniz et al. [2].

5. The link state probabilities originating from the two-ball approximation in Table 13.1 provide useful guidelines on how to choose the average cell radius of mmWave systems. Radii of the order of 5 are expected to guarantee a very good connectivity, at the cost of a denser deployment. Radii larger than 200 m, on the other hand, are expected to be too big for establishing a sufficiently reliable connection between BS and MT.

6. Table 13.1 shows that the connectivity properties of mmWave networks operating at 28 and 73 GHz are similar. This is in agreement with the conclusions drawn in by Rappaport et al. [1] and Akdeniz et al. [2].

13.3.2 COMMUNICATION BLOCKAGE PROBABILITY

As mentioned in Remark 13.1, the peculiarity of the three-state link model in Section 13.2.3 is the presence of communication blockages if no BSs are available for serving the MT. The following lemma provides the probability that this event occurs.

Lemma 13.4. *The probability* $\mathcal{P}_{\text{blockage}} = \Pr\{\Psi_{\text{LOS}} = \emptyset \cap \Psi_{\text{NLOS}} = \emptyset\}$ *that a communication blockage occurs is* $\mathcal{P}_{\text{blockage}} = \exp\left(-\Lambda_{\text{blockage}}\right)$, *where:*

$$
\begin{aligned}
\Lambda_{\text{blockage}} = {} & \pi\lambda\left(\delta_{\text{OUT}}^{-1}\ln\left(\gamma_{\text{OUT}}\right)\right)^2 \\
& + 2\pi\lambda\delta_{\text{OUT}}^{-2}\gamma_{\text{OUT}}\left(\gamma_{\text{OUT}}^{-1} + \gamma_{\text{OUT}}^{-1}\ln\left(\gamma_{\text{OUT}}\right)\right)
\end{aligned}
\tag{13.15}
$$

Proof. See Appendix 13.A. □

The communication blockage probability $\mathcal{P}_{\text{blockage}} = 0$ if $\delta_{\text{OUT}} = 0$, that is, $p_{\text{OUT}}(r) = 0$ in Eq. (13.3). In practice, based on Table 13.1, this operating regime emerges if the average cell radius is of the order of 50 m, that is, if the cellular network is sufficiently dense. In general, thus, let $\mathcal{P}_{\text{LOS}}$ and $\mathcal{P}_{\text{NLOS}}$ be the probabilities that the MT is served by an LOS and an NLOS BS, respectively, we have $\mathcal{P}_{\text{LOS}} + \mathcal{P}_{\text{NLOS}} + \mathcal{P}_{\text{blockage}} = 1$ and $\mathcal{P}_{\text{LOS}} + \mathcal{P}_{\text{NLOS}} \leq 1$. This implies that the coverage probability may be zero even for $T = 0$. This occurs if $\mathcal{P}_{\text{LOS}} = \mathcal{P}_{\text{NLOS}} = 0$ and $\mathcal{P}_{\text{blockage}} = 1$. A similar comment applies to the average rate. By direct inspection of Eq. (13.15), this occurs if $\delta_{\text{OUT}} \to +\infty$, which corresponds to $p_{\text{LOS}}(r) = p_{\text{NLOS}}(r) = 0$ and $p_{\text{OUT}}(r) = 1$ in Eq. (13.3). As discussed in Section 13.3.1, there is a critical distance where this operating regime emerges, which corresponds to 200 m for the considered mmWave channel model.

13.4 MODELING COVERAGE AND RATE: NOISE-LIMITED APPROXIMATION

In this section, we introduce a mathematical framework for computing coverage and rate of mmWave cellular networks by assuming that the other-cell interference is negligible compared to the noise. In this case, in particular, $\text{SINR} \approx \text{SNR} = U^{(0)}/\sigma_N^2$. As remarked in Akdeniz et al., Singh et al., and Thomas and Vook [2, 20, 21], the noise-limited approximation is expected to hold for well-designed mmWave cellular networks, which use a transmission bandwidth of the order of gigahertz.

From Eq. (13.5), $U^{(0)} = PG^{(0)}\left|h_s^{(0)}\right|^2/L^{(0)}$, where $s = \text{LOS}$ or $s = \text{NLOS}$ if the MT is served by an LOS or an NLOS BS, respectively, and $U^{(0)} = 0$ if a communication blockage occurs. By assuming that the network is noise-limited, the $\text{SINR} \approx \text{SNR}$ can be formulated as follows:

$$
\text{SNR} \stackrel{(a)}{=} \frac{PG^{(0)}\left|h_{\text{LOS}}^{(0)}\right|^2}{\sigma_N^2 L^{(0)}}\delta\left\{L^{(0)} - L_{\text{LOS}}^{(0)}\right\} + \frac{PG^{(0)}\left|h_{\text{NLOS}}^{(0)}\right|^2}{\sigma_N^2 L^{(0)}}\delta\left\{L^{(0)} - L_{\text{NLOS}}^{(0)}\right\}
\tag{13.16}
$$

where (a) takes into account that the distribution of LOS and NLOS links is different.

Proposition 13.1. *Let the SNR in Eq. (13.16). The coverage probability in Eq. (13.6) can be formulated as follows:*

$$P_N^{(cov)}(T) = P_{LOS}^{(cov)}(T) + P_{NLOS}^{(cov)}(T)$$

$$P_s^{(cov)}(T) = \frac{1}{2}\int_0^{+\infty} \text{erfc}\left(\frac{\ln(Tx/\gamma^{(0)}) - \mu_s}{\sqrt{2}\sigma_s}\right)\dot{\Lambda}_s([0,x))\exp(-\Lambda_\Phi([0,x)))dx$$

$$\approx \frac{1}{2}\int_0^{+\infty} \text{erfc}\left(\frac{\ln(Tx/\gamma^{(0)}) - \mu_s}{\sqrt{2}\sigma_s}\right)\dot{\Lambda}_s^{(approx)}([0,x))\exp\left(-\Lambda_\Phi^{(approx)}([0,x))\right)dx$$

(13.17)

where $s = \{LOS, NLOS\}$, $\gamma^{(0)} = PG^{(0)}/\sigma_N^2$, $\mu_s = \mu_s^{dB}\ln(10)/10$, $\sigma = \sigma_s^{dB}\ln(10)/10$, *and* $\Lambda_\Phi^{(approx)}([0,x)) = \Lambda_{LOS}^{(approx)}([0,x)) + \Lambda_{NLOS}^{(approx)}([0,x))$, *with* $\Lambda_{LOS}^{(approx)}(\cdot)$, $\Lambda_{NLOS}^{(approx)}(\cdot)$ *being defined in Lemma 13.3*, $\dot{\Lambda}_s^{(approx)}([0,x))$ *is the first derivative of* $\Lambda_s^{(approx)}([0,x))$ *which can be formulated as follows:*

$$\dot{\Lambda}_s^{(approx)}(x) = \frac{d}{dx}\Lambda_s^{(approx)}([0,x))$$

$$= \frac{2}{\beta_s}\mathcal{G}_s^{(1)}x^{2/\beta_s-1}\bar{\mathcal{H}}\left(x - (\kappa_s D_1)^{\beta_s}\right) + \frac{2}{\beta_s}\mathcal{G}_s^{(2)}x^{2/\beta_s-1}\bar{\mathcal{H}}\left(x - (\kappa_s D_2)^{\beta_s}\right)$$

(13.18)

$$+ \frac{2}{\beta_s}\mathcal{G}_s^{(6)}x^{2/\beta_s-1}\mathcal{H}\left(x - (\kappa_s D_2)^{\beta_s}\right)$$

Proof. See Appendix 13.B. □

Proposition 13.1 provides an exact single-integral expression of the coverage. In particular, the two-ball approximation in Section 13.3.1, which is obtained by replacing $\Lambda_{LOS}(\cdot)$ and $\Lambda_{NLOS}(\cdot)$ with $\Lambda_{LOS}^{(approx)}(\cdot)$ and $\Lambda_{NLOS}^{(approx)}(\cdot)$, respectively, is used. The rate can be computed from Eq. (13.7), for example, by using the GCQ formulation.

Remark 13.3. By direct inspection of, for example, Eq. (13.17), it follows that coverage and rate increase as P, $G^{(0)}$ and λ increase. They decrease, on the other hand, as σ_N^2 increases. In the noise-limited regime, thus, the performance of mmWave cellular networks improves by increasing the transmit power, the directivity gain of the intended link and the density of BSs.

13.5 MODELING COVERAGE AND RATE: ACCURATE MODELING OF THE OTHER-CELL INTERFERENCE

In Section 13.4, the analytical formulation of coverage and rate is based on a noise-limited approximation for mmWave cellular networks. This approximation is expected to be accurate for well-designed mmWave cellular networks, which capitalize on a large transmission bandwidth and on a high directional beamforming gain. In general, however, the noise-limited approximation may be sensitive to the choice of the parameters. The generalized mathematical approach proposed in this section is applicable to those system setups where the aggregate other-cell interference may not be neglected, for example, for cell radii of the order of 50 m or less. The price to be paid is, however, a higher numerical complexity of the resulting mathematical framework. We show, nevertheless, that the resulting framework is still tractable.

Let the same assumption as in Section 13.4 hold. The SINR, $\text{SINR} = U^{(0)}/(\sigma_N^2 + I_{\text{agg}})$, can be formulated as follows:

$$
\begin{aligned}
\text{SINR} &= \frac{PG^{(0)}\left|h_{\text{LOS}}^{(0)}\right|^2}{\sigma_N^2 L^{(0)} + I_{\text{agg}}\left(L^{(0)}\right)}\delta\left\{L^{(0)} - L_{\text{LOS}}^{(0)}\right\} \\
&+ \frac{PG^{(0)}\left|h_{\text{NLOS}}^{(0)}\right|^2}{\sigma_N^2 L^{(0)} + I_{\text{agg}}\left(L^{(0)}\right)}\delta\left\{L^{(0)} - L_{\text{NLOS}}^{(0)}\right\}
\end{aligned}
\tag{13.19}
$$

where the aggregate other-cell interference can be expressed as follows:

$$
I_{\text{agg}}\left(L^{(0)}\right) = \sum_{s\in\{\text{LOS},\text{NLOS}\}}\sum_{i\in\Psi^{(\setminus 0)}}\left(PG^{(i)}\left|h_s^{(i)}\right|^2/L_s^{(i)}\right)\mathbf{1}\left(L_s^{(i)} > L^{(0)}\right)
\tag{13.20}
$$

where $\mathbf{1}(\cdot)$ is the indicator function.

The computation of the coverage probability of the SINR in Eq. (13.19) is obtained by taking the other-cell interference into account with the aid of the characteristic function (CF) of the interference, which is provided in the following lemma.

Lemma 13.5. *Let BSs be distributed according to a PPP and the MT be associated to the BS providing the smallest path-loss. The CF of the aggregate other-cell interference in Eq. (13.20) conditioned on the path-loss of the intended link, $L^{(0)}$, has closed-form expression as follows:*

$$
T(\omega, L^{(0)}) = \prod_{s\in\{\text{LOS},\text{NLOS}\}}\exp\left(\overline{\Theta}_s\left(\omega, L^{(0)}\right)\right)
\tag{13.21}
$$

where,

$$
\begin{aligned}
\overline{\Theta}_s(\omega, x) &= \frac{\omega_{\text{BS}}}{2\pi}\frac{\omega_{\text{MT}}}{2\pi}\Theta_s\left(\omega, x; G_{\text{BS}}^{(\max)}G_{\text{MT}}^{(\max)}\right) \\
&+ \frac{\omega_{\text{BS}}}{2\pi}\left(1 - \frac{\omega_{\text{MT}}}{2\pi}\right)\Theta_s\left(\omega, x; G_{\text{BS}}^{(\max)}G_{\text{MT}}^{(\min)}\right) \\
&+ \left(1 - \frac{\omega_{\text{BS}}}{2\pi}\right)\frac{\omega_{\text{MT}}}{2\pi}\Theta_s\left(\omega, x; G_{\text{BS}}^{(\min)}G_{\text{MT}}^{(\max)}\right) \\
&+ \left(1 - \frac{\omega_{\text{BS}}}{2\pi}\right)\left(1 - \frac{\omega_{\text{MT}}}{2\pi}\right)\Theta_s\left(\omega, x; G_{\text{BS}}^{(\min)}G_{\text{MT}}^{(\min)}\right)
\end{aligned}
\tag{13.22}
$$

$$\Theta_s(\omega,x;g) = \mathcal{G}_s^{(1)} F_{\beta_s}(\omega \mathcal{P}g,x)\bar{\mathcal{H}}\left(x - (\kappa_s D_1)^{\beta_s}\right)$$

$$- \mathcal{G}_s^{(1)} F_{\beta_s}\left(\omega \mathcal{P}g,(\kappa_s D_1)^{\beta_s}\right)\bar{\mathcal{H}}\left(x - (\kappa_s D_1)^{\beta_s}\right)$$

$$+ \mathcal{G}_s^{(2)} F_{\beta_s}(\omega \mathcal{P}g,x)\bar{\mathcal{H}}\left(x - (\kappa_s D_2)^{\beta_s}\right) \tag{13.23}$$

$$- \mathcal{G}_s^{(2)} F_{\beta_s}\left(\omega \mathcal{P}g,(\kappa_s D_2)^{\beta_s}\right)\bar{\mathcal{H}}\left(x - (\kappa_s D_2)^{\beta_s}\right)$$

$$+ \mathcal{G}_s^{(6)} F_{\beta_s}\left(\omega \mathcal{P}g, \max\left\{x,(\kappa_s D_2)^{\beta_s}\right\}\right)$$

$$F_{\beta_s}(B,A) = A^{2/\beta_s}\left[1 - \frac{1}{\sqrt{\pi}}\sum_{k=1}^{N_{\text{GHQ}}}\tilde{\omega}_{k1} F_1\left(-\frac{2}{\beta_s}, 1 - \frac{2}{\beta_s}, j\frac{B}{A}10^{\frac{\sqrt{2}\sigma_s \tilde{s}_k + \mu_s}{10}}\right)\right] \tag{13.24}$$

and $_1F_1(\cdot;\cdot;\cdot)$ be the confluent hypergeometric function, and $\tilde{\omega}_k$ and $\tilde{s}_k$ are the weights and abscissas factors of N_{GHQ}th order Hermite polynomial.

Proof. See Appendix 13.C. □

From the CF of the other-cell interference, the coverage probability in the presence of interference is formulated in the following theorem.

Theorem 13.1. *Consider a cell association based on the smallest path-loss. Let the SINR be formulated as in Eq. (13.19). Let $j = \sqrt{-1}$ be the imaginary unit and $\text{Im}\{\cdot\}$ be the imaginary part operator. By using the two-ball approximation of Section 13.3.1, the coverage probability, that is, $P^{(\text{cov})}(T) = \Pr\{\text{SINR} > T\}$ can be formulated as follows:*

$$P^{(\text{cov})}(T) \approx P_N^{(\text{cov})}(T) + P_{NI}^{(\text{cov})}(T) \tag{13.25}$$

where $P_N^{(\text{cov})}(T)$ is the coverage probability in the noise-limited regime, which is given in Proposition 13.1, *and $P_{NI}^{(\text{cov})}$, which provides the impact of the other-cell interference, is defined as follows:*

$$P_{NI}^{(\text{cov})}(T) = \sum_{s\in\{\text{LOS},\text{NLOS}\}}\frac{1}{\pi}\int_0^\infty\int_0^\infty\frac{1}{\omega}\text{Im}\left\{\exp\left(j\omega\sigma_N^2\right)\Theta_s(T,\omega,x)(1 - T(\omega,x))\right\}$$
$$\times \dot{\Lambda}_s^{(\text{approx})}([0,x))\exp\left(-\Lambda_\Phi^{(\text{approx})}([0,x))\right)dx \tag{13.26}$$

where the following definitions hold:

$$\Theta_s(T,\omega,x) = \frac{1}{\sqrt{\pi}}\sum_{k=1}^{N_{\text{GHQ}}}\tilde{\omega}_k\exp\left(-j\omega\frac{\mathcal{P}G^{(0)}}{Tx}10^{\frac{\sqrt{2}\sigma_s \tilde{s}_k + \mu_s}{10}}\right) \tag{13.27}$$

The coverage probability in Theorem 13.1 is formulated as the summation of the coverage probability in the noise-limited scenario and a correction term that depends on the aggregate other-cell interference. As result, (1) Theorem 13.1 reduces to Proposition 13.1 if the noise-limited approximation is invoked, that is, $T(\omega,x)=1$, and (2) $P_{NI}^{(\text{cov})}(\cdot)$ is expected to be small in mmWave cellular networks, except in dense network deployments.

13.6 NUMERICAL AND SIMULATION RESULTS

In this section, we illustrate some numerical examples to validate the accuracy of the proposed mathematical frameworks and to compare mmWave and μWave cellular networks. The frameworks are substantiated with the aid of Monte Carlo simulations, where some modeling assumptions used for analytical tractability are not enforced in the system simulator. In Section 13.4, notably, coverage and rate are computed under the noise-limited assumption. This approximation is *not* retained in the system simulator, in order to show to what extent the noise-limited assumption holds for mmWave systems. In Section 13.6.1, in particular, the accuracy of the PPP-based approach for application to mmWave cellular networks is studied by taking into account actual BSs locations taken from the OFCOM database [28] and the footprints of the buildings taken from the ordnance survey (OS) database [29], corresponding to a dense urban environment (downtown London). In Section 13.6.2, in addition, the accuracy of the mathematical frameworks in Section 13.4 and 13.5 based on the two-ball approximation is verified against the empirical link state modeling given in Eq. (13.3).

13.6.1 EXPERIMENTAL VALIDATION OF PPP-BASED MODELING

In order to test the accuracy of the PPP-based model for the locations of the BSs, we use experimental data from the O2 telecommunication operator in the UK. Similar to Lu and Di Renzo [30], these data are obtained from OFCOM [28]. More specifically, 183 BSs are located in an area of 4 km², as illustrated in Fig. 13.1. Sixty-two out of 183 BSs lie inside a geographical region where a building is located. They are referred to as "rooftop BSs." One-hundred twenty-one of 183 BSs lie in a geographical region where no buildings are located. They are referred to as "outdoor BSs." Let $\mathcal{A}$ and $\mathcal{N}$ denote the area of the geographical region under analysis and the number of BSs in that area, respectively. The density of BSs can be computed as $\lambda_{\text{BS}} = \mathcal{N}/\mathcal{A}$.

To take realistic blockages into account, that is, LOS and NLOS links originating from the locations and the shapes of the buildings, we use experimental data related to the actual deployment of buildings that corresponds to the metropolitan areas of Fig. 13.1. Similar to Lu and Di Renzo [30], these data are obtained from OS [29]. The consistency of the data from the two independent web sites of OFCOM and OS has been verified with the aid of Google maps. In our study, the elevation of the buildings is not considered, since these data are not available from the database.

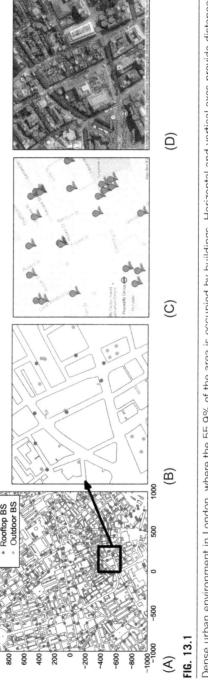

FIG. 13.1

Dense urban environment in London, where the 55.9% of the area is occupied by buildings. Horizontal and vertical axes provide distances expressed in meters. The points show the positions of BSs from all telecom operators in London. (A) Entire region under analysis. (B) Magnification of a smaller region. (C) Google map view of (B). (D) Satellite view of (B).

The presence of buildings in dense urban environments constitutes an inherent source of blockages, which results in LOS and NLOS links. Based on the locations of the BSs and on the locations and shapes of the buildings, respectively, LOS and NLOS links can be empirically identified. To this end, the following criterion is used. Consider a generic "outdoor BS" and a generic MT. The related link is in LOS if no building is intersected by connecting the BS and the MT with a straight line. Otherwise, the link is in NLOS. Consider a generic "rooftop BS." The BS-to-MT links are assumed to be in NLOS. This is a simplifying assumption, used by other researchers as well [22], which seems to be acceptable if no information on the elevation of the buildings is available.

Channel model, directional beamforming, and the cell association criterion are assumed to be those introduced in Section 13.2. The following setup is considered:

- The transmission bandwidth is $B_W = 2$ GHz. The noise power is $\sigma_N^2 = -174 + 10 \log_{10}(B_W) + \mathcal{F}_{dB}$, where $\mathcal{F}_{dB} = 10$ dB is the noise figure.
- The directional beamforming model is as follows: $G_{BS}^{(max)} = G_{MT}^{(max)} = 20$ dB, $G_{BS}^{(min)} = G_{MT}^{(min)} = -10$ dB, and $\phi_{BS} = \phi_{MT} = 30$ degrees.
- The shadowing model is as follows [2, Table I]: $\sigma^{(LOS)} = 5.8$ dB, $\sigma^{(NLOS)} = 8.7$ dB if $F_c = 28$ GHz and $\sigma^{(LOS)} = 5.8$ dB, $\sigma^{(LOS)} = 7.7$ dB if $F_c = 73$ GHz. On the other hand, $\mu^{(S)}$ is assumed to be equal to 0 dB for both LOS and NLOS scenarios.
- The path-loss constants, $\kappa_s = 10^a \alpha_s/(10\beta_s)$, and the path-loss exponents, β_s, for $s \in \{LOS, NLOS\}$ are those reported in Table 13.1 for transmission at 28 GHz and 73 GHz.

In addition, the performance of μWave and mmWave cellular networks are compared as well. To this end, the following setup for μWave cellular networks is considered [2]: $F_c = 2.5$ GHz, $B_W = 40$ MHz, $G_{MT}^{(max)} = G_{MT}^{(min)} = 0$ dB. The channel model is chosen as in Akdeniz et al. [2, Eq. (11)], that is, $l(r)^{(dB)} = 22.7 + 36.7 \log_{10}(r) + 26 \log_{10}(2.5)$. All the channels are assumed to be in NLOS, with a shadowing standard deviation equal to $\sigma^{(NLOS)} = 4$ dB. The rest of the parameters are the same as for mmWave cellular networks.

In Fig. 13.2, we observe that the PPP-based abstraction model is quite accurate for modeling practical BSs deployments in dense urban areas. The figure also shows that mmWave cellular networks operating at $F_c = 28$ GHz slightly outperform their counterparts operating at $F_c = 73$ GHz due to the smaller path-loss.

In Fig. 13.3, mmWave and μWave cellular networks are compared. This figure shows that mmWave systems are capable of outperforming μWave systems in the considered dense urban area of London, without the need for an increase in the density of BSs.

13.6.2 VALIDATION OF THE NOISE-LIMITED APPROXIMATION

In this section, we illustrate some numerical examples to validate the accuracy of the mathematical frameworks in Sections 13.4 and 13.5, and to compare mmWave and μWave cellular networks. The frameworks are substantiated with the aid of Monte

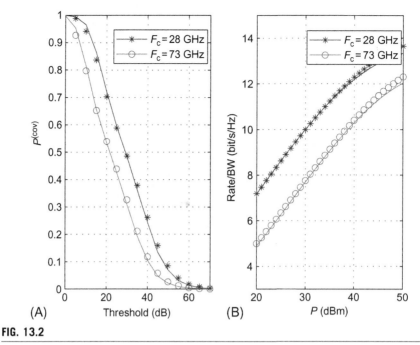

FIG. 13.2

(A) Coverage probability of a mmWave cellular network. (B) Average rate of a mmWave cellular network normalized by the transmission bandwidth. Markers show Monte Carlo simulations for empirical BSs locations (OFCOM) and a practical blockage model (OS). Solid lines show the simulations where the location of BSs follow a PPP with the same density and the blockage model is still obtained from OS. The transmit power of the BSs is $P = 30$ dBm.

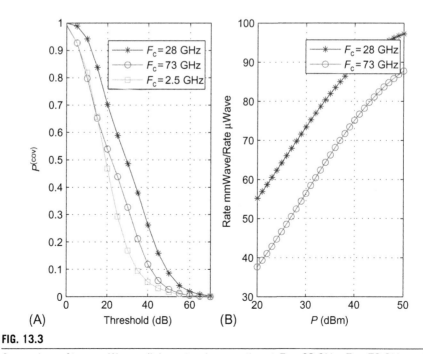

FIG. 13.3

Comparison of two mmWave cellular networks operating at $F_c = 28$ GHz, $F_c = 73$ GHz and a μWave cellular network operating at $F_c = 2.5$ GHz: (A) Coverage probability. (B) Ratio of the average rates of two mmWave networks and of the μWave network. The transmit power of the BSs is $P = 30$ dBm.

Carlo simulations, which are obtained by using the system simulator described in Di Renzo et al. (Jul. 2013), Di Renzo and Lu (May 2014, Jan. 2015, Mar. 2015), Di Renzo and Guan (Aug. 2014, Sep. 2014) [8–13].

The considered setup is the same as that in Section 13.6.1, except for the blockage model that is obtained from Akdeniz et al. [2, Table I]. In particular, we have: $\delta_{LOS} = 1/67.1$, $\gamma_{LOS} = 1$, $\delta_{OUT} = 1/30$, $\gamma_{OUT} = \exp(5.2)$, for both $F_c = 28$ GHz and $F_c = 73$ GHz. Also, the BSs are assumed to be distributed according to a PPP.

As far as the mathematical frameworks are concerned, the frameworks in Proposition 13.1 (noise-limited approximation) and Theorem 13.1 are used. In particular, all curves are generated by using the two-ball approximation.

Selected numerical results are illustrated in Figs. 13.4–13.8. We observe that the noise-limited approximation is quite accurate for the considered setup. If $R_c \geq$ 100 m, in particular, we observe that mmWave cellular networks can be assumed to be noise-limited. If the density of BSs increases, on the other hand, this approximation may no longer hold. The performance gap compared to Monte Carlo simulations is, however, tolerable and this shows that, in any case, mmWave cellular networks are likely not to be interference-limited. This finding is in agreement with recently published papers that consider a simplified blockage model [20]. The figures also show

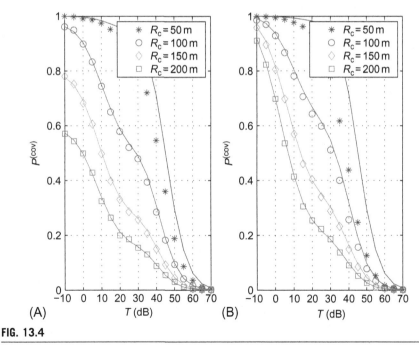

FIG. 13.4

Coverage probability of a mmWave cellular network at $F_c = 28$ GHz. (A) $p_{OUT}(\cdot)$ in Eq. (13.3). (B) $p_{OUT}(r) = 0$. *Solid lines*: mathematical framework based on the noise-limited approximation. *Markers*: Monte Carlo simulations.

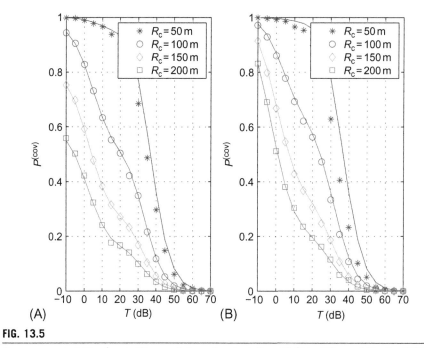

FIG. 13.5

Coverage probability of a mmWave cellular network at $F_c = 73$ GHz. (A) $p_{OUT}(\cdot)$ in Eq. (13.3). (B) $p_{OUT}(r) = 0$. *Solid lines*: mathematical framework based on the noise-limited approximation. *Markers*: Monte Carlo simulations.

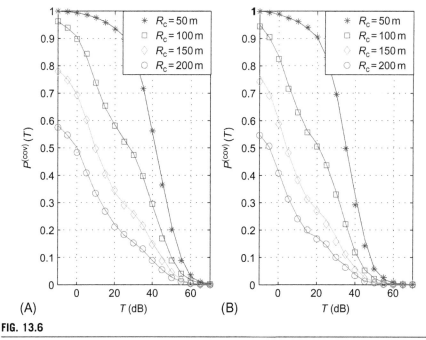

FIG. 13.6

Coverage probability of a mmWave cellular network at $F_c = 28$ GHz (A) and $F_c = 73$ GHz (B). $p_{OUT}(\cdot)$ in Eq. (13.3) is used. *Solid lines*: mathematical framework that takes the other-cell interference into account. *Markers*: Monte Carlo simulations.

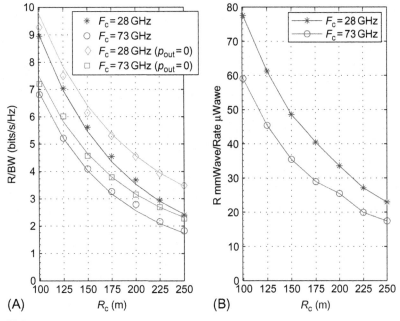

FIG. 13.7

Average rate of a mmWave cellular network at $F_c = 28$ GHz and $F_c = 73$ GHz. (A) The normalized rate R/B_W is shown. (B) Ratio of the average rates of two mmWave networks at $F_c = 28$ GHz and $F_c = 73$ GHz and of a μWave network at $F_c = 2.5$ GHz.

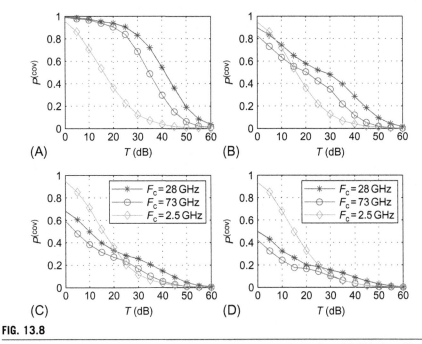

FIG. 13.8

Coverage probability of mmWave and μWave cellular networks at $F_c = 28$ GHz (mmWave), $F_c = 73$ GHz (mmWave) and $F_c = 2.5$ GHz (μWave). As for mmWave networks, $p_{OUT}(\cdot)$ in Eq. (13.3) is used. As for the μWave cellular network, $p_{OUT}(r) = 0$. (A) $R_c = 50$ m. (B) $R_c = 100$ m. (C) $R_c = 150$ m. (D) $R_c = 200$ m.

that, in general, the presence of an outage state reduces the coverage. This is noticeable, in particular, for small values of the reliability threshold T. Furthermore, as expected, the performance gets better as the average cell radius R_c decreases, that is, for denser network deployments. Some figures deserve additional comments.

In Fig. 13.7 the rate of mmWave and μWave networks is compared. The figures confirm that mmWave networks are capable of significantly enhancing the average rate. This is mainly due to the larger transmission bandwidth, which is 50 times larger, in the considered setup, for mmWave systems. The figures show, however, that the gain can be larger than the ratio of the bandwidths, especially for medium/dense cellular deployments.

In Fig. 13.8 mmWave and μWave cellular networks are compared in terms of the coverage probability. This figure shows that mmWave systems have the potential of providing a better coverage than μWave systems, provided that the network density is sufficiently high. Otherwise, μWave systems are still to be preferred, especially for small values of the reliability threshold T. As expected, mmWave transmission at $F_c = 28$ GHz slightly outperforms its counterpart at $F_c = 73$ GHz due to the smaller path-loss.

13.7 CONCLUSION

In this chapter, a new analytical framework for computing coverage and rate of mmWave cellular networks has been proposed. Its novelty lies in taking into account realistic channel and blockage models for mmWave propagation, which are based on empirical data. A two-ball approximation for modeling the link-state of mmWave communications is introduced, which is based on matching the intensity measures of the PPPs of empirical three-state and approximated two-ball link models. The proposed mathematical methodology based on the noise-limited approximation of mmWave cellular systems has been shown to be sufficiently accurate for typical densities of BSs and for transmission bandwidths of the order of gigahertz. For those system setups where the aggregate other-cell interference may not be totally neglected, for example, for cell radii of the order of 50 m or less, the mathematical approach has been generalized by taking into account the distribution of other-cell interference. Numerical examples have confirmed that sufficiently dense mmWave cellular networks have the capability of outperforming their μWave counterpart. These findings have been validated with the aid of experimental data for the locations of cellular BSs and blockages in a dense urban environment.

APPENDIX

13.A PROOFS OF THE RESULTS IN SECTION 13.3

Proof of Lemma 13.1

The proof follows by using a methodology similar to Blaszczyszyn et al. [27, Section II-A]. In particular, by invoking the displacement theorem of PPPs [23, Theorem 1.10], the process of the propagation losses $\Phi = \left\{ l\left(r^{(n)}\right), n \in \Psi \right\}$ can be interpreted as a transformation of Ψ, which is still a PPP on $\mathbb{R}^+$. From Section 13.2.3, we know that $\Psi = \Psi_{LOS} \cup \Psi_{NLOS} \cup \Psi_{OUT}$. Since Ψ_{LOS}, Ψ_{NLOS} and Ψ_{OUT} are independent, the density (or intensity), $\Lambda_\Phi(\cdot)$, of $\Phi = \left\{ l\left(r^{(n)}\right), n \in \Psi \right\}$ is equal to the summation of the intensities of Φ_{LOS}, Φ_{NLOS} and Φ_{OUT}. Since the path-loss of the links in outage is infinite, its intensity, by definition, is equal to zero. The intensities, $\Lambda_{LOS}(\cdot)$ and $\Lambda_{NLOS}(\cdot)$ of Φ_{LOS} and Φ_{NLOS}, respectively, on the other hand, can be computed by using mathematical steps similar to the proof of Blaszczyszyn et al. [27, Lemma 1]. More specifically:

$$\Lambda_s([0,x)) = 2\pi\lambda \int_0^{+\infty} \mathcal{H}\left(x - (\kappa_s r)^{\beta_s}\right) p_s(r) r dr \qquad (13.28)$$

where $p_s(\cdot)$ for $s = \{LOS, NLOS\}$ is defined in Eq. (13.3).

Eq. (13.8) follows by inserting $p_s(\cdot)$ of Eq. (13.3) in Eq. (13.28) and by computing the integrals with the aid of the notable result $\int_a^b e^{-cr} r dr = (1/c^2)$ $\left(e^{-ca} + ae^{-ca} - e^{-cb} - be^{-cb}\right)$.

Proof of Lemma 13.4

Since Ψ_{LOS} and Ψ_{NLOS} are independent, the following equality holds:

$$\mathcal{P}_{blockage} = \Pr\{\Psi_{LOS} = \emptyset \cap \Psi_{NLOS} = \emptyset\} = \Pr\{\Psi_{LOS} = \emptyset\} \Pr\{\Psi_{NLOS} = \emptyset\} \qquad (13.29)$$

From the void probability theorem of PPPs [23], we have,

$$\Pr\{\Psi_s = \emptyset\} = \exp\left(-2\pi\lambda \int_0^{+\infty} p_s(r) r dr\right) \qquad (13.30)$$

for $s = \{LOS, NLOS\}$, where $p_s(\cdot)$ is defined in Eq. (13.3). The integral in Eq. (13.30) can be computed in closed-form from Eq. (13.28) by letting $x \to +\infty$. The proof follows with the aid of some simplifications. Alternatively, the proof may be obtained directly from Lemma 13.1. By definition of communication blockage, in fact, the equalities $\mathcal{P}_{blockage} = \Pr\left\{\Phi^{(0)} = +\infty\right\} = \Pr\left\{L^{(0)} \geq +\infty\right\} = 1 - F_{L^{(0)}}(x \to +\infty)$ hold.

13.B PROOFS OF THE RESULTS IN SECTION 13.4

Proof of Proposition 13.1

From Eqs. (13.6) and (13.16), the coverage can be formulated, by definition, as follows:

$$P^{(cov)}(T) = \mathbb{E}_{L_{LOS}^{(0)}}\left\{ \Pr\left\{ \frac{PG^{(0)}\left|h_{LOS}^{(0)}\right|^2}{\sigma_N^2 L_{LOS}^{(0)}} > T \middle| L_{LOS}^{(0)} \right\} \Pr\left\{ L_{NLOS}^{(0)} > L_{LOS}^{(0)} \middle| L_{LOS}^{(0)} \right\} \right\}$$

$$+ \mathbb{E}_{L_{NLOS}^{(0)}}\left\{ \Pr\left\{ \frac{PG^{(0)}\left|h_{NLOS}^{(0)}\right|^2}{\sigma_N^2 L_{NLOS}^{(0)}} > T \middle| L_{NLOS}^{(0)} \right\} \Pr\left\{ L_{LOS}^{(0)} > L_{NLOS}^{(0)} \middle| L_{NLOS}^{(0)} \right\} \right\}$$

$$(13.31)$$

Denote the first and second addends in Eq. (13.31) by $P_s^{(\text{cov})}(\cdot)$, where $s = \text{LOS}$ and $s = \text{NLOS}$, respectively. Both addends can be computed by using the following notable results:

$$\Pr\left\{\left|h_s^{(0)}\right|^2 > L_s^{(0)}T/\gamma^{(0)}\Big|L_s^{(0)}\right\}$$

$$\overset{(a)}{=} 1/2 - (1/2)\text{erf}\left(\left(\ln\left(L_s^{(0)}T/\gamma^{(0)}\right) - \mu_s\right)/\left(\sqrt{2}\sigma_s\right)\right)$$

$$\Pr\left\{L_{\text{NLOS}}^{(0)} > L_{\text{LOS}}^{(0)}\Big|L_{\text{LOS}}^{(0)}\right\} \overset{(b)}{=} \exp\left(-\Lambda_{\text{NLOS}}\left(\left[0,L_{\text{LOS}}^{(0)}\right)\right)\right)$$

$$\Pr\left\{L_{\text{LOS}}^{(0)} > L_{\text{NLOS}}^{(0)}\Big|L_{\text{NLOS}}^{(0)}\right\} \overset{(c)}{=} \exp\left(-\Lambda_{\text{LOS}}\left(\left[0,L_{\text{NLOS}}^{(0)}\right)\right)\right) \qquad (13.32)$$

where (a) follows from the CDF of lognormal distribution, and (b) and (c) follow from Lemma 13.1, Lemma 13.2, since $L^{(0)} = \Phi^{(0)} = \min\left\{L_{\text{LOS}}^{(0)}, L_{\text{NLOS}}^{(0)}\right\}$, $L_{\text{LOS}}^{(0)} = \Phi_{\text{LOS}}^{(0)}$ and $L_{\text{NLOS}}^{(0)} = \Phi_{\text{NLOS}}^{(0)}$. The proof follows by writing the expectation with respect to $L_{\text{LOS}}^{(0)}$ and $L_{\text{NLOS}}^{(0)}$ in terms of their PDFs, which is, similar to (b) and (c), equal to $f_{L_s^{(0)}}(\xi) = d\Pr\left\{L_s^{(0)} < \xi\right\}/d\xi = \dot{\Lambda}_s([0,\xi))\exp\left(-\Lambda_s([0,\xi))\right)$, since $\Pr\left\{L_s^{(0)} < \xi\right\} = \exp\left(-\Lambda_s([0,\xi))\right)$. The rest of the proof follows by representing $\dot{\Lambda}_s([0,\xi))$ and $\Lambda_s([0,\xi))$ with $\dot{\Lambda}_s^{(\text{approx})}([0,\xi))$ and $\Lambda_s^{(\text{approx})}([0,\xi))$, respectively.

13.C PROOFS OF THE RESULTS IN SECTION 13.5

Proof of Lemma 13.5

The conditional CF of the aggregate other-cell interference is computed from its definition as follows:

$$CF(\omega, L^{(0)}) = \mathbb{E}\left\{\exp\left(j\omega I_{\text{agg}}\left(L^{(0)}\right)\right)\right\}$$

$$\overset{(a)}{=} \prod_{s\in\{\text{LOS},\text{NLOS}\}} \mathbb{E}\left[\exp\left(j\omega \sum_{i\in\Phi_s^{(0)}} \frac{PG^{(i)}h_s^{(i)}}{L^{(i)}}\mathbf{1}\left(L^{(i)} > L^{(0)}\right)\right)\right] \qquad (13.33)$$

$$\overset{(b)}{=} \prod_{s\in\{\text{LOS},\text{NLOS}\}} \exp\left(\mathbb{E}\left\{\int_{L^{(0)}}^{\infty}\left(e^{j\omega\frac{PG^{(i)}h_s^{(i)}}{x}} - 1\right)\dot{\Lambda}_s(x)dx\right\}\right)$$

where (a) is obtained from the independence assumption of LOS and NLOS links, (b) follows from the Probability Generating Functional (PGFL) theorem of PPPs [23, Proposition 1.2.2], and $\dot{\Lambda}_s^{\text{approx}}(x)$ is the first-order derivative of the intensity $\Lambda_s^{\text{approx}}(x)$ given in Eq. (13.18).

The rest of the proof follows by computing the integral in Eq. (13.33) with the aid of the notable result as follows:

$$\int_A^{\infty}\left(\exp\left(-jKx^{-1} - 1\right)\right)x^{2/\beta-1}dx = (\beta/2)A^{2/\beta}\left(1 - {}_1F_1(-2/\beta, 1 - 2/\beta, jK/A)\right) \qquad (13.34)$$

as well as by computing the expectation with respect to the shadowing with the aid of Gaussian-Hermite Quadrature rule, and the expectation with respect to the antenna array gain by using Eq. (13.2).

Proof of Theorem 13.1

From the SINR in Eq. (13.19) and from the total probability theorem, the coverage probability in the presence of other-cell interference can be formulated as follows:

$$
P^{(\mathrm{cov})}(T) = \mathbb{E}_{L_{\mathrm{LOS}}^{(0)}} \left\{ \Pr\left\{ \frac{PG^{(0)}\left|h_{\mathrm{LOS}}^{(0)}\right|^2}{(\sigma_N^2 + I_{\mathrm{agg}})L_{\mathrm{LOS}}^{(0)}} > T \,\middle|\, L_{\mathrm{LOS}}^{(0)} \right\} \Pr\left\{ L^{(0)} = L_{\mathrm{LOS}}^{(0)} \,\middle|\, L_{\mathrm{LOS}}^{(0)} \right\} \right\}
$$
$$
+ \mathbb{E}_{L_{\mathrm{NLOS}}^{(0)}} \left\{ \Pr\left\{ \frac{PG^{(0)}\left|h_{\mathrm{NLOS}}^{(0)}\right|^2}{(\sigma_N^2 + I_{\mathrm{agg}})L_{\mathrm{NLOS}}^{(0)}} > T \,\middle|\, L_{\mathrm{NLOS}}^{(0)} \right\} \Pr\left\{ L^{(0)} = L_{\mathrm{NLOS}}^{(0)} \,\middle|\, L_{\mathrm{NLOS}}^{(0)} \right\} \right\}
\tag{13.35}
$$

The conditional coverage probability in Eq. (13.35) can be expressed in terms of the CDF of the other-cell interference, as follows, $\Pr\left\{ I_{\mathrm{agg}}\left(L_s^{(0)}\right) \le PG^{(0)}h_s^{(0)} / \left(L_s^{(0)}T\right) - \sigma_N^2 \right\}$, for $s \in \{\mathrm{LOS}, \mathrm{NLOS}\}$. By invoking the Gil-Pelaez theorem [31], it can be computed from the CF of the other-cell interference given in Lemma 13.5. Moreover, the probability that the intended link is in state s, $\Pr\left\{ L^{(0)} = L_s^{(0)} \right\}$, can be formulated as follows:

$$
\Pr\left\{ L^{(0)} = L_s^{(0)} \right\} = 1 - \Pr\left\{ L_{\hat{s}}^{(0)} \le L_s^{(0)} \right\} \stackrel{(a)}{=} \exp\left(-\Lambda_{\hat{s}}\left(\left[0, L_s^{(0)}\right]\right) \right)
\tag{13.36}
$$

where (a) is obtained from the void probability theorem of $\Phi_{\hat{s}}$ [23]. The expectation with respect to $L_s^{(0)}$ can be computed using the probability density function of $L_s^{(0)}$, which is obtained by calculating the derivative of the CDF of $L_s^{(0)}$, which, from the void probability theorem of PPPs, is $1 - \exp\left(-\Lambda_{\hat{s}}([0,x])\right)$. Finally, Eq. (13.25) follows by noting that the coverage can be split into two parts by using the equality $e^{j\omega\sigma_N^2}\mathcal{T}(\omega,x) = e^{j\omega\sigma_N^2} - e^{j\omega\sigma_N^2}(1 - \mathcal{T}(\omega,x))$. In particular, $P_{\mathrm{N}}^{(\mathrm{cov})}(T)$ and $P_{\mathrm{NI}}^{(\mathrm{cov})}(T)$ correspond to the first and second addends of this equality, respectively. The expression of $P_{\mathrm{N}}^{(\mathrm{cov})}(T)$ follows from Eq. (13.17).

REFERENCES

[1] T.S. Rappaport, S. Sun, R. Mayzus, H. Zhao, Y. Azar, K. Wang, G.N. Wong, J.K. Schultz, M. Samimi, F. Gutierrez, Millimeter wave mobile communications for 5G cellular: it will work! IEEE Access 1 (2013) 335–349.

[2] M.R. Akdeniz, Y. Liu, M.K. Samimi, S. Sun, S. Rangan, T.S. Rappaport, E. Erkip, Millimeter wave channel modeling and cellular capacity evaluation, IEEE J. Sel. Areas Commun. 32 (6) (2014) 1164–1179.

[3] S. Rangan, T. Rappaport, E. Erkip, Millimeter-wave cellular wireless networks: potentials and challenges, Proc. IEEE 102 (3) (2014) 366–385.

[4] A. Ghosh, T.A. Thomas, M.C. Cudak, R. Ratasuk, P. Moorut, F.W. Vook, T.S. Rappaport, G.R. MacCartney Jr., S. Sun, S. Nie, Millimeter wave enhanced local area systems: a high data rate approach for future wireless networks, IEEE J. Sel. Areas Commun. 32 (6) (2014) 1152–1163.

[5] J.G. Andrews, F. Baccelli, R.K. Ganti, A tractable approach to coverage and rate in cellular networks, IEEE Trans. Commun. 59 (11) (2011) 3122–3134.

[6] H. ElSawy, E. Hossain, M. Haenggi, Stochastic geometry for modeling, analysis, and design of multi-tier and cognitive cellular wireless networks: a survey, IEEE Commun. Surveys Tut. 15 (3) (2013) 996–1019.

[7] M. Di Renzo, C. Merola, A. Guidotti, F. Santucci, G.E. Corazza, Error performance of multi-antenna receivers in a Poisson field of interferers: a stochastic geometry approach, IEEE Trans. Commun. 61 (5) (2013) 2025–2047.

[8] M. Di Renzo, A. Guidotti, G.E. Corazza, Average rate of downlink heterogeneous cellular networks over generalized fading channels—a stochastic geometry approach, IEEE Trans. Commun. 61 (7) (2013) 3050–3071.

[9] M. Di Renzo, Lu. Wei, The equivalent-in-distribution (EiD)-based approach: on the analysis of cellular networks using stochastic geometry, IEEE Commun. Lett. 18 (5) (2014) 761–764.

[10] M. Di Renzo, P. Guan, A mathematical framework to the computation of the error probability of downlink MIMO cellular networks by using stochastic geometry, IEEE Trans. Commun. 62 (8) (2014) 2860–2879.

[11] M. Di Renzo, P. Guan, Stochastic geometry modeling of coverage and rate of cellular networks using the Gil-Pelaez inversion theorem, IEEE Commun. Lett. 18 (9) (2014) 1575–1578.

[12] M. Di Renzo, W. Lu, End-to-end error probability and diversity analysis of AF-based dual-hop cooperative relaying in a Poisson field of interferers at the destination, IEEE Trans. Wirel. Commun. 14 (1) (2015) 15–32.

[13] M. Di Renzo, W. Lu, Stochastic geometry modeling and performance evaluation of MIMO cellular networks using the equivalent-in-distribution (EiD)-based approach, IEEE Trans. Commun. 63 (3) (2015) 977–996.

[14] M. Di Renzo, W. Lu, On the diversity order of selection combining dual-branch dual-hop AF relaying in a Poisson field of interferers at the destination, IEEE Trans. Veh. Technol. 64 (4) (2015) 1620–1628.

[15] M. Di Renzo, Stochastic geometry modeling and analysis of multi-tier millimeter wave cellular networks, IEEE Trans. Wirel. Commun. 14 (9) (2015) 5038–5057.

[16] W. Lu, M. Di Renzo, Stochastic geometry modeling and system-level analysis & optimization of relay-aided downlink cellular networks, IEEE Trans. Wirel. Commun. 63 (11) (2015) 4063–4085.

[17] T. Bai, A. Alkhateeb, R.W. Heath Jr., Coverage and capacity of millimeter-wave cellular networks, IEEE Commun. Mag. 52 (9) (2014) 70–77.

[18] T. Bai, R. Vaze, R.W. Heath, Analysis of blockage effects on urban cellular networks, IEEE Trans. Wirel. Commun. 13 (9) (2014) 5070–5083.

[19] T. Bai, R.W. Heath Jr., Coverage and rate analysis for millimeter wave cellular networks, IEEE Trans. Wirel. Commun. 14 (2) (2015) 1100–1114.

[20] S. Singh, M.N. Kulkarni, A. Ghosh, J.G. Andrews, Tractable model for rate in self-backhauled millimeter wave cellular networks, IEEE J. Sel. Areas Commun. 33 (10) (2015) 2196–2211.

[21] T.A. Thomas, F.W. Vook, System level modeling and performance of an outdoor mmWave local area access system, in: Proceedings of IEEE Personal, Indoor, and Mobile Radio Communication (PIMRC), 2014, pp. 108–112.

[22] M.N. Kulkarni, S. Singh, J.G. Andrews, Coverage and rate trends in dense urban mmWave cellular networks, in: Proceedings of IEEE Global Communications Conference, 2014, pp. 3809–3814.

[23] F. Baccelli, B. Blaszczyszyn, Stochastic Geometry and Wireless Networks, Part I: Theory, Now Publishers, Delft, The Netherlands, 2009.

[24] G.R. MacCartney Jr., J. Zhang, S. Nie, T.S. Rappaport, Path loss models for 5G millimeter wave propagation channels in urban microcells, in: Proceedings of IEEE Global Communications Conference, 2013, pp. 3948–3953.

[25] G.R. MacCartney Jr., M.K. Samimi, T.S. Rappaport, Omnidirectional path loss models in New York City at 28 GHz and 73 GHz, in: Proceedings of IEEE International Symposium on Personal Indoor and Mobile Radio Communications, 2014, pp. 303–307.

[26] M. Abramowitz, I.A. Stegun, Handbook of Mathematical Functions With Formulas, Graphs, and Mathematical Tables, ninth edition, Dover Publications, New York, 1972.

[27] B. Blaszczyszyn, M.K. Karray, H.P. Keeler, Using Poisson processes to model lattice cellular networks, in: Proceedings of IEEE International Conference on Computer Communications, 2013, pp. 773–781.

[28] OFCOM, Sitefinder Database [Online], Available at: http://stakeholders.ofcom.org.uk/ sitefinder/sitefinder-dataset/.

[29] Ordnance Survey and Britain's Mapping Agency, OS Open Data [Online], Available at: https://www.ordnancesurvey.co.uk/opendatadownload/products.html.

[30] W. Lu, M. Di Renzo, Stochastic geometry modeling of cellular networks: analysis, simulation and experimental validation, in: Proceedings of ACM International Conference on Modeling, Analysis and Simulation of Wireless and Mobile Systems, 2015, pp. 179–188.

[31] J. Gil-Pelaez, Note on the inversion theorem, Biometrika 38 (1951) 481–482.

Index

Note: Page numbers followed by *b* indicate boxes, *f* indicate figures and *t* indicate tables.

Printed in the United States
By Bookmasters